Technological Change and Employment

Editors: Ronald Schettkat and Michael Wagner

The meta-study *The Effects of Modern Technologies on the Labor Market,* a joint research project for the investigation of the relationships between technological change, employment structures and unemployment, was commissioned by the West German Federal Ministry of Scientific Research and Technology and carried out by the nine institutes listed below. The research teams analyzed innovations in the West German economy from different perspectives, using various methods and taking into account micro and macro-interdependencies. The result is a detailed and comprehensive study.

Deutsches Institut für Wirtschaftsforschung Berlin,
Projectdirector: Dr. Frieder Meyer-Krahmer

Ifo-Institut für Wirtschaftsforschung München,
Projectdirector: Dr. Lothar Scholz

Institut für Stadtforschung und Strukturpolitik GmbH Berlin,
Projectdirector: Prof. Dr. Hans-Jürgen Ewers

Infratest Sozialforschung GmbH München,
Projectdirector: Lisa Höflich-Häberlein

Institut für Sozialforschung und Gesellschaftspolitik e. V. Köln,
Projectdirector: Dr. Werner Friedrich

Institut für Wirtschafts- und Sozialforschung Wien,
Projectdirector: Dr. Michael Wagner

Wissenschaftszentrum Berlin für Sozialforschung/Forschungsschwerpunkt Arbeitsmarkt und Beschäftigung,
Projectdirector: Dr. Ronald Schettkat

Basler Arbeitsgruppe für Konjunkturforschung/Forschungsstelle für Arbeitsmarkt- und Industrieökonomik der Universität Basel,
Projectdirector: Prof. Dr. Peter Kugler

Technische Universität Berlin/Heinrich-Hertz-Institut,
Projectdirector: Prof. Dr. Gernot Weißhuhn

Coordinating team: Prof. Dr. Egon Matzner (Wissenschaftszentrum Berlin für Sozialforschung/Forschungsschwerpunkt Arbeitsmarkt und Beschäftigung), Dr. Ronald Schettkat (Wissenschaftszentrum Berlin für Sozialforschung/Forschungsschwerpunkt Arbeitsmarkt und Beschäftigung), Dr. Michael Wagner (Institut für Wirtschafts- und Sozialforschung Wien)

Technological Change and Employment

Innovation in the German Economy

Editors: Ronald Schettkat and Michael Wagner

Translated by Andrew Watt
Layout and editing: Barbara Schulz

Walter de Gruyter · Berlin · New York 1990

German edition published by Walter de Gruyter & Co., Berlin 1989.

Dr. Ronald Schettkat, Wissenschaftszentrum Berlin für Sozialforschung/Forschungsbereich Arbeitsmarkt und Beschäftigung

Dr. Michael Wagner, Institut für Wirtschafts- und Sozialforschung Wien

This book includes 29 figures and 97 tables

Library of Congress Cataloging-in-Publication Data

Technologischer Wandel und Beschäftigung. English.
Technological change and employment : innovation in the German economy / editors. Ronald Schettkat and Michael Wagner : translated by Andrew Watt.
Translation of : Arbeitsmarktwirkungen moderner Technologien. 1, Technologischer Wandel und Beschäftigung.
Includes bibliographical references (p.)
ISBN 0-89925-682-1 (U.S.)
1. Labor supply--Germany (West)--Effect of technological innovations on. 2. Unemployment, Technological--Germany (West) I. Schettkat. Ronald. II. Wagner, Michael. III. Title.
HD6331.2.G28T4313 1990
331.12'0943--dc20 90-2841
CIP

Deutsche Bibliothek Cataloging in Publication Data

Technological change and employment : innovation in the German economy / ed.: Ronald Schettkat and Michael Wagner. Transl. by Andrew Watt. - Berlin ; New York : de Gruyter, 1990
Dt. Ausg. u.d.T.: Technologischer Wandel und Beschäftigung
ISBN 3-11-012427-0
NE: Schettkat, Ronald [Hrsg.]

Printing: Kupijai & Prochnow Buch- und Offsetdruckerei, Berlin
Binding: Lüderitz & Bauer GmbH, Berlin
Cover design: Johannes Rother, Berlin

Preface

The labor-market effects of modern technologies have always been a subject of controversial discussion in economics and the social sciences. While the tension between conflicting perspectives, analytical methods and empirical findings has been a powerful motor behind scientific progress in this area, the resultant thematic and analytical confusion has been an obstacle to the requirement of established academic research for clear criteria with which to judge the success or failure of specific analytical approaches.

Given this situation any large-scale research project has to opt for a pragmatic approach: while tried and tested methods must be applied to recent data to obtain information on the most recent developments, it is also necessary to attempt to test new methods in order to gain access to significant functional relationships which have proved impenetrable to traditional methods of analysis.

From the very beginning this pragmatic attitude has characterized the Meta Study - an investigation of the labor-market effects of modern technologies commissioned by the West German Minister for Research and Technology, whose findings are presented in this volume. This can be seen in the composition of the Advisory Board which included representatives from different disciplines and specific "schools" within these disciplines, in the heuristic formulation of specific questions from the problematic as a whole, in the selection of the project teams to be entrusted with individual research projects and in the creation of interfaces between the individual - and scientifically autonomous - teams.

This presentation of the central findings of the Meta Study is also to be understood against the background of the organizational form in which the research was undertaken. The degree of complementarity between the content of the individual contributions presented here is higher than may be apparent at first sight due to the stylistic and conceptual differences between the project teams involved.

The editors of this summary presentation can do little to alter this fact, at least to the extent that they avoid the temptation to offer an overall evaluation. This temptation was resisted, if only for the simple reason that both were involved in one of the project teams. The necessary self-restraint notwithstanding, in the introduction the editors have attempted to provide a guide to the development and structure of the research conducted within the framework of the Meta Study. This, it is hoped, will help the reader to identify the relationships between the individual projects and their place within the Meta Study as a whole.

This book would not have been possible without the cooperation of a large number of people. In particular the project teams were kind enough to summarize their original reports to enable their inclusion in this edition.

Ronald Schettkat/Michael Wagner San Francisco/Berlin, May 1989

Contents

Part II: Industries

Tables and Figures

Employment Effects of Modern Technology
A Wide Range of Empirical Findings and Competing Approaches of Analytical Integration (Schettkat/Wagner):

The Effects of the Use of Computer-Aided Technology in Industrial Enterprises.
It's the Context that Counts (Ewers/Becker/Fritsch):

Innovation, Growth and Employment. Innovative Activity at Plant, Sectoral and Intersectoral Level and its Effects on the West German Economy in the 1980s (Scholz/Penzkofer/Schmalholz/Beutel):

TANDEM: Simulations within the Innovation-Growth-Employment-Circuit (Frühstück/Wagner):

Perspectives for Macro-Economic Development at Different Rates of Innovative Activity (Blazejczak):

Employment Effects of Modern Technology

A Wide Range of Empirical Findings and Competing Approaches of Analytical Integration

Ronald Schettkat and Michael Wagner

Technical progress influences a wide range of factors on which economists base their explanations of economic processes. New products are developed, opening additional possibilities to the consumer, the production of goods and services can be more effectively organized leading to higher incomes and shorter working hours. Finally, new skill requirements develop, while traditional skills and qualifications may lose some of their former value.

Just as controversial as the effects of technical progress are the factors which drive it forwards. Which factors lead firms to introduce new technology? What are its effects at enterprise level? Does technical progress lead to deskilling or reprofessionalization of the work force? Is the level of employment reduced as a result of the use of new technology, or is technical progress a necessary condition for increasing employment?

These are just some of the many questions which play an important part in the debate on the labor market effects of modern technology and which are dealt with in this book.

1. Technology Indicators

Economic research is often obliged to treat technical progress as a "black box" (Rosenberg 1982). At times it has been included in econometric models as a trend value, at others it was simply defined as a residual category. Starting from the premise that firms combine the factors of production, capital and labor, in such a way as to minimize costs, and therefore that the mutual substitution of the factors is possible, technical progress is taken to be that part of factor substitution which has been empirically observed but which cannot be explained by factor price variations. It is thus a "measure of ignorance" (Abramowitz 1956), since to define technical progress as a residual value is to make no attempt at a positive measure of one of the central variables of economic development. Depending on whether this residual value leads to greater, lesser or unchanged use of a factor, technical progress is considered to be factor-using, factor-saving or factor-neutral (Hicks 1932).

At the same time the prices for the different factors of production are considered to provide the signals for the direction of technical progress, as the latter is determined by the attempt to replace the relatively more expensive (i.e. the scarcer) factor (cf. Kugler et al. in this volume). However, doubts have been cast by other writers on this view of technical progress as an element steering the economy towards equilibrium (Rosenberg 1976, David 1975). Schumpeter (1926), for example, viewed disequilibrium, the opening of new markets through new products and the high profits which this makes possible, as the motor behind technical progress.

Different types of innovation are considered central in the two theoretical approaches mentioned: (1) process innovation and (2) product innovation. Process innovation leads to the rationalization and more efficient organization of the production process, while product innovation is seen as vital to open up new markets and thus as the motor behind economic growth. However, in the real world "pure" forms of the two types of innovation are seldom to be found because the introduction of a new product is usually accompanied by changes in the production process, while the introduction of new plant automatically results in changes in product quality (Ewers/Becker/Fritsch and Höflich-Häberlein/Häbler both in this volume).

Modern economies are characterized by a highly specialized division of labor. For this reason the character of innovation changes as it moves up or down the vertical production chain: product innovations in one industry (e.g. engineering) become process innovations in another (e.g. the automobile industry). Thus whether an innovation is to be considered a process or a product innovation depends on one's perspective. At the macro level, however, these differences become completely blurred.

1.1 Measuring Technical Progress

Two principal means of measuring technical progress can be distinguished: indicators based on concrete, specific innovations and those based on indicators using value units for measurement purchases (Hansen 1986).

Often individual technologies - such as industrial robots or computers - are selected in order to analyze the effects of technical progress. While such an approach allows detailed studies to be made, it also entails a number of problems. Even where specific technologies are considered, as the research goes into greater detail differences become apparent so that it becomes impossible to make generalized statements. Industrial robots, for example, can be used in many different ways - in the paint shop, for assembly, welding etc. (cf. Edler et al. in this volume). Depending on the way in which a given technology is used its effects on skill requirements, working conditions and the level of employment may be very different.

Depending on the interest of the analysis, the researcher may have to differentiate not only between the different uses of a given technology but also between "machine generations". Such a demand for detail may run counter to the other aims of the analysis. To put it another way, how are more general relationships and effects to be isolated if the specific is put before the general? Specific indicators are of course a useful measure of the diffusion or effects of a single technology, such as are to be found in case studies and surveys, but at the macro level there is the danger that excessive attention will be paid to particularly spectacular innovations although their quantitative importance may be negligible. It has been shown (Fishlow 1966), for example, that progress in rail transport was only marginally due to the actual invention of the railway. Productivity gains were due rather to improvements in details such as the reduction in the weight of the wagons, the use of steel in place of iron and the refinement of the locomotive. In such a case it was less the spectacular technical inventions which led to improvements in the transport system than the numerous subsequent innovations *en détail.*

1.2 Value Units as Indicators

At branch or macro level, or where more than one technology is being considered, such specific (physical) indicators cannot be used because there are no criteria by which the results can be normed, and so comparison is impossible. Indicators can be used which are non-specific as regards the technology in question and which can provide a measure of technical progress. Such value indicators (e.g. expenditure on innovation) or the number of employees engaged in research and development (R&D) are used as a measure of innovative activity. In addition to their degree of "specificity", indicators can be differentiated according to whether they measure the input, output or throughput of innovations (Scholz et al. in this volume).

Expenditure on R&D or the number of employees (physicists, chemists, engineers etc.) working in this field are purely measures of input. They do not provide information on the success of innovative activity. The number of patents registered by a firm tells us only how many inventions have occurred within that firm, and nothing about their economic success or otherwise. This indicator is a measure of throughput. A measure of output, i.e. of the success of innovative activity, would be the share of new products in the total sales of a company.

Each indicator of innovation has its specific advantages and disadvantages. An additional problem for empirical analysis is that the required information is often not available, a particularly thorny problem for longitudinal studies covering an extended period of time. It is clear that a universal measure of technical progress, equally suited to all levels and aims of analysis, is not at hand. Rather it is necessary to apply different indicators in different situations. The indicators used in the Meta Study to cover the various aims, levels, analytical instruments and observation periods are set out below.

1.3 Innovation Indicators Used in the Meta Study

Within the framework of the Meta Study specific technology indicators were used for the case studies and surveys. The IfS analyzed the rate and level of diffusion of computer-aided technology in various branches (including, in four selected branches, diffusion within the enterprise, cf. Ewers/Becker/Fritsch in this volume). The DIW studied the effects of the use of industrial robots, taking into account intersectoral exchange relationships with the help of input-output analyses (Edler et al. in this volume). These studies were all concentrated on specific technologies. Since 1979 the Ifo-Institute has collected data on innovative activity in West German manufacturing industry based on company surveys. A broadly defined concept of innovation is used, on the basis of which the companies taking part in the survey classify themselves as innovators or non-innovators. In addition data is collected on certain specific forms of innovation and firms' expenditure on them. The Ifo-Institute's survey generates information on new products or considerable improvements to existing ones (product innovation) and on changes in assembly and production methods (process innovation). This survey enables the level of innovative activity in the different branches of manufacturing industry in the Federal Republic of Germany to be determined (Scholz et al. in this volume).

The innovation value indicators collected by the Ifo-Institute provided the data base for several of the project teams involved in the Meta Study (Warnken/Ronning, Schettkat/Bangel and Frühstück/Wagner in this volume). The values for expenditure on innovation take account not only of the expenditure carried out by the firm itself (direct innovation expenditure) but also innovations purchased from other branches (indirect innovation expenditure). It then proved possible, by integrating the data from the innovation test into an input-output framework, to model the intersectoral exchange relationships for innovations (Erber/Horn and Scholz et al. in this volume).

The innovation expenditure data gained in this way can be differentiated into product and process innovation and direct and indirect innovation expenditure.

The Meta Study also applied time-series data on R&D expenditure made available by the "Stifterverband der Deutschen Wissenschaft", on the basis of which the DIW constructed a R&D capital stock matrix with which the level of R&D expenditure in the different branches can be determined. The data were then used as innovation indicators in model simulations (Erber/Horn in this volume). With the help of a technology flow matrix, R&D carried out within the firm can be differentiated from that which is bought in from other branches.

Overview 1.1 provides an overview of the concepts of innovation, the data bases and the survey periods used in the Meta Study. Due to the different degrees of aggregation and the various analytical approaches it was necessary to use different innovation indicators for each of the analytical levels used in the Meta Study. While this is not to be considered a disadvantage, as it is not possible at present to construct an innovation indicator which is fully suited to all situations, at the same time such differences are to be borne in mind in interpreting the results of the analyses.

2. The Effects of Modern Technology on the Level of Employment

As an aid to the evaluation of the effects of new technology on employment a distinction can be drawn between the direct or primary effects resulting from the use of new technology itself, and the indirect or secondary effects which are mediated through other mechanisms (Hagemann 1985).

It is also important to distinguish between different dimensions - the time, space and organizational dimensions - when analyzing the effects of new technology on employment. What may appear to be a severe cut in employment in a single plant may in fact lead to employment gains at the macro level if additional jobs are created in other enterprises, be they competitors or firms from other branches at different stage in the same production chain. Equally, the observation of positive employment effects in one enterprise as a result of the introduction of new technology does not necessarily mean an overall increase in employment at the macro level. The opposite may well be the case where the gains in one firm occur at the expense of employment in competitor firms or those delivering to or purchasing from the firm in question (Edler et al. in this volume).

Table 1.1: Innovation Concepts, Survey and Analysis Levels

research project (authors)	innovation concept	survey/ sources	survey interval (period)	survey levels of the data used	levels of analysis
DIW-Find (Erber/Horn)	expenditures on R&D, capital stock of R&D	Stifterverband, own calculations	annually 1971-85	firm/sector	sector
DIW-Langfrist (Blazejczak)	effects of innovations on the macro level	micro level results		firm/sector	total economy (scenarios)
DIW-IO-Analyse (Edler et al.)	changes of inputs, changes of demand, selected technologies: industrial robots, CNC;	DIW and IfS surveys, expert interviews	1987	firm, department	sector
	R&D matrix	own calculations		sector	
DIW micro electronics	specific technologies	DIW survey	1985	firm	firm/sector
BAK/FAI (Kugler/ Müller/Sheldon)	change of factor inputs, change in factor scarcity	Federal Statistical Office	annually	enterprise	sector
Ifo (Scholz et al.)	direct expenditures on innovation	innovation, business cycle and investment test,	annually 1979-1985	firm	firm/sector
	indirect expenditures on innovation	expert interview and Infratest enterprise interviews	1980, 1986		functional area/ firms
IfS (Ewers/Becker/Fritsch)	use of computer-aided technology	postal survey and enterprise interviews	1986 1987	firm	sectors of manufacturing industry
Infratest (Höflich-Häberlein/Häbler)	use of selected technologies or technological investment	COMTEC panel, case studies, enterprise interviews	annually (1982-85) 1987	firm	service sector industries
ISG (Warnken/Ronning)	direct expenditure on innovation, indirect expenditure on innovation	Ifo tests Ifo innovation matrix	annually (1979-85)	firm	sector
	innovative investment	Ifo investment matrices	annually	sector	
IWS (Frühstück/Wagner)	direct expenditure on innovation, R&D capital stock	Ifo Test, Stifterverband		firm	sector total economy
WZB (Schettkat/Bangel)	share of innovative firms,	Ifo Test	annually (1979-85)	firm	sector
	direct expenditure on innovation, indirect expenditure on innovation	Ifo Tests Ifo innovation matrix	annually (1979-85)	firm	sector
	innovative investment	Ifo investment matrices	annually	sector	
	specific technologies	BIBB/IAB	1979	person	total economy

In the analysis of employment effects it is important to distinguish between the *producers* who bring an innovative product onto the market (product innovators) and the *users* of new technologies who introduce such products into their production processes (process innovators). Both types can experience employment effects compared with the use or production of traditional technology because:

- the amount of labor per unit of output is reduced (increased productivity) and/or
- the level of output changes.

Certain effects are then passed on from both the users and producers of new technology to the firms above and below them in the production chain. New products and production processes generally require changes in the structure of inputs, such as the increased use of electronic components. In some cases they enable the vertical integration of production processes which used to be performed outside the firm (Höflich-Häberlein/Häbler).

As far as employment is concerned, the primary effect of process innovation is usually considered to be one of rationalization, whereas product innovation tends to be create employment. However, even in the case of product innovations it cannot automatically be assumed that they will lead to market expansion. New products will to some extent at least merely replace those already on the market because they perform the same or similar functions or because incomes are limited (Mettelsieffen/ Bahrens 1987). It is rare to find a product which meets needs previously not catered for at all; usually they enter into competition, at least partially, with products already available on the market.

Although the primary effect of process innovation is to reduce the amount of labor required for a given level of output these primary effects may be offset by compensatory effects in other markets or through income or demand effects.

While primary employment effects can be observed relatively easily at plant level, it is obviously somewhat difficult to determine the secondary effects occurring in other firms (suppliers, purchasers and competitors) in such a way. New technology may be labor-saving yet it may not lead to overall job losses if the primary rationalization effects is compensated by secondary effects such as increasing sales. Although the secondary effects are methodologically difficult to determine, in the final analysis it is they which are all important in evaluating the total employment effects of new technology.

The primary effects of process innovations can be compensated at both higher and lower stages of the vertically integrated production chain. Process innovation requires investment which creates demand and so employment in capital good industries. Changes also occur in the intermediate products (such as energy) bought from other firms. In some cases the production of such inputs may be integrated into the firm, while in others their production may be contracted out. If production costs can be reduced, price cuts may lead to higher sales.

The rationalization effects of process innovations and the consequent increase in productivity can result in the form of higher wages, shorter working time or increased profits. If total wages rise so will the demand for consumer goods and services. The compensation effects will thus occur not only in the innovative sector but also, indeed primarily, in the consumer-good sectors. An increase in profits can be viewed in a similar way: to the extent that higher profits are invested they create additional demand and thus jobs in the capital-good sectors. However, if the increase in profits is not invested it will 'leak out' of aggregate demand, at least for a time, and so reduce the impact of compensatory effects. It has also been suggested (Kugler/Müller/Sheldon in this volume) that the costs of capital and labor could adjust to the new situation brought about by technical progress, such that as a reaction to rationalization the price of labor (i.e. wages) would fall, in order that rationalization comes to a halt or even be reversed.

The Meta Study employed a variety of methods in order to determine the extent of the direct and secondary employment effects of modern technology at different levels. In this regard the most urgent requirement was to construct a "bridge" between the effects at the micro level, i.e. in the individual plant, and those at meso (branch) and macro (national) level. It is only by considering the effects at all levels that a reliable estimate of the relative importance of labor-saving and compensation effects of the use of new technology can be made.

In the enterprise-specific analyses carried out by the IfS, Infratest, Ifo and the DIW, changes in the level of employment in firms using new technology were contrasted with firms in which new technology had not been introduced. The influence of innovative activity on changes in the level of employment was then determined using analytical models. In the analyses conducted at branch level (Kugler/Müller/Sheldon and Erber/Horn in this volume) both the direct effects of technical progress in the firm and the compensation effects due to the stimulation of demand in the same sector were analyzed.

Empirically-based input-output analyses including the dynamic input-output analyses developed by Leontief and Duchin (1986) were conducted for the use of industrial robots (Edler et al. in this volume) in order to take the movement of innovative products between branches into account. Income effects and the influence of foreign trade were considered in scenarios based on the DIW's longitudinal model (Blazejczak in this volume).

3. Changes in Skill Profiles

In the wake of the diffusion of modern technologies throughout the production process the skills required of the work force are undergoing constant changes. However, the direction of these changes is not so much determined by technology as by a wide range of social processes with a large number of degrees of freedom (Doeringer/Piore 1971, Lutz 1986, Kern/Schumann 1985). Indeed the freedom to determine the effects on skills and qualifications are likely to increase with the intensification of the use of micro-electronics, as their application itself leads to a considerable increase in the level of flexibility (Sorge et al. 1982).

In addition to knowledge of their immediate task a production process based on an highly specialized division of labor requires of workers an awareness of the entire production process, a willingness to cooperate and the ability to communicate. Distinctions can be drawn between functional and extra-functional (Dahrendorf 1963), formal and informal skills (Lutz 1969) and those linked directly and or merely indirectly to actual production (Kern/Schumann 1970). The threat of devaluation is greatest for skills directly tied to the production process as they are immediately affected by process innovations. Skills and qualification are thus to be considered as a multi-dimensional variable with the result that it is difficult to find a suitable empirical indicator.

For the purposes of empirical study skills are usually measured by the following criteria: formal level of education (final school certificate, vocational training, university or polytechnic degree); occupational classification (printer, fitter, mechanic, chemist etc.); job status (blue-collar workers, foreman, white-collar worker); and the field of activity (in the production process itself, in an activity complementary to production). For specific analyses, therefore, the effort has been made to develop criteria based on a combination of factors (Stooß 1984). Every indicator of skills and qualifications has its specific advantages and disadvantages which can only be evaluated in the context of the specific analytical purpose. The final choice is often determined by the availability of data. A variety of different skill indicators were used in the context of the Meta Study (Ewers/Becker/Fritsch, Höflich-Häberlein/ Häbler, Warnken/Ronning, Kugler/Müller/Sheldon, Schettkat/Bangel, König/Weißhuhn in this volume).

Changes in skill requirements induce adjustment processes on the labor market. While analyses based on stocks are perfectly suitable for studying the effects of new technology on the level and structure of employment (de Neubourg 1987), this approach is less suited to the analysis of labor market *processes* because stocks cannot be used to draw conclusions as to the adjustment *processes* taking place on the labor market. The difference between two stock values is merely the result of such processes and does not reveal the flows which lie behind them (Freiburghaus 1978). The data base for the analysis of labor market flows in the Federal Republic is very limited. For the macro level values are on the flows between labor market aggregates, available from the IAB (Reyer/Bach 1986), but they cannot be differentiated by branch.

A constituent part of the adjustment process to technological and structural changes is mobility, either between jobs within an enterprise or branch or between different industries. Moreover, mobility can go beyond the employment system itself and result in unemployment or non-participation in the labor force (early retirement, return to the role of housewife, education and training etc.). There is of course also a flow into the employment system and also the possibility of direct exchange between unemployment and non-participation. Furthermore, the active working population of a country can experience both quantitative and qualitative change through migration.

Skill flexibility is also partly due to the fact that there is a constant flow from the educational system into the employment system. Clearly, the most painful form of adjustment for the individual worker occurs in the form of mobility out of the employment system. The interests of workers are better served by internal training and education within the enterprise or branch. This explains the particular relevance of the question as to whether adjustment to technological and structural change can be made via internal flexibility (Sengenberger 1987) or whether external flexibility - and thus a high level of labor market mobility - is a necessary feature of the process of innovation.

The relationship between innovative activity and labor market mobility out of the employment system (i.e. into unemployment and non-participation) was analyzed for the Meta Study by the WZB/AMB (Schettkat/Bangel in this volume). This analysis, which was oriented towards developments at branch level, was supplemented by detailed studies at the micro level conducted for manufacturing industry by the IfS (Ewers/Becker/Fritsch in this volume) and for the service sector by Infratest (Höflich-Häberlein/Häbler in this volume).

4. Functional Relationships between Various Areas of Analysis

Initially the group responsible for the design and coordination of the Meta Study (the "Wissenschaftliche Beirat") adopted a modular perspective, which may serve as a basis for a better understanding of the various contributions presented in this volume. However, during the course of research a second analytical framework proved to be useful as it helped to define potential interfaces between various projects.

4.1 Modular Perspective

Based on a pilot study into the current state of research in the field (Friedrich/ Ronning 1985) the project as a whole was divided into four modules, each with a specific set of questions to be researched.

Module 1: An analysis of the speed and extent of the diffusion of product and process innovation, the conditions under which they are used and their effects within the enterprise (current level and course of the spread of selected new technologies; factors promoting/impeding their use in the enterprise; quantitative and qualitative effects on employment).

Module 2: Analysis of the determinants and extent of factor substitution at the enterprise level (changes in the use of factors of production within the enterprise; economic and other variables influencing such changes; studies based on aggregate data of the determinants of the demand for labor differentiated by branch and skill level; econometric determination of the elasticities of substitution).

Module 3: An evaluation of available data sets (the data on which official statistics are based contain information which provide considerable insights into the labor market and employment effects of modern technologies, but often have not been appropriately evaluated; concepts are to be developed and the necessary steps taken to enable such sources to be used for disaggregated analyses of the employment structure and the way it changes over time).

Module 4: Combination of the evaluation of the consequences of technology (Technikfolgenabschätzung) with developments at the macro level and trends in the different branches (to take in to account structural conditions at the macro level and the complex relationship between economic and technological development; to develop plausible assumptions for model scenarios; to provide simulation exercises).

From this summary of the various aspects of the Meta Study it should have become clear that what the Federal Ministry for Research and Technology, which commissioned the study, wanted was a whole range of analyses and scientific studies which, both conceptually and methodologically, had to be carried out at different levels of inquiry.

The heterogeneous nature of the approaches becomes quite clear from a glance over the contributions to this volume: module 1 consists of the contributions from IfS, Infratest and the DIW; for model 2 the analyses are carried out by Ifo,

BAK/FAI, ISG, WZB/AMB and the DIW; the themes in module 3 were taken up by the Ifo and the ISG and also by the Technical University Berlin/HHI while the macroeconomic orientation of module 4 was included in the studies by the DIW and IWS.

In spite of the modular nature of the way in which the problematic was approached, it was clear from the very beginning that the Meta Study should attempt to construct a "bridge" between the micro and macro levels. This made it necessary to develop a framework providing interfaces between the four modules. Its structure is described in the next section.

4.2 An Analytical Framework

The functional framework described below is intended to show how the various analytical areas of the Meta Study are related. Figure 1.1 provides an outline of this analytical framework. The ten areas for study are linked with arrows indicating functional relationships. This analytical model is an open functional circuit which includes circular feedback effects between the different areas. For each of the arrows in the diagram it proved possible to determine a quantitatively specified functional relationship within the Meta Study. A central feature of this analytical model is that it lays great emphasis on the link between the micro and macro levels, one of the major aims of the Meta Study.

Area 1: External Conditions at Branch and Macro Level

To a considerable extent the introduction of new technology is dependent on external economic conditions. This applies both to "active" innovative activity and the "passive" adoption of new technology in the form of goods exchange between branches. Generally speaking there is a positive relationship between those factors which favor overall investment and those conducive to the level of innovative activity.

There is also a clearly identifiable relationship in the opposite direction, i.e. the introduction of new technology exerts an influence on external economic conditions. For example, the change in the productivity of the factors of production provoke reactions in the collective bargaining process, the effects of which are felt in all major areas of the economy.

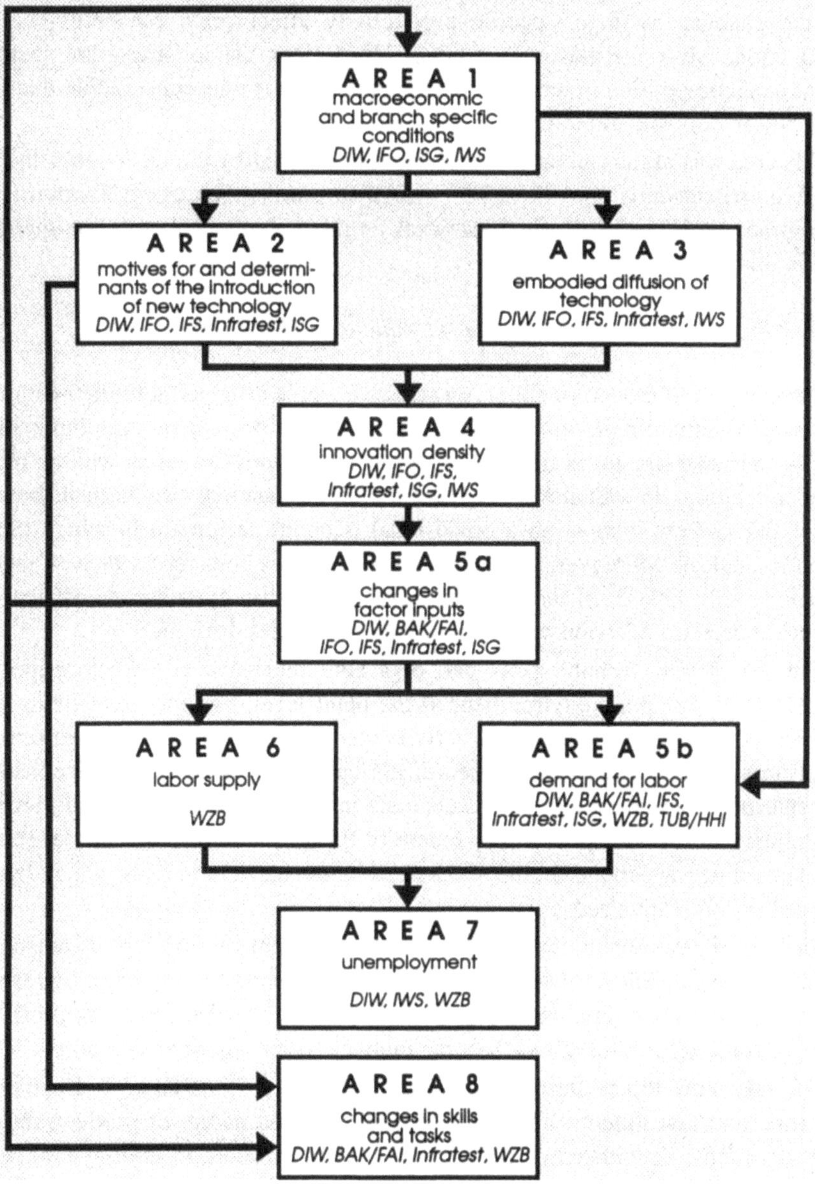

Figure 4.1: A Survey of the Meta Study's Analytical Areas

In this context foreign trade relations are of particular importance. On the one hand the changes in factor-specific productivity affect price competitiveness in foreign trade, while on the other product innovations are an important factor in international competition in terms of product quality. It is thus conceivable that these effects pull in opposite directions.

This area was at the center of the analyses conducted by the DIW (with the help of its Langfristmodell) and the IWS (with their simulations using Tandem). The contributions by Erber/Horn and Blazejczak in this volume deal with the questions raised in this area.

Area 2: Aims and Determinants of the Introduction of New Technology

The introduction of modern technology affects a whole range of activities within the enterprise. A distinction is usually made between product and process innovations, whereby the latter are to be defined as to include organizational as well as merely technical changes. In whichever way the attempt is made to distinguish between product and process innovation a broad band is bound to remain in which the definition is unclear. Moreover, it is usually the case that both forms of new technology are closely linked. In the service sector especially, even the theoretical distinction between product and process innovation becomes difficult.

The use of new technology is part of a comprehensive attempt to cope with changes in the economic environment at the plant level. For the great majority of enterprises the impulse to innovative activity comes direct from the product or factor markets in which the enterprise is operating. Changes in product markets constantly force enterprises to adopt qualitative changes in their products and/or to a competitive adjustment of their prices. The extent of this pressure to innovate depends on market structures which are in turn influenced by the barriers to entry which may be increased by innovative activity.

The use of new technology requires constant investment; it is for this reason that it is heavily dependent on the same factors which determine enterprise investment decisions. At the same time, factors of a more general, social nature (social orientation, organizational culture) also affect the innovative activity of an economy.

The following topics are of interest in this context: the aims of and barriers to innovation and investment; the relative proportions and nature of product and process innovations, anticipated or actual effects of innovation on employment, input structure, market opportunities, corporate culture, enterprise organization; additional investment planned.

In the Meta Study these topics were dealt with in the following ways:
- case studies conducted by the IfS (in manufacturing industry),
- case studies and an enterprise survey conducted by Infratest (in the service sector),
- special results from the Ifo Innovation and Investment Test,
- results of the DIW's project "Enterprise and Innovation".

Further details on the data sources, methods and results of these studies are to be found in Ewers/Becker/Fritsch, Höflich-Häberlein/Häbler and Scholz et al. in this volume.

Area 3: The Diffusion of Embodied Technology

The spread of modern technology can be viewed under two aspects. Firms may attempt to develop new technology themselves, or they may "import" know-how developed in other firms into their sphere of production. It is therefore possible to distinguish between the density of innovation on the side of the "developers" and that of the "users" in an economy. Such a distinction is particularly useful in capital-good sectors where the product innovation of the producers of, say, machinery becomes process innovation in the firms using the new machines. The more rapid the "active" progress of the producers, the more embodied innovation will be "passively" absorbed by user industries. The medium of diffusion in such cases is the purchase of investment goods by individual industries. On this view, two determinants of the density of innovation can be discerned: the extent of the innovation potential created and the intensity of use in the form of purchases of investment goods. The density of innovation of an economy may, for example, decrease if there is a slowdown of new investment although the expenditure on innovation in the capital-good sectors may be on the increase.

This basic idea of technological diffusion via the exchange of goods between sectors can be extended beyond the capital-good sectors, although in some cases the principle may be less clear-cut. An example is innovation in the industries producing material inputs. The higher the number of linkages between branches in an economy, the broader are the diffusion effects of new technology, whereby in general there are considerable differences between the density of innovation among the developers and the users.

Suitable indicators for measuring the density of innovation are based on innovation statistics which are founded on the structure of goods exchange within the economy (or the way it changes over time). Concrete attempts have been made to analyze the diffusion of new technology using an input or investment-good

matrix. The following projects were set up to tackle such problems within the framework of the Meta Study:
- the innovation-exchange matrix constructed by the Ifo (linking the data from the innovation test with the input-exchange matrix),
- the ISG's innovation-exchange matrix (quantifying the diffusion of "innovative" groups of products with the help of the investment matrix),
- the DIW's R&D capital stocks.

A selection of the results achieved in this area is to be found in the contributions by Erber/Horn and Scholz et al. in this volume.

Area 4: The Density of Innovation and the Speed of its Diffusion

The introduction of new technology is fraught with considerable risks, both at branch and the enterprise level. For this reason it is usually only a small number of enterprises in a branch which make the initial decision to use a specific new technology. At first these "pioneers" limit themselves to an experimental approach. If the innovation in question turns out to be successful, the incentive is there for its introduction throughout the enterprise or the industry. However, this does not occur simply through imitation, as new technology is continually being developed, improved and adapted.

The more obvious the advantages of a new technology, the more rapidly the willingness to innovate spreads among departments and enterprises. However, the introduction of new technology comes up against constraints and limitations, of which the most important ones are financial and organizational. The latter arise because the introduction of new technology affects the interests of the work force, in many cases in a fundamental way, as certain tasks and departments lose or gain status. If additional skills are required from the work force there is always the danger that existing groups of workers will be displaced by newcomers.

This struggle between the incentives and the barriers to the use of modern technology generates a pattern of innovation density which is specific to each branch. The dynamic process which this produces and the extent of the diffusion of innovative activity within a branch (or of a certain technology) can be illustrated with the help of quantitative indicators of the following types:
- indicators which measure the spread (i.e. the use) of a given technology,
- indicators which also measure innovative activity which does not necessarily have a direct impact, e.g. the level of expenditure on R&D.

This area was covered by the Meta Study in several ways: The longitudinal/ cross-sectional surveys conducted by Ifo make it possible to quantify:

- the proportions of innovators to non-innovators, and product to process innovations,
- the relationship between expenditure on innovation and turnover or employment,
- the structure of expenditure on innovation,
- and the effects of investment cycles.

The surveys conducted by the IfS indicate the number of adopters of specific technologies as a share of the enterprises of West German manufacturing industry over time, providing information on the past and future course of the diffusion process. The evaluations performed by Infratest enable the spread of certain technologies in the service sector to be quantified. The input-output analyses of the DIW provide information on the innovation dynamic in selected sectors with additional information available from the branch studies conducted by the ISG in its comprehensive project report.

A selection of the findings reached in this area is to be found in the papers by Höflich-Häberlein/Häbler, Frühstück/Wagner, Ewers/Becker/Fritsch and Edler et al. in this volume.

Area 5: Changes in Input Factor Proportions

The introduction of modern technology leads to changes in the structure of production at plant level, particularly in the proportions in which the different factors of production are combined. Indeed in neoclassical micro-economics the extent and direction of technical progress is usually characterized by its effects on the demand functions for the factors of production. The influence of technical progress on the price elasticity of factor demand is thus of considerable interest.

The heterogeneous nature of the factors of production suggests that their differentiation should go beyond the traditional division into "intermediary goods", "capital" and "labor". It is apparent, for example, that the introduction of new technology does not have the same effects on all types of labor, as its use often requires additional highly skilled staff whereas in certain other areas a reduction in labor input is possible.

A further important distinction is that between the demand for working hours and that for workers. Enterprises have a wide range of strategies at their disposal in order to adjust to changes in the demand for labor time while keeping the level of employment constant. This means that the effects of technical progress on the demand for working time are not automatically reflected in the firm's behavior *vis a vis* the level of employment.

In the Meta Study the factors of production were divided into sub-categories and analyses were conducted at both branch and macro level as follows:

- the sectoral longitudinal study of the BAK/FAI (Kugler/Müller/Sheldon in this volume),
- the FIND model of the DIW (Erber/Horn in this volume),
- the input-output analysis of the DIW (Edler et al. in this volume) and
- the estimates made by the IWS for Tandem (Frühstück/Wagner in this volume).

Area 6: The Supply of Labor

The implementation of new technology can provoke reactions in the supply of labor if, for example, skills and qualifications are devalued or the level of employment falls. The age and gender structure of the work force is of particular importance if new technologies demand new skills which are not evenly distributed between age and gender groups. In addition to the flow from employment into unemployment the age and gender-specific flow out of employment into non-participation is central to the evaluation of branch and group-specific labor market chances in the face of the increasing use of new technology and the changes in skill requirements which accompany it.

It is difficult to determine the overall maximum potential supply of labor, influenced as it is by a whole range of cultural, social, political and economic factors. The actual supply of labor, on the other hand, can be analyzed comparatively easily and at a high level of differentiation on the basis of regularly compiled and published employment statistics. Some conclusions can be drawn, for example, about the extent of the "discouraged worker" effect. The flows out of employment mentioned above can be analyzed using available statistics and, in conjunction with figures for, say, age structure in different branches, some conclusions can be drawn.

Such an approach was used by the WZB/AMB, covering the following topics:

- branch-specific flow out of employment into retirement,
- branch-specific flow out of employment into other forms of non-participation in the labor force,
- age and gender-specific changes in participation rates.

Detailed results of these studies are documented in Schettkat/Bangel in this volume.

Area 7: Unemployment

The discussion on the diffusion of modern technology in the production process is usually directly linked to that on unemployment. The link is provided by changes in both the level and the structure of employment. In the first case, changes in the level of employment and unemployment due to the use of technology require analysis, whereas in the second case, in addition to the changes in the structure of employment and unemployment, structural changes must be analyzed as a process, thereby illuminating the form and extent of external labor market adjustment.

A glance behind the stock figures for unemployment usually used to describe unemployment, reveals a considerable dynamic reflected in the flows into and out of unemployment and its average duration. The flow into unemployment relative to the number of those potentially affected by it (wage and salary earners) is an indicator for the risk of becoming unemployed. Its size varies between branches and is dependent on both the degree of external adjustment of the work force to new skill requirements and the branch-specific employment trend. The duration of unemployment in a given branch, on the other hand, is only loosely related to innovative activity in that branch, simply because persons made unemployed from one industry may reenter the employment system in another or may leave the system altogether (early retirement, non-registered unemployment).

It is therefore necessary to study not only the stock values of unemployment but also labor market dynamics in order to take the processes into account which lie behind changes in stock values. The most obvious differentiation of labor market indicators is by branch; a further differentiation by gender was undertaken to identify gender-specific differences and their causes in background variables.

The central themes in this area as far as the Meta Study is concerned were:

- changes in the (stock values of the) level of unemployment,
- the risk of becoming and remaining unemployed differentiated by branch (calculated by the WZB/AMB on the basis of data provided by the Federal Employment Agency),
- the analytical link between labor market indicators and innovative activity (also conducted by the WZB).

The main results of these studies are to be found in the contributions by Blazejczak, Frühstück/Wagner and Schettkat/Bangel in this volume.

Area 8: Changes in Skills and Qualifications

Changes in skills and qualifications are of central importance for the evaluation of the labor market effects of modern technology. The (supposed) direction of changes in skill requirements is of major importance for the debate on the requirements and quality of work. Moreover, changes in skills and qualifications are among the causes for subsequent adjustment processes in the labor market.

Beside the skills and qualifications which are required directly at the place of work, qualities must also be considered which are used only indirectly, such as the ability for social interaction, or "structural knowledge" both of which are of increasing significance in highly complex production processes.

Multidimensional phenomena are a problem in that it is difficult to identify and quantify them adequately, so that theoretical and empirical analyses tend to concentrate on certain aspects only.

Occupational classifications, for example, are an indicator which is often used to illustrate changes in skills and qualifications. However, such an indicator does not only describe qualifications and types of work but also social value judgements. Moreover, such classifications themselves do not remain homogeneous over time with respect to skill and activity profiles, but are themselves subject to constant change. Every conceivable indicator has its specific advantages and disadvantages and no one indicator is able to represent the full complexity of skills and qualifications. Despite these difficulties some of the aspects of changes in skills and qualifications were analyzed in detail within the framework of the Meta Study.

The employment statistics were analyzed by the ISG, using a specially constructed skill index based on occupational position and educational and training variables. The DIW determined changes in occupational structure with the help of its input-output analysis, a method which takes both the direct and indirect effects of new technology into account. The study conducted by the DIW into "Micro-electronics and Employee Qualification" provides information on the qualification process. The ISG, BAK/FAI and the IWS worked with different groups of blue and white-collar workers, classified according to relevant data from the Federal Statistical Office. The ISG and the WZB/AMB used occupational classifications to analyze changes in skills and qualifications.

The contributions by Warnken/Ronning and Schettkat/Bangel provide an overview of the results achieved in this the final area.

5. Outlook

The contributions of the research teams presented here provide a snapshot of the rapidly developing field of economic research into the consequences of the use of modern technology. In many respects this snapshot is rather selective. It presents a small set out of a rich variety of research in this field for the Federal Republic of Germany. During the conduct of the Meta-Study several contacts were made to other research teams, but this is scarcely reflected in the research of this volume.

Finally it would appear worthwhile to sketch out some ideas as to the potential for the future development of the ground which the Meta Study has prepared. These, the (entirely subjective) views of the editors, can be grouped under the four headings: "concepts", "analytical fields", "models" and "data".

As far as concepts are concerned, economic research into innovation still has a huge task ahead of it. It is not a question of replacing the (fruitful) pluralism of concepts (through a single flash of genius) by a single consistent framework or taxonomy, but rather of providing answers to certain central conceptual questions. These include: How does new knowledge "replace" existing information? Can innovation be interpreted as a factor of production in its own right? By which media (information, investment goods, human capital) are new technologies transmitted? Under what conditions can a variety of technological changes be expressed in terms of an "amount" of technical progress (measured in prices)?

As regards analytical fields, the contribution made by the Meta Study requires further development. This is particularly true of the fields working time, the supply of labor and working conditions. The fact that these areas were scarcely mentioned in the original commission for the project was felt to be a severe limitation, especially as the employment effects of new technology are visible not least in terms of changes in working time. Within the framework provided by the following contributions there are a wide range of possibilities to link up with research into working time as well as the supply of labor and working conditions. Only when these areas together with a more intensive analysis of adjustment processes are included in the analysis can we really speak of the "labor market effects" of new technology.

The work on models to illustrate the effects of technical progress was pushed so far within the Meta Study that it is already possible to see the direction in which the next steps will lead. The construction of embodied innovation flow matrices was entirely successful and this "bridge" merely requires further refinement. However, an open question remains in the form of the endogenization of innovative activity within the framework of disaggregated macro-economic models. It is only when price and quantity effects, including feedback effects at the macro level, are simultaneously and explicitly modelled that economic research into innovation can claim to have reached the current level of macro-econometric model construction.

As far as data are concerned, the Meta Study confirms the experience made elsewhere that it is partly the lack of data but also partly the variety of heterogeneous information which pose the genuine challenge to the analytical capabilities of social scientists. Against such a background interested public bodies would be well advised to continue to support those activities in which an attempt is being made to construct coherent data bases on technical progress and labor market processes in the Federal Republic of Germany. For the development of German social science as a whole, the development and maintenance of such an infrastructure will be of increasing importance in the years ahead.

PART I

ENTERPRISE LEVEL

The Effects of the Use of Computer-Aided Technology in Industrial Enterprises: It's the Context that Counts

Hans-Jürgen Ewers, Carsten Becker, Michael Fritsch

1. Introduction

The following is a summary of the results of our study into the effects of the introduction of computer-aided technology in industry. Our results are not such as to enable us either to confirm or to refute the 'job-killer hypothesis' once and for all. We consider this debate to be as futile as those on whether the innovative activity of an economy is 'science-pushed' or 'demand-pulled' or whether large firms are more innovative than small or medium-sized ones (or vice versa). In view of the complexity of the real world such 'all-or-nothing' questions are facile and mitigate against deepening our understanding of the differentiated way in which economic processes of adjustment and innovation occur. The value of our conclusions is to be seen far more in the fact that they make it possible to distinguish more clearly between the various constellations which give rise to the application of new technological opportunities and the different patterns of effects they produce. This should contribute to establishing points of departure for the regulation of the effects of new technology which goes beyond undifferentiated hypotheses.

In the discussion on the effects of new forms of technology it is now generally recognized that technology is not something which suddenly descends on an economy, destroying established structures and steering developments in a new and predetermined direction. The numerous studies which have been made into the effects of new technology in general and of computers in particular have provided ample evidence that technological determinism in the sense just described does not exist (cf. Lutz 1987). The extent of the effects of a new technology and the way in which such effects develop over time depend not only on the diversity of potential applications which it opens but also on the problems and constraints which it has been designed to help overcome. This is because new technologies are applied primarily as a means to overcome bottlenecks within the framework of already existing structures. To this extent they merely reinforce already established trends, a fact which by no means precludes the possibility that, over time, a recognizable pattern of effects specific to that technology may develop. In the course of its application - while the primary aim remains to overcome existing constraints - the search is on for further ways in which the new technology can be profitably applied: such possibilities may be discovered more or less accidentally. In other words, it is only after time-lag that the special 'talents' of a technology are revealed, inducing applications and structural changes which can then be interpreted as 'technology-specific' effects. The identification of such effects thus requires a longitudinal analysis over an extended period. In the case of computer-aided technology it is still too early to identify these effects because in most cases its application is not yet beyond the initial stage of overcoming constraints.

Despite the great diversity of the effects of technology described below, two general conclusions would appear to be of major importance for economic policies aimed at influencing the application of new technology:

- Firstly, it is apparent that such policies must take into consideration the 'environment' in which technology is applied. This is because the effects of the use of technology follow the logic of the situation in which its users find themselves, or, alternatively, the aims which they hope to achieve through its application. It is therefore not surprising that the pattern of technological impacts has certain recognizable characteristics, at least in the initial stages of its application. It reflects (and reinforces) the effects of those external conditions (such as changes in the international division of labor, the trend towards the tertiarization of industrial production and the diversification of supply) which in all probability would have determined the development of the economy even without the influence of new technology. Seen from this point of view, discussions as to whether or not new technology is a 'job killer' miss the point completely: if certain effects of new technology are particularly desirable (or undesirable) then it is necessary to begin with those factors which, at the micro level, induce firms and enterprises to apply technology in a certain way.
- It is equally apparent that, despite the diversity of effects induced by various technologies, almost all forms of their application lead to a considerable acceleration of the selection and filter processes already at work on the labor market. This occurs at various levels. In the case of technology-induced redundancies, unskilled workers are affected to a much greater extent than more highly skilled workers. Equally, technology-induced hiring tends to favor better qualified skill-groups. Technology-induced retraining at plant level also follows the same pattern: a far lower proportion of unskilled and semi-skilled workers participate in retraining measures than do skilled workers, and the extent of such training (in terms of its duration) is much less.

This means that the almost certain increase in the use of computer-aided technology cannot be expected to make a positive contribution to the most urgent labor market problem at present, namely the long-term unemployment of unskilled workers. This is true even if the balance of the short-term, direct employment effects were to prove positive. Indeed it seems likely that, at least, during a transition period the increased use of computer-aided technology will aggravate the problem still further. It is possible that increased demand leading to a more complete utilization of the potential labor force (aided by demographic factors) may then provide for a gradual reintegration of this particularly hard-hit section of the population into the employment system. It is clear, however, that labor market policy cannot remain inactive until then. The new jobs which will be created during the 'transition period' require skills and qualifications which those currently unemployed - and those who will become unemployed - cannot match. There is therefore a need for policies to compensate those who, in the transition to a new production structure, will remain 'at the back of the queue' for the foreseeable future.

In this summary we have avoided making detailed references to the longer, published version of our report and to our sources in order to enhance its readability and limit its size. By comparing the titles of the following sections with the 'contents' of the long version of our study the reader should have no trouble finding any details of further interest.

We proceed in the next section to a short description of the data base which provided the foundation on which our study was built. In section 3 we discuss the diffusion of computer-aided technology in the manufacturing industry of the Federal Republic of Germany, including both the determinants of and motives for the application of new technology and its diffusion within the enterprise. Section 4 deals with the overall effects of computer-aided technology at plant level, which is further developed in section 5, the analysis of the direct employment effects. Section 6 discusses the extent of retraining as a result of the application of computer-aided technology at plant level. Finally, in section 7, the effects of computer-aided technology on employment at plant level are summarized.

2. The Data Base

Our analyses were based on two sources of data. The first consisted of a mailed questionnaire entitled 'Development Problems of Industrial Enterprises in the Context of Structural Change' conducted in selected regions of the Federal Republic of Germany. The questionnaire identified the use of computer-aided technology disaggregated to department level in more than 3,300 industrial enterprises (for further details cf. Fritsch 1989). Building on the information provided by the survey the effects of the use of computer-aided technology were studied in detail by means of

274 interviews, some of which took the form of case studies, with enterprises in selected industries.

The target group for the postal survey were industrial enterprises with at least 10 employees and on-site production; plants concerned solely with distribution and administration were not considered. The questionnaire aimed primarily to establish the conditions influencing technological developments at plant level. It encompassed both turnover and employment trends together with a series of enterprise characteristics such as resources, product range, methods of production etc. In order to take account of the full diversity of on-site conditions, regions were selected so as to take into account different geographical factors. Because such site conditions as local labor market characteristics, the availability of services, infrastructure and information tend to vary with the degree of industrial 'density', the selection of the regions was guided by the physical density of both industry and housing. Within each 'density-group' regions were chosen in which in the manufacturing sector exhibited either a pronounced positive or clearly negative employment trend.

It is difficult to evaluate the proportion of completed questionnaires received and the representativity of the sample, as precise information on the complete set of firms considered relevant for the questionnaire (industrial enterprises with on-site production) was not available. An analysis of those firms which declined to complete the questionnaire suggests that new enterprises and enterprises which are part of a larger company and have little local autonomy are underrepresented in the sample. Based on the total number of firms we contacted, we estimate that 25% of the firms who were in a position to supply information did in fact provide responses which were suitable for evaluation. Enterprises with less than 50 employees were somewhat underrepresented while those with more than 500 employees were probably slightly overrepresented.

The postal survey identified the extent of the use of computer-aided technology with the question: "Do you employ computer-aided technology in the following departments?". The possible responses were: "administration", "machine workshop (CNC)", "production planning" and "design (CAD)". If an affirmative response was given, firms were asked to state the year in which the technology was introduced in the relevant department. The question: "Does your enterprise use one of the following forms of telecommunications technology?" was aimed at establishing the extent of the use of new telecommunications technologies. Possible responses were "Telefax", "Teletex", "Datex" and "Bildschirmtext" (Videotext). As these technologies are still in the early stages of diffusion there was no question as to their year of introduction.

In the case studies/interviews with users of computer-aided technology the aim was to investigate the effects of the application of new technology with particular emphasis on the "administration/production planning", "machine workshop" and "design" departments. The effects in various other departments - "assembly", "quality control", "storage" - were also studied, but the use of computer-aided technology in these departments is as yet fairly restricted, so that the sample contains a relatively small number of cases. The interviewees consisted of enterprises which had indicated in the postal survey that they used computer-aided technology in machine workshops and/or design. In order to ensure a degree of homogeneity regarding conditions of production and sector-specific features of the use of technology, interviews were restricted to enterprises in certain branches. The main criterion for the selection of the branches was the trend level of demand in these sectors since the mid-1970s. By concentrating on branches with clear (positive or negative) demand trends it should be possible to include the effects of market situation and the overall branch 'environment' in the analysis.

The engineering and electrical industries which have been marked by a relatively positive demand trend since the mid-1970s were selected, together with the wood-processing and textile industries which have been faced with falling demand over the last ten years. The emphasis of the 274 interviews/case studies was on the engineering (46.7%) and electrical industries (34.3%). 11.7% of the interviewees were enterprises in the wood-processing industry and 7.3% in the textile industry. As large firms often make extensive use of computer-aided technology while smaller firms tend to be 'late adopters', large firms are somewhat overrepresented in the sample. Nevertheless, some 30% (72.6%) of the enterprises interviewed employ less than 50 (200) workers. Thus small and medium-sized enterprises, which are often largely ignored in studies into the effects of computer-aided technology, are comparatively well represented.

Within the framework of the interviews/case studies an attempt was made to study the effects of the use of computer-aided technology at both the (micro) level of the enterprise as a whole and the ('micro-micro') level of the various departments, even down to the level of the individual work station or machine. Our point of departure was the place in which new technology was first employed in an enterprise, the place, so to speak, where the 'stone hits the water's surface', and from where the effects emanating from the impact can be relatively reliably and accurately identified. However, in enterprises which have made intensive use of new technology over a long period the limitations of this approach, due in particular to its retrospective nature, rapidly become apparent. Here, the study of the uses and effects of computer-aided technology at the level of the enterprise as a whole complements the analysis at the 'micro-micro' level of individual work stations or departments.

Furthermore, using this approach it is also possible to consider indirect effects of the use of new technology which cannot be analyzed at work-place level.

3. The Diffusion of Computer-Aided Technology

3.1 The Diffusion of Computer-Aided Technology in Manufacturing Industry in the Federal Republic of Germany

Although 80% of the West German manufacturing enterprises in our postal survey already make use of computer-aided technology in at least one department, the degree of diffusion (measured by the proportion of enterprises) and the depth of diffusion (i.e. the intensity of its use, cf. section 3.3.) in the majority of departments is still relatively low. As shown in table 3.1, the majority of enterprises (75.8%) in the postal survey apply computer-aided technology in administration, while the degree of diffusion in other departments is much lower. On average only a quarter of the enterprises (26.8%) use CNC machines in production, a fifth (22.0%) use computers for production planning and only one in ten enterprises (10.1%) employ them in design and in development. Slightly less than a third (32.9%) of the enterprises in the survey stated that they used at least one of the 'new' forms of telecommunications technology, with Telefax and Teletex as the clear favorites.

If the degree of diffusion in the various departments is differentiated by branch, as in table 3.1, the variance between the different branches is seen to be very high, especially in "design". These differences are a clear indication that the proportion of enterprises which are able to make profitable use of computer-aided technology in "design" varies from branch to branch. The potential for users of computer-aided technology is restricted to those firms for which, under the primary consideration of profitability (price against performance of the technology), such an acquirement is worth while. Similarly, the relatively small variance in branch-specific adoption rates for computer-aided technology in the field of administration indicates that enterprises in all industries can be considered potential adopters of office-oriented, computer-aided technology.

The variance in adoption rates between the different branches may also reflect differences in the branch-specific size structure of enterprises. If the adoption rates are differentiated by enterprise size, the pattern established in numerous other studies on diffusion is confirmed: adoption rates clearly increase with enterprise size (number of workers). The differences between enterprises of different sizes is particularly pronounced in the departments "production planning" and "design".

Table 3.1: Adoption Rates for Computer-Aided Technologies by Branches of Industry (%)

	administration	production planning	CNC	CAD	percentage of firms using EDP	new telecommunic. technologies
all industries	75.8	22.0	26.8	10.1	80.0	32.9
crude oil**	100.0	16.7	33.3	0.0	100.0	66.7
stone and clay	75.0	20.6	22.2	4.4	77.0	16.5
iron producing**	92.3	30.8	46.2	7.7	92.3	69.2
non-ferrous metals	86.7	33.3	46.7	20.0	93.3	60.0
foundries	76.5	27.5	23.7	3.9	78.4	37.3
drawing etc.	68.4	17.3	32.7	4.1	75.5	30.6
steel and light metal	72.5	10.0	18.8	8.8	75.0	35.0
machinery	75.7	23.9	48.9	19.4	82.7	46.6
road vehicles	81.2	33.3	42.0	8.8	82.6	52.2
ship building	87.5	50.0	31.1	37.5	87.5	56.3
aircrafts**	92.9	64.3	57.1	57.1	92.9	78.6
electronics	84.8	30.4	31.6	29.4	89.6	43.4
precision engineering	83.9	30.1	44.1	16.1	92.5	44.1
iron/sheet metal/metal goods	70.3	21.8	36.2	7.0	79.5	28.8
musical instruments, toys	70.5	7.7	19.2	1.3	74.4	17.9
chemistry	85.8	21.6	10.4	6.0	86.6	42.5
office machinery, computers	75.0	50.0	25.0	54.2	75.0	62.5
precision ceramics	83.3	29.2	12.5	0.0	83.3	33.3
glass	78.4	21.6	29.7	16.2	78.4	37.8
wood working	52.9	10.3	14.9	1.1	60.9	13.6
wood processing	72.7	22.5	19.8	4.3	75.9	20.3
pulp**	71.4	14.5	42.9	0.0	78.6	64.3
paper and paper products	80.7	15.8	10.5	5.3	82.5	29.8
printing	66.2	13.2	22.1	5.9	72.1	33.8
plastics	72.5	19.8	32.9	3.9	75.4	36.2
rubber	94.1	23.5	17.6	5.9	94.1	47.1
leather	85.7	14.3	3.6	3.6	85.7	14.3
textiles	72.6	20.4	16.1	3.8	74.7	28.0
clothing	69.7	17.4	5.3	4.5	70.5	18.2
food	79.4	19.7	15.4	0.6	82.6	21.7
tobacco**	85.7	14.3	0.0	0.0	85.7	42.9

* adopters as a percentage of total enterprises questioned in each industry
** less than 15 cases

Enterprise size as a determinant of the application of new technology conceals a variety of related factors: large enterprises are usually marked by large, standardized flows of information to a much greater extent than smaller ones, a factor which remains an important precondition for the profitable use of computer-aided technologies. Their staff is usually well qualified and they have large investment budgets so they can afford to 'try out' new technologies with less risk than smaller enterprises. Last but not least, large firms invest in new equipment more frequently enabling them to acquire innovative products shortly after they become available.

Further analysis of the differences between adopters and non-adopters, based on the postal survey, shows that the results of such comparisons are very much dependent on where the line is drawn between potential and non-potential adopters. Different results are gained depending on how restrictive the conception of 'potential' is defined. If firms are included which, in view of the objective profitability of the application of computer-aided technology, are not really potential adopters, the determinants of the objective profitability of the use of computer-aided technology tend to be dominant in the comparison between adopters and non-adopters. If, on the other hand, the group of potential adopters is kept very restrictive, other socio-economic characteristics, independent of objective profitability, exert greater influence on the decision whether or not to introduce new technology. Using variously restrictive conceptions of potential users, our analyses show that the differences between adopters and non-adopters of computer-aided technology can be interpreted in nearly all cases as being determinants of the objective profitability of the introduction of the new technology. Despite the relatively diverse information about the enterprises gained from the postal survey it was not possible to identify determinants which clearly were not related to objective profitability. If such factors do exist they are either of secondary importance and/or they were such as to escape identification in our study. The fact that other, similar analyses (e.g. Thwaites/Edwards/Gibbs 1982, Müdespacher 1987) come to the same conclusion indicates that standardized postal questionnaires are obviously unsuited to identify the influence of factors such as management attitudes, organizational structure etc., on adoption behavior, always assuming, of course, that these factors do in fact play a significant role.

Comparing industrial enterprises which introduced computer-aided technology in 1985 or 1986 with those which by the end of 1986 were still non-users, we can summarize as follows: Users

- are generally larger in terms of both work force and turnover
- grew more rapidly between 1980 and 1985
- are more heavily involved in research and development
- have a larger proportion of graduates in their work force

- more frequently purchase services from other companies
- have a more highly differentiated product range and offer more supplementary services
- tend to be producing more products in the early stages of their 'life-cycle' as a proportion of turnover and are more clearly oriented towards new products and improving quality
- are more heavily involved in export markets
- have increased the vertical integration of production and its flexibility more significantly in recent years
- make greater use of fixed, written planning
- see bottlenecks primarily in a shortage of skilled and managerial staff and in internal organizational deficits
- view the future development of employment in the enterprise considerably more positively.

From our survey the adopters of computer-aided technology appear to constitute the type of enterprise which has reacted positively to the current wave of structural change and, in terms of employment and turnover, has indeed benefitted from it.

3.2 Motives for the Introduction of Computer-Aided Technology

The interviews/case studies with user enterprises included a standardized question ("What were the most important reasons for the introduction of computer-aided technology in this enterprise?") aimed at establishing the motives for the introduction of new technology. Included in the question were 13 preformulated answer categories with the additional possibility of naming "other reasons". Figure 3.1 illustrates the relative importance of the different answer categories. The columns in each line indicate the factors to which no statistically significant difference can be established. The answer categories to the left (right) of these columns are significantly more (less) important than the answer categories of the respective line.

As can be seen from figure 3.1, reducing unit production costs and improving product quality are top of the list of motives; no significant difference could be established between them. Equally important was the desire to shorten production times. The relative significance attached to the motives "positive sales expectations", "improvement of production organization " and "capacity bottlenecks" together with the last place accorded to "decreasing sales or market share" clearly indicate that computer-aided technologies are frequently introduced with the aim of expanding capacity and seldom in the context of a "defensive rationalization strategy", i.e. as a reaction to demand constraints.

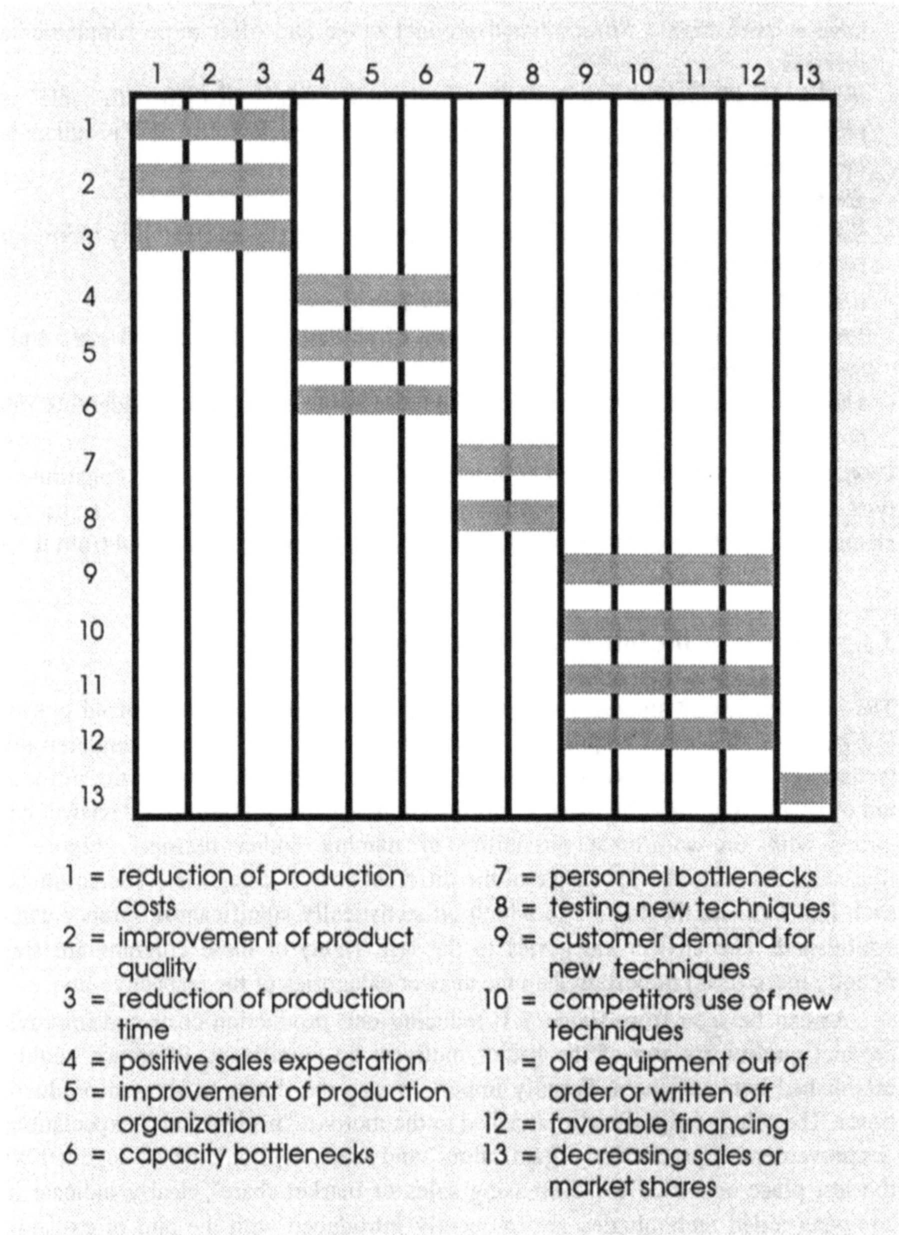

Figure 3.1: Rank Order of Motives for the Introduction of Computer-Aided Technology

The 7th place for "personnel bottlenecks" is a further indication of the significance of expansion as a motive. The "testing of new techniques" manages only place 8, from which it can be concluded that 'technological curiosity' and the love of experimentation is at best of relatively minor importance in the decision to introduce computer-aided technology. The following categories also appear to be relatively insignificant: "customers demanding new process technology", "competitors use of new technology", "old equipment out of order or written off", "favorable financing" (cf. figure 3.1).

Correlation and factor analyses produce four different motive-groups for the introduction of computer-aided technology, each of which can be interpreted as a response to a particular constraint. The most important of these is probably the 'expansion motive', i.e. new technology is introduced in the context of expansionary investment, primarily in order to overcome capacity constraints (particularly of labor). This is followed by the 'replacement and rationalization motive' in which the reduction of unit production costs and/or the replacement of old or defective plant are the most important motives. In the group ranked third in importance, 'the competitive pressure motive', the constraint to be overcome is less one of cost than of product quality. A fourth constellation of constraints lies behind the final group, which can be termed the 'experimentation motive' for the introduction of computer-aided technology: given positive sales expectations and favorable terms for financing investment, enterprises experiment with new processes in the hope of improving the organization of production, increasing the product quality and reducing their production times. The desire to reduce costs is of secondary importance in this context.

It is apparent that the motives for the introduction of computer-aided technology are influenced to a considerable extent by the external environment in which the enterprise finds itself. Thus enterprises which are exposed to tough price competition emphasize the reduction of unit production costs as a motive for adopting new technology. The greater the importance of product quality for the market position of an enterprise, the greater the weight attached to "improving product quality". Similarly the importance of "short delivery times" for the success of the enterprise is reflected in an emphasis on 'reducing process/production time' as the aim of adopting new technology.

The question as to why an enterprise does not make still greater use of computer-aided technology can to some extent be conceived of as the reverse of that as to the motives for its introduction. It also provides important information on the constraints limiting adoption. As table 3.2 shows, the majority of the enterprises interviewed (54.1%) indicated that they did not make greater use of computer-aided technology because, given their existing product range, it would not have been profitable to do so. 33.7% of the enterprises stated "insufficient liquidity" as a constraint, while 31.7% indicated that they were unsure whether new plant would be sufficiently utilized (constraint: insufficient demand).

Table 3.2: Answers to the Question "Why didn't you introduce more computer-aided technology?" (in %)

would not be profitable within the given product range	54.1
insufficient liquidity	33.7
unsure whether utilization of new plant would be sufficient	31.1
lack of suitably qualified personnel	30.0
insufficient management capacity	29.3
available equipment not yet sufficiently developed	26.7
existing machine-stock not yet sufficiently amortized	18.1
employees' resistance	5.6

more than one answer possible

In 30% of the enterprises interviewed the more intensive use of computer-aided technology is hindered by the lack of suitably qualified employees and in almost the same number (29.3%) insufficient "management capacity" - which was reported to be "fully occupied with other projects". 26.7% of those interviewed consider the equipment currently available on the market to be not yet fully developed, while 18.1% see an important constraint in the fact that the existing plant had not depreciated sufficiently; in many cases this can be taken to be the reason for the lack of liquidity. Only a very small percentage (5.6%) reported "employee resistance" as limiting the more intensive use of computer-aided technology, with only two enterprises (0.7%) listing this as the major constraint. In conclusion, the large number of enterprises which indicate that computer-aided technology could not be profitably introduced within the framework of the current product range and that the equipment currently available on the market was not yet not sufficiently developed, suggests that a further differentiation of the supply of the relevant investment goods could effect a not inconsiderable increase in the use of computer-aided technology in manufacturing industry.

3.3 The Diffusion of Computer-Aided Technology within the Enterprise

In most enterprises computer-aided technologies are usually first used in "administration/production planning". Our survey shows that this is not due simply to the fact that suitable office equipment has been available since the 50s and 60s and that the process of diffusion into other applications began much later. An examination of the enterprises which first introduced computer-aided technology between the start of 1985 and the postal survey (Summer/Autumn 1986) reveals that even now 91.1% of enterprises begin the application of computer-aided technology in "administration/production planning".

While the most intensive use of computer-aided technology continues to be in "administration/production planning", information provided by the enterprises on their plans for the future indicate that the other departments will begin to catch up soon. Irrespective of branch and enterprise size the increase in the intensity of use is greatest in those departments in which intensity of use is currently low. This applies in particular to "assembly", "quality control", "storage" and "transport". A qualitative change is foreseen in IT networking within departments, between departments and between enterprises and other market actors. At present the level of IT networking in the overwhelming majority of enterprises is still very low although in a number of enterprises considerable progress has been made in this field. The extent of IT networking between the administration and the production planning departments is particularly surprising, a development which appears to have been overlooked in the general euphoria over the realization of Computer Integrated Manufacturing (CIM) concepts. According to our survey Production and Planning Systems (PPS) are networked more frequently with the administrative than the direct production departments. The same applies to the networking of Computer Aided Planning (CAP) systems.

An unsurprising result of our survey was that the overall density of computer-aided technology (measured in terms of the absolute value of the plant, the number of terminals or of employees working with computer-aided technology) correlates closely to enterprise size. However the correlation between enterprise size and the relative intensity of use in the different departments (measured in terms of the proportion of employees or the investment budget linked to computer-aided technology) is very diffuse. The intensity of use of computer-aided technology in "administration/production planning" is positively correlated to enterprise size (measured by the investment share of that department in total investment in computer-aided technology), while in "design" and "assembly" computer-aided technology is used less intensively in larger enterprises.

A provisional explanation for these results may be provided by the very clear correlation between the time of 'first use' of computer-aided technology in a department and current intensity of use: the earlier an enterprise introduces computer-aided technology in a particular department, the higher the current relative intensity of use in that department. This result, which applies to all departments and for all sizes of enterprise, is a clear reflection of the potential for rationalization and increasing profitability arising out of the extensive use of computer-aided technology in a department. As the first adopters of computer-aided technology tended to be the larger enterprises and its first application tended to be in "administration/ production planning", it is understandable that large enterprises still tend to concentrate their investment in computer-aided technology in this department. Clearly the incentives (or indeed the compulsion) to create continuous data flows within a department are very great once parts of it have been computerized.

A second and equally important explanation for the intensive use of computer-aided technology in "administration/production planning" lies in a behavior pattern which can also be identified for all departments and enterprise sizes and which is, moreover, also of importance for the diffusion of computer-aided technology between enterprises: the tendency of enterprises to give priority in the introduction of computer-aided technology to those departments which are considered to be a constraint on enterprise development. This is shown by the high degree of correspondence between the motives for the introduction of the new technology and its department-specific intensity of use. For example, the intensity of use of computer-aided technology in "administration/production planning" is relatively low in enterprises which indicated that their primary motives were "reducing production costs", "capacity constraints", "improving product quality", "competitors introducing new technology" and/or "customers demanding new process technology". Such motives indicate that investment in computer-aided technology is likely to be concentrated in the technical production departments. The intensity of use of computer-aided technology is, on the other hand, significantly greater where "extent of product range" or the quality of "counseling and service" or the "nature and intensity of sales promotion" are particularly important aspects of the competitive position of the enterprise. Of the enterprises interviewed, those which identified the above aspects as being of great significance were mostly large plants which were clearly attempting to exert greater control over their complex relationships with other market actors through the use of computer-aided technology.

In contrast to the diffusion of computer-aided technology in the "administration/ production planning" department, the relative intensity of use of computer-aided technology in "production" and "assembly" is most closely linked to turnover growth. To this extent it is not surprising that it is small and/or young enterprises which tend to exhibit a higher than average intensity of use in these departments; enterprises in this category are typically characterized by rapid growth. Enterprises which apply CAD and/or CNC relatively intensively are typified by a string of characteristics which fit the cliché of the sort of firm which has adapted successfully to the trends of current structural change: a large proportion of their turnover consists of products in the "introductory" or "growth" phases of the product 'life-cycle'; production is one-off or small-batch production; their work force has above-average skills and qualifications; and finally they emphasize the importance of "flexibility in meeting specific customer demands" for their competitive position.

There can be little doubt here as to the identity of cause and effect: the strategic position of such enterprises entails turnover growth and enables them to introduce technology rapidly and on a broad front. Indeed, it might be said that it forces them to do so, as new technology is often the only way to overcome the constraints caused by rapid growth. New technology then enables the enterprise to improve its competitive position creating favorable conditions for further expansion.

4. The Effects of Computer-Aided Technology

This section is concerned with the overall effects of the application of computer-aided technology in the enterprises interviewed, with the exception of the direct employment effects which are the subject of section 5. First, the general effects of computer-aided technology at enterprise level are discussed, followed by the specific effects of the use of certain types of computer-aided technology (CAD systems and CNC machines) at the level of the individual department.

4.1 The Overall Effects of Computer-Aided Technology at Enterprise Level

The question: "What effects have occurred or are expected to occur following the introduction of computer-aided technology?" was included in our survey in order to ascertain the relative importance of the various conceivable effects of the application of computer-aided technology at enterprise level. We were particularly keen to discover when the various impacts took effect, and which were expected to occur in the future.

As can be seen from table 4.1, the effects most commonly reported were such as to be relatively difficult to quantify - "reduction of routine tasks in favor of creative activities" (82.5%), "improved product quality" (81.0%), "improved basis for calculation" (79.2%) and "improved deadline compliance" (76.8%). The percentage of enterprises which reported that "personnel reduction" was a consequence of the introduction of computer-aided technology was, at 62.7%, substantially lower. By far the least common effect reported was that of a change in forms of wage payment: only 19.9% had made changes in this area.

The time lags before the different developments take effect (which can also be seen from table 4.1) illustrate the extent to which certain effects are conditional on various educational and adaptation processes having taken place in the enterprise:

- 61% of those plants which indicated that "improved product quality" was (or was expected to be) an effect reported that such effects made themselves felt in the installation phase, i.e. right at the very beginning of the technological process. Obviously the introduction of computer-aided technology leads to an increase in product quality without the need for far-reaching (and so time-consuming) adjustment measures. A good example of this is the use of CNC machines which, due to their great accuracy in parts' production, usually bring about an immediate improvement in quality.
- Some effects, on the other hand, require the integration of new technology into the production process. For example, for 77.1% of those who claimed "improved deadline compliance" as a consequence of the use of computer-aided technology, this effect occurred only after a certain period of time or was still expected. The same applies to the effects; "increased division of labor within the plant", "higher degree of formalization of production procedures", "improved basis for calculation" and "reduction of routine tasks in favor of creative activities". In the case of "personnel reduction", 68.9% reported that this took effect over a period of time or that job losses were expected in the future.

Our survey indicates that the use of computer-aided technology tends to lead to a process of organizational differentiation at enterprise level (see table 4.2):

- 2.2% of the enterprises stated that the number of departments had been reduced following the introduction of computer-aided technology while 29.5% reported an increase.
- In as far as the distribution of decision-making responsibilities was affected by the use of computer-aided technology, it led on balance to a decentralization: 7.1% of those interviewed reported a centralization whereas 16.1% stated that decision-making powers had been decentralized.

Almost half of the enterprises reported changes in the organization of the production process. Major changes were made to the distribution of tasks between different workers and departments.

Table 4.1: Answers to the Question "Which effects of the introduction of computer aided technologies have been observed or are expected in this plant?"*

	occurred from the beginning	occurred with a time lag	expected in the future	not observed or expected
personnel reduction	19.5	29.9	13.3	37.3 (50.6)
greater control and monitoring possibilities	26.2	31.0	10.0	32.8 (42.8)
reduction of routine tasks in favor of creative activities	30.9	40.5	11.2	17.4 (28.6)
change in forms of wage payment (hourly, bonus payment)	3.7	5.6	8.9	81.8 (90.7)
services previously purchased from external sources now performed within the plant	15.6	12.6	3.7	68.1 (71.8)
improved basis for calculation	26.8	36.1	16.4	20.7 (37.1)
reduction of inventory	7.8	27.2	20.5	44.5 (65.0)
improved deadline compliance	17.6	40.8	18.4	23.2 (41.6)
improved product quality	49.4	27.1	4.5	19.0 (23.5)
diversification of product range	18.6	18.2	8.2	55.0 (63.2)
increased division of labour within the plant	8.5	20.7	9.3	61.5 (70.8)
higher degree of formalization of production procedure	18.8	29.5	14.3	37.4 (51.7)

* Percentage of the response of all plants in the study (sum of a line = 100%). In brackets: percentage of responses "expected in the future" and "not observed or expected".

There is a clear correlation between changes in structure and organization and enterprise size. In enterprises with less than 20 (50) employees only 24.1% (37.8%) made changes the distribution of tasks, while in those with more than 500 (1000) employees the figure was 71.4% (84.1%). Similarly, changes in organizational structure - (de)centralization of decision-making powers; increase/decrease in the number of departments - were made significantly more often by larger enterprises.

Table 4.2: Answers to the Question "Did the use of computer-aided technology lead to changes in plant-organization?" (%)

	no	already occurred	expected in the future
number of departments has increased	67.9	29.5	2.6
number of departments has decreased	97.4	2.2	0.4
centralization of decision-making responsibilities	91.0	7.1	1.9
decentralization of decision-making responsibilities	82.8	16.1	1.1
other changes in plant-organization	96.3	3.7	0.0

Disadvantages of the use of computer-aided technology were seen primarily in "increased dependence on technology" (21.0%) and in an "increased disruption of the overall production process in the case of equipment failure" (22.5%) (cf. table 4.3). 16% mention an "increased dependence on certain employees", while only 10% thought that the use of computer-aided technology had led to "losses due to organizational friction" and a mere 7.9% reported "longer stoppages" as a negative consequence. It is interesting to note that the "increased dependence on technology" was mentioned mainly by small enterprises whereas the "increased dependence on certain employees" would appear to be a problem for larger firms. This may be a reflection of the greater extent of the division of labor and the consequently greater number of 'specialists' in large enterprises. The other potential disadvantages listed in table 4.3, however, exhibited no such correlation with enterprise size.

The search for the determinants of the various effects of the use of computer-aided technology produced further evidence that, in the main, technology is introduced to overcome various constraints. For example, a series of statistically significant relationships were identified between the motives behind the introduction of the new technology and its effects (and/or between the aspects of the firm's competitive position which it considered paramount and the effects). The effects reported by the enterprises vary according to the constraint which initiated introduction and are consistent both with the motives for introduction and with the improvement in those aspects considered important for competitive position. Some examples should help to illustrate this:

- The effect "personnel reduction" is significantly and positively correlated with the importance of the motive "reduction of production costs" and the competitive aspect "price".
- There is a positive relationship between the effects "improved product quality" and "diversification of product range" and the motive "improvement of product quality", and also between the effect "improved product quality" and the competitive aspect "product quality".
- Finally, there is a positive relationship between the effect "improved deadline compliance" and the importance of the two motives "reducing production times" and "improvement of production organization". The same applies to the relation between this effect and the aspect of competition "deadline compliance".

Table 4.3: Answers to the Question: "What do you see as the major disadvantages resulting from the introduction of computer-aided technology?"

	percentage of responses in %
increased disruption of the overall production process in the case of equipment failure	22.5
increased dependence on technology	21.0
increased dependence on certain employees	16.6
losses due to organizational friction	9.7
longer down-times	7.9

more than one answer possible

4.2 The Effects of the Use of Computer-Aided Technology in Individual Departments

Detailed analyses of the entire spectrum of effects of computer-aided technology were conducted for the departments "design" and "parts' production". Our interest centered on the effects on productivity (in terms of total time expended on a given activity) but we were also concerned with a differentiated bundle of effects on productive activity in the 'innovative' department and in those departments indirectly affected by the innovation.

Table 4.4: Answers to the Question "How did the time required for certain design-activities change due to the use of CAD compared to conventional design methods?"
(change of time in %)

	Percentiles												
	5	10	20	25	30	40	50	60	70	75	80	90	95
overall change	-69	-60	-50	-35	-30	-20	0	0	+20	+25	+43	+60	+76
completion of the first drawing	-71	-50	-25	-18	-10	0	0	0	0	+10	+20	+50	+56
alterations of existing drawings	-90	-86	-64	-50	-50	-20	0	0	+18	+45	+50	+76	+90
calculations	-75	-50	-11	0	0	0	0	0	0	0	+14	+47	+89
provision of lists of components/work-schedules	-80	-50	-11	0	0	0	0	0	0	0	+14	+50	+76

4.2.1 Effects of the Use of CAD Technology

The spectrum of effects of CAD technology, both the changes in productivity in the construction department and the other effects associated with its use in the enterprise is so broad that it is well-nigh impossible to generalize about 'the' effects of the use of CAD technology. This is reflected, as table 4.4 shows, in the data on changes in the total time required for design resulting from the use of computer-aided technology. The figures for the overall change in time spent on design indicate that approximately 40% of users achieved time savings, and some 20% little or no change while the remaining 40% experienced an increase compared with the initial position. The broad spectrum of effects is evidently a reflection of the fact that, while CAD systems offer the possibility of rationalizing design, they also open new opportunities in this department which had previously been closed to the enterprise. For example, where the introduction of CAD induces alterations to the product range, its use can lead to an overall increase in design activity, which must be regarded as technologically induced.

The results obtained on the other effects of the use of CAD technology (cf. table 4.5) emphasize further the diversity of the spectrum of effects. Surprisingly, the effect most frequently mentioned is not felt in the department where CAD is directly employed, namely design, but rather in other departments: 48.1% of the users of CAD technology give "improved basis for work in other departments" as an effect, of which for 75% it was the most or second most important effect of all.

Table 4.5: Effects of the Use of Computer-Aided Design (answers in %)

improved basis for work in other departments	48.1
more complicated construction tasks taken on	29.0
improved deadline compliance	27.6
improvement in parts' quality	23.6
trend to standard solutions	23.6
fewer model-constructions	21.5
reduced number of part-types	18.9
widening of product-range	17.4

more than one answer possible

Table 4.6: Answers to the Question "How did the time needed for production change due to use of CNC-equipment" (change of time in %)

	Percentiles												
	5	10	20	25	30	40	50	60	70	75	80	90	95
overall production time	-70	-60	-50	-50	-40	-30	-25	-5	0	0	+20	+50	+70
transport and storage time	-70	-50	-20	-10	0	0	0	0	0	0	0	0	0
checking time	-50	-50	-17	-10	-5	0	0	0	0	0	0	0	0
setting up time	-62	-50	-38	-30	-20	-10	0	0	0	0	+4	+27	+50
processing time	-80	-75	-60	-54	-50	-45	-32	-25	-30	-2	0	0	+14

Almost one third (29%) of the enterprises employing CAD technology stated that they had "taken on more complicated design tasks". The importance attributed to this effect indicates that CAD technology is only partly a tool for rationalization - qualitative changes are at least as important.

4.2.2 The Effects of the Use of CNC Technology

The spectrum of effects of the use of CNC technology for "parts' production" proved to be as broad as that for CAD in construction. As can be seen from table 4.6, the introduction of CNC machinery by no means automatically results in a shortening of overall production time, or in its individual components (transport, storage, checking, setting-up, processing). Some 40% of users stated that total production time had not changed or had even increased. The proportion of CNC users reporting a relative reduction in the time required for the various components of total production time was even lower: only 25% of users stated that transport and storage times had been shortened, and in only 30% of cases was it possible to reduce checking times.

By contrast, the effects of the introduction of CNC technology on product quality are clear and positive: 89.2% of enterprises stated that product quality had increased as a result of the use of CNC machinery; 44.5% considered this improvement to be "considerable", while just over 10% classified it as "minor".

A trend towards smaller production batches, which is frequently assumed to result from the use of CNC machinery, cannot be confirmed by our study. In about 50% of cases, batch size had not changed, more than 21% had increased batch size while 27.9% of the users interviewed reported a reduction. A similarly diffuse picture emerges from the response to the question as to the batch size which - in contrast to conventional plant - is now produced primarily with CNC machines: slightly less than half of the enterprises produce larger batches, while a marginally smaller proportion tended to produce smaller batches on the new machines.

The answers to the question regarding the extent to which the range of parts produced had changed following the introduction of CNC machines show that the potential for flexibility offered by CNC machinery does not necessarily lead to product diversification. Only 52.2% of users had increased their product range, while for 44.5% no (noticeable) change had occurred and 3.3% stated that the range of parts produced had actually narrowed.

Neither did we find any evidence of the reduction in storage and warehousing so often claimed as a consequence of CNC technology. For 70.1% of CNC users stocks had remained unchanged on balance, while the proportion of enterprises which reported a reduction of stocks was, at 16.1%, to say the least modest, and, above all, only slightly larger than the proportion reporting an increase (12.8).

Table 4.7 provides a summary of the responses to the question on the further effects of the introduction of CNC. Surprisingly, the most frequently mentioned impacts take effect only in the later stages of the production process: 54.3% give "reduced need for further work on parts" as the most important effect. The exactitude and homogeneity of production ("reduced quality fluctuations") is accorded only second place, at 43.6%. The "improved deadline compliance" was also quoted relatively frequently (36.9%). Somewhat less than a third stated that they were now producing parts which had previously been manufactured outside the plant (29.9%). This can be considered as evidence that the (negative) consequences of rationalization for employment are transferred across enterprise 'boundaries'.

Table 4.7: Effects of CNC Use (answers in %)

reduced need for further work on parts	54.3
reduced quality fluctuations	43.6
improved deadline compliance	36.9
production of parts previously manufactured by other firms	29.9
reduced need for specialized machinery	25.7
extension of product-range	25.2
wider range of design possibilities	23.0
reduction of material input	21.0
design of parts more CNC-compatible	15.7
other	5.6

more than one answer possible

4.3 Determinants of the Effects of Computer-Aided Technology

4.3.1 Determinants of the Effects of CAD Technology

An investigation of the determinants of the concrete effects of the use of CAD technology (in particular productivity changes, in the sense of relative changes in the time required for construction) in each individual case confirms our claim that both the specific way in which the technology is used and the enterprise environment into which the technology is introduced play a decisive role. The use of CAD technology leads to time savings in those enterprises in which design is characterized by efforts at rationalization and/or standardization and where the number of different parts produced is reduced ('off the shelf' production). The time reduction is much lower where major "changes in product quality" were achieved, "more complicated construction tasks" were taken on and/or the product range was increased.

The concrete use made of CAD technology - and so indirectly changes in productivity - are influenced to a great extent by conditions specific to the individual enterprise. In order to illustrate this, the situative context for the application of CAD will now be described. In the following, the role played by various market conditions is presented together with the different competitive strategies pursued by CAD users (for a treatment of other factors cf. Ewers/Becker/Fritsch 1988).

Relatively large CAD users emphasize "reduction of number of parts", "increased punctuality of delivery", "increased product range" and "change in product quality" as being the most important effects of the introduction of the new technology. The constraints faced by all these enterprises lie primarily in increased competition combined with stagnant or negative employment and/or sales trends. Depending on certain other conditions there appear to be several ways in which the use of CAD technology can help overcome these constraints, each of which will have specific effects on productivity. In the case of some of the larger users the market exerts no pressure to diversify production; neither an extended range of products nor high product quality are required, with the result that the rationalization of the construction process (reduction of number of parts) is the major aim of the new technology. It is the qualitative aspects of the use of CAD technology, i.e. improving product quality and/or increasing product range, together with a better relationship with customers (this is indicated by the great significance attached to "customer service and counselling") which play the most important role.

The decision whether to increase the quality of existing products or rather to concentrate on an extension of the product range appears to be influenced by enterprise-specific conditions. In particular, the older enterprises among the larger users of CAD technology tend to be subject to tough price competition. For some of these users the extension of their product range is apparently an attempt to circumvent this price competition. The specific way in which CAD technology is used in such enterprises explains why this group of adopters have not realized time savings compared with more traditional methods of production.

For CAD adopters whose output is based largely on assembly-line production - in our sample this was concomitant with a considerable proportion of turnover in stagnant markets - such tough price competition is not typical. In such cases, enterprises see the best chance for success in qualitative improvements in the existing product range (increasing product quality; taking on more complicated construction tasks). Given that this applies mainly to large enterprises which, furthermore, tend to change their product range less frequently than smaller firms, it is a plausible supposition that on balance time savings will be made due to the benefits of economies of scale and/or economies of scope.

As our correlation analyses show, CAD users which concentrate production on medium-sized series are not faced with the problem of selling products on stagnating markets. They are therefore not forced to make such far-reaching changes to their existing product range; instead they tend to be making great efforts to improve sales promotion. In view of such a market situation it seems plausible that changes in construction will be oriented largely towards rationalization: such enterprises tend to realize time savings in all aspects of design.

In contrast to these large enterprises, small firms, especially those concentrating on one-off or small-batch production, face entirely different market conditions. Given that "flexibility in meeting special customer demands" is a quality required of enterprises of this type and that therefore new and changing design tasks must be continually performed, the increase in time expended on the various aspects of the construction process is not particularly surprising. Finally, many CAD users now perform design tasks formerly carried out externally, a trend which also provides an explanation for the increase in time expenditure associated with computer-aided design.

4.3.2 Determinants of the Effects of CNC Technology

The importance of the role played by the context in which new technology is used in determining the effects of that technology is also well illustrated by the determinants of the effects of the use of CNC machines. This can be shown in terms both of the technologically induced change in productivity (changes in total production times and those of its component parts - transport, storage, checking, setting up and actual production time) and of changes in batch size, product range and stock levels.

An examination of the differences between the various users of CNC machinery with regard to changes in overall production time reveals that there is no direct correlation with enterprise size, nor with current size of production runs (one-off, serial or mass production), nor with the absolute number of CNC machines used. The correlations which do exist are to be found in the specific conditions under which the new technology is introduced and applied, as illustrated by the following examples:

- The higher the degree of automation of production the greater the reduction in overall production times, whereby the reductions were achieved primarily in the transport and storage departments.
- The individual adjustment of new plant to the existing conditions of the production process and the overall standard of the capital equipment are two of the factors which result in considerable reductions in overall production times.
- The change in production times is relatively small (indeed in some case an increase was registered) if the organizational form chosen is such that machines are programmed at workshop level.

These correlations with the specific conditions under which the technology is operated can also be applied to the individual components of total production time. The effect of such specific conditions is particularly striking in the case of transport, storage and also setting up time. CNC users whose production is concentrated in one-off and small-batch production (usually relatively small enterprises) reported only minor time savings or even increases in transport and storage times. CNC users whose production consists to a relatively large extent of medium-sized to large series (mostly larger enterprises) achieved considerable time savings in these areas. Enterprises producing one-off products tend to experience an increase in setting-up time due to the use of CNC machines.

If we examine the specific conditions faced by CNC users in the group of enterprises concentrating on one-off production, an explanation can be offered for the "deviations" from the effects usually to be found in the literature. The most important characteristic of the competitive position of such enterprises is the requirement for a relatively high degree of "flexibility in meeting special customer demands". The frequent product changes which this implies lead to correspondingly frequent alterations in setting-up. Such users typically program their machines in the workshop, even in the case of relatively complicated products. The time expended on programming the machines will have been included in setting-up time in most cases so that an increase in this time component as a consequence of the use of CNC machines would appear to be a plausible result. As enterprises where a large proportion of total production is of one-off products also tend to have increased average batch size following the introduction of CNC machines, a further consequence is the increase of storage times at the point of production.

Similarly, variations in 'context' lie behind the differences in the technologically induced changes in the average size of production runs.

- A tendency towards a reduction of the size of production runs was identified, especially in the case of larger enterprises which conventionally produce relatively large series. With larger production runs the exploitation of the potential flexibility gains even from just one aspect, such as the reduction in tied-up capital through the reduction in stocks, proves lucrative. At the same time, the fact that no definitive correlation could be established between the importance of mass-production in an enterprise and changes in the size of production runs indicates that such opportunities are not always fully exploited, even by the larger enterprises.

- In view of what has already been said about the changes in overall production times in the case of smaller users of CAD technology, the fact that this group with its comparatively large share of one-off production tended to increase average production runs seems completely compatible with profitability considerations. Moreover, it was not to be expected that small production runs would automatically decrease in size as a result of the introduction of new technology.

In most of the small enterprises applying CNC machines with their comparatively large share of one-off production the introduction of the new technology has not led to changes in product range. For the (usually larger) enterprises with a large proportion of mass and/or serial production there was on average an extension of product ranges.

There is, finally, a clear correlation between changes in stock levels and changes in product range and/or with changes in batch size. Where batch sizes fell as a result of the use of CNC machines, stock levels also fell, while an increase in product range tended to be accompanied by increases in stocks. If this is linked with the reactions of the different users described above it can be seen that on balance those users with a relatively large share of one-off production increased their stock levels, whereas larger firms and/or those for which large series production is relatively important tended to experience a fall in stock levels. For mass production-oriented enterprises no clear trend could be identified, due primarily to the fact that there is no uniform tendency in such enterprises regarding changes in average production runs as a consequence of the introduction of CNC technology.

5. Direct Quantitative Employment Effects of Computer-Aided Technology

This section dealing with the direct quantitative employment effects of the use of computer-aided technology is so structured as to allow a distinction to be made between the structure and the level of the effects, By the 'structure' of employment effects we understand the mixture of different employment changes brought about by the use of new technology. These effects may run contrary to one another; they include hirings, the redeployment of labor to work with computer-aided plant, displacement of labor into other departments within the enterprise and the exit of labor from the enterprise ("exits"). The 'level' of the employment effects describes the absolute (or sometimes relative, i.e. measured against total employment in an enterprise or department) extent of the various movements of labor induced by the new technology.

5.1 The Structure of the Employment Effects of the Use of Computer-Aided Technology

Slightly less than half (48.7%) of the users of computer-aided technology we interviewed reported having hired new workers in connection with the introduction of new technology (cf. table 5.1). At the same time a mere 3.9% said that the requirement for additional labor to operate the new plant was covered exclusively by hirings. Clearly the redeployment of labor within the enterprise plays an important role: 30% of users interviewed covered their requirements exclusively in this manner.

Table 5.1: Structure of the Employment Effects of the Use of Computer-Aided Technology at Plant Level (Percentage of Plants)

only hiring of new personnel	3.9
only redeployment of labor to work at computer-aided equipment	30.0
hiring of new personnel <u>and</u> labor redeployment	26.5
redeployment <u>and</u> exits	21.4
hiring of new personnel <u>and</u> exits	1.2
hiring of new personnel, redeployment <u>and</u> exits	17.0

An examination of labor flows at department rather than enterprise level reveals that the labor required to operate computer-aided plant in "construction" was covered to a large extent by hirings: 35% took on some new workers in this department and 8.4% covered their requirements exclusively from the external labor market.

At plant level 39.7% of enterprises indicated that workers had been displaced from at least one department, i.e. labor had been transferred within the enterprise and/or workers had left the enterprise. This figure varies greatly if we examine the different departments separately. Only 10% of CAD users reported labor displacement from "design" as a result of the introduction of new techniques whereas every single user in our survey reported labor force reductions in the "stockkeeping" department, The figures for the other departments were as follows: "administration/ production planning" 24.1%; "parts' production" 33.6%; "assembly" 46.7%.

Table 5.2: Structure of Exits due to the Introduction of Computer-Aided Equipment (percentage of plants, exits in percent)

type of exit	exclusively	with other types
redundancies	23.2	51.8
voluntary resignation	19.6	51.8
early retirement	14.3	39.4
other	7.1	12.5

In enterprises where computer-aided technology displaced labor, transfers within the enterprise play at least as important a role as redundancies. At plant level 45.1% of users compensated for labor released exclusively by intra-enterprise transfer; in 74.5% of cases this was a partial solution. In only 25.5% of cases did all labor released by the introduction of computer-aided technology leave the enterprise. As can be seen from table 5.2, where new technology results in labor leaving an enterprise, voluntary resignations are at least as important as redundancies. Early retirement also account for a large proportion of exits.

5.2 The Level of the Direct Employment Effects of Computer-Aided Technology

Table 5.3 provides a summary of the level of direct, technologically induced employment effects at plant level in both absolute and relative terms. By far the largest block is represented by the transfer of labor to operate computer-aided technology followed by 'exits' and 'deplacements'. The level of exits is consistently higher than that of hirings.

Nevertheless, the net movement of labor, i.e. the balance of hirings and exits was zero in 45.6% of the enterprise in our survey (cf. table 5.4). At department level the percentage of such 'labor-market neutral' cases is even higher, with values bunching between 60% and 70%. The percentage of enterprises reporting net hirings was (except for the department "assembly") twice as high as that for which net exits were registered.

Table 5.3: Gross-Labor Movement due to the Introduction of Computer-Aided Technology at Plant Level

number of workers

	Percentiles										
	10	20	25	30	40	50	60	70	75	80	90
hiring of new workers	1	1	1	1	2	2	3	4	5	5	14
redeployments	1	1	2	2	2	3	5	8	10	17	30
exits	1	1	1	2	3	4	6	8	10	20	42
transfer of labor to work at computer-aided equipment	2	3	4	5	7	10	15	23	30	40	83

percentage of all workers in the plant

	Percentiles										
	10	20	25	30	40	50	60	70	75	80	90
hiring of new workers	0.4	0.8	0.9	1.1	2.0	3.3	4.4	5.8	6.5	8.6	16.7
redeployments	0.5	1.0	1.4	1.7	2.2	2.7	3.5	4.5	6.2	6.9	12.6
exits	0.3	0.9	1.2	1.3	2.3	4.0	6.2	7.9	10.0	11.0	17.6
transfer of labor to work at computer-aided equipment	2.6	5.4	6.5	7.5	10.2	13.4	17.6	22.5	25.0	27.6	39.6

Table 5.4: Structure of Technology Linked Labor Movements due to the Introduction of Computer-Aided Technologies in Different Departments of German Manufacturing Plants (percentages of all cases).

	net exit from the department (plant)	boundary passing net movement equals zero	net entrance into the (plant)
administration and production planning	12.1	63.2	24.7
design	2.1	68.8	29.1
production	9.1	67.7	23.2
assembly	18.9	62.2	18.9
quality control	6.9	77.2	15.9
plant total	16.4	45.6	38.0

If the sum of net hirings and exits are compared, considerable differences emerge between the different departments. In "administration/production planning" and "assembly" net exits exceed hirings by a factor of 2.5 and 12 respectively. Although "parts' production" also shows a clear excess of exits over hirings, this result has been heavily influenced by one enterprise with a particularly high level of exits. If this case is removed from the calculation net exits and hirings balance almost exactly. In the departments "design" and "quality control" on the other hand, net hirings exceed exits by 14.5 and 2.5 times respectively. At enterprise level net exits are 2.6 times as numerous as net hirings. However this result is similarly distorted by two large plants with exceptionally high exits figures. Without these plants both net movements are approximately equal.

Clearly the use of computer-aided technology has labor displacing effects primarily in departments typically characterized by simple manual work; in departments in which creative (brain)work is required (e.g. "design") the net employment effects tend to be positive.

5.3 Determinants of the Direct Quantitative Employment Effects

The results of our analysis of the determinants of the relative extent of technologically induced displacement of labor (i.e. intra-enterprise redeployment and/or exits as a proportion of total employment in an enterprise) are relatively unambiguous. Ignoring for a moment the (relatively minor) differences between the various types of computer-aided technology, the relative extent of displacements increase with both enterprise size and the (absolute and relative) intensity of use of the technology. It is the market context which decides whether labor released from departments where new technology has been introduced is merely redeployed or actually results in exits from the enterprise: it appears that enterprises a large proportion of whose products are sold on stagnant or shrinking markets (with corresponding sales and employment trends) are much less in a position to compensate for employment losses resulting from the use of computer-aided technology within the enterprise than firms operating in the context of dynamic markets. Moreover, enterprises with a high proportion of exits tend to have a more 'anonymous' relationship with the outside world (in terms of the secondary role played by such aspects as "counselling/ service" and "intensity of sales promotion". It is to a large extent the older and larger adopters of the new technology which tend to experience a high proportion of exits.

The relative extent of hirings correlates positively with the following factors: the relative intensity of use of the technology, the trend rate of expansion sales and employment in the enterprise and the importance attached to "positive sales expectations" as a motive for the introduction of the technology. Above average rates of hirings also occur in relatively young enterprises which, as we have seen, tend to be expanding rapidly.

In addition to the expansion of the enterprise as a whole following the introduction of computer-aided technology, other factors at enterprise level may also play a role in increasing hiring rates. For example, (larger) CAD and CNC users which diversified their product range and/or improved product quality frequently took on new workers.

6. Retraining as a Consequence of the Use of Computer-Aided Technology and its Problems

A necessary condition for the use of new technology by an enterprise is that at least some of its employees are in a position to operate it. If the existing work force lacks the requisite know-how this can be rectified a) by hiring workers from the external labor market with the required skill profiles (in some cases this will also imply simultaneous redundancy for workers with obsolete qualifications) and/or b) by retraining the existing work force.

Following the introduction of computer-aided technology 74.8% of the enterprises in our survey carried out some form of retraining over and above 'training on the job'. 10.8% of the total workforce in the enterprises in our survey, underwent retraining. Measured against the number of those employees working directly with the computer-aided plant, the retraining quota was 41.6%.

As can be seen from table 6.1 this quota varies greatly between the different departments. While in "design" 75.1% of those directly employed with the new technology had taken part in some form of retraining, the figures for "assembly" and "administration" are 56.9% and a mere 31.5% respectively. In the case of "parts' production" the figure of 44.7% is probably as low as it is because not only machine operatives and programmers are included but also those who maintain and set up the machines. The 15.3% of enterprises which reported that they had neither carried out special retraining measures nor hired new workers to operate the machines are mainly relatively small plants and/or those with a relatively low intensity of use.

The employees selected for further training and the duration of training follow a clearly recognizable pattern: it is primarily employees who already have above-average qualifications who benefit from further training, and their training is much more intensive (measured by the amount of time spent acquiring the new qualification) than that of less skilled employees. As table 6.1 shows, the proportion of those undergoing retraining who were unskilled or semi-skilled was only 19.6%, whereas such workers make up some 38% of the total work force. On the other hand, skilled workers (workers with a college/polytechnic certificate) made up 68.1% (10.3%) of those retrained compared with a share of the workforce of under 55% (5.1%).

Table 6.1: Proportion of Retrained Employees According to Initial Qualifications and Duration of Retraining to the Total of all Retrained Employees (by initial qualification)

	total	of which:			
a b		university degree	poly-technic	skilled workers	unskilled workers
total	10.8 -	11.8 2.1	11.8 10.3	11.3 68.1	6.8 19.6
administration	7.7 -	9.5 1.3	9.0 10.9	7.6 74.3	6.5 13.5
design	14.2 -	12.7 10.4	16.2 39.0	15.7 49.9	16.5 0.7
production	11.6 -	6.7 0.2	11.7 1.9	14.9 76.4	6.4 21.5
assembly	14.7 -	- -	20.0 4.1	18.3 38.2	12.1 57.7
quality control	7.4 -	8.1 1.7	9.0 10.2	10.0 25.1	6.1 63.1
stockkeeping	6.9 -	- -	10.0 4.3	8.4 69.7	2.2 26.1

a: duration of retraining (days per retrained employee)
b: proportion of retrained employees to all retrained employees in that department by initial qualification

The time spent on retraining each employee can be taken as a proxy for the intensity of the training. Taking all departments and skill groups together the average time spent on retraining measures amounted to 10.8 person-days (PD). While there was no substantial difference between the average time spent retraining personnel with a university or polytechnic degree and other skilled workers - 11.8 PD and 11.3 PD -, the average figure for unskilled participants in retraining schemes was only 6.8 PD (cf. table 6.1). Thus, unskilled and semi-skilled workers are not only underrepresented in terms of participating in retraining measures; when they do participate, the intensity of retraining is comparatively low. Average time spent on retraining is lowest in the departments "administration/production planning", "quality control", and "stockkeeping", and highest in "design" and "assembly". With the exception of retraining in "design", employees with a university degree are also underrepresented. This is most probably due to the fact that such employees are frequently in managerial positions and as such do not use computer-aided equipment in such a way as to require explicit further training in order to enable them to perform their functions.

During our in-depth interviews other selection criteria were mentioned beyond the employee's present qualifications: these included the employee's age, motivation and his or her own desire to undergo retraining. Often management saw the opportunity for retraining as a 'mark of distinction' with which it reacted to an articulated interest on the part of individual workers in learning how to come to terms with new technology.

Table 6.2: Problems Experienced in Retraining Personnel (% of plants with retraining activities mentioning the problem)

insufficient initial qualification of workers	30.9
facilities not available locally and/or difficult to get to	19.6
lack of willingness on the part of workers	18.6
organizational problems in incorporating retraining measures into existing schedules	18.6
desired subjects/courses were not offered	9.3
offered retraining programs are too expensive	7.8

More than half of the enterprises which had carried out retraining measures in connection with the use of computer-aided technology pointed to problems experienced in retraining personnel. This applied mainly to large plants which made relatively intensive use of computer-aided technology, so that the existence of problems and constraints involved in retraining can be at least partly explained in terms of the greater absolute scale of the measures required. As can be seen from table 6.2, the most common single problem (mentioned by 30.9%) was the insufficient initial qualification of the workers. 19.6% see a barrier to the further training of their staff in the fact that facilities are not available locally and/or are difficult to get to. 18.6% complain of a lack of willingness on the part of workers while the same number see organizational problems in incorporating such measures into existing schedules. By comparison, only a relatively small number of enterprises which have carried out retraining programmes complained that the "desired subjects/courses were not offered". The cost of such measures does not seem to have called retraining programs into question to any great extent: only 7.8% mention this as a constraint and only 1.5% (4 enterprises) considered this the most important problem.

In more than half the enterprises (54.4%) in which retraining measures had been undertaken employees experienced wage and salary increases as a direct result of their improved qualifications. In fact this is almost certainly an understatement of the effects of higher qualifications on pay because, especially in larger enterprises, better qualifications often lead to more rapid promotion: the higher payment which normally accompanies promotion is then attributed to other factors than the improved qualifications. In approximately half of the cases in which improved qualifications resulted in wage and salary increases, the raise was by up to 10%; an increase of 20% and above occurred in only 10% of cases, with increases of more than 30% as a rare exception. In 40% of all cases the wage/salary increase was between 10% and 20%. If the attempt to retrain existing employees to operate computer-aided equipment does not appear to offer reasonable prospect of success, or if sufficient personnel are already available, most enterprises are very concerned to find alternative employment within the enterprise for workers whose previous work has become obsolete due to the introduction of computer-aided technology.

7. On the Effects of the Use of Computer-Aided Technology on Employment Trends at Enterprise Level

Although we are in a position to describe relatively exactly the direct employment effects of the use of computer-aided technology in the enterprises with which we conducted interviews, it is difficult to generalize our results and so estimate the effects of the new technology on the employment trend of enterprises in the Federal Republic of Germany as a whole. There are several reasons for this. Firstly, our sample is distorted by the so-called 'survivor bias', i.e. it contains only enterprises which have 'survived' in the market with the consequence that the employment trends in our sample are naturally more positive than would actually the case for the overall economy (on the effects of this bias cf. Fritsch/Hull 1987). A further distortion arises from the fact that we, consciously and for good reasons, restricted our interviews to the users of computer-aided technology. In as far as these enterprises achieve greater competitiveness through their use of computer-aided technology - and our analysis of the effects of the use of the new technology points clearly in this direction - this may lead to employment losses and closures among the 'losers' in the competitive process. It seems plausible to suppose that the 'losers' would be largely firms which had adopted the new technology only at a very late stage or not at all.

After completion of the installation phase (which often requires considerable additional labor) the direct effects of the use of computer-aided technology must be seen as 'labor saving'. With the same input of labor it is possible to produce a higher quality (and so more expensive) product, with the additional possibility of extending the product range. Thus, at a constant level of output a reduction in labor input is to be expected at the point of use of computer-aided technology.

In reality, however, a quantitative determination of the extent of the induced rationalization effects is complicated by the fact that the new plant usually means an increase in productive capacity. This is possible, for example, due to the shortening of the various components of total production time, and enterprises are naturally keen to utilize this additional capacity in order to benefit from the advantages of diminishing marginal costs. It is relatively often the case that, while the number of operatives remains constant, the output of computer-aided machinery is higher than that of conventional equipment. Since computer-aided technology is often acquired with the aim of expanding output (as was shown under point 2.2), it is possible for employment to increase if the new technology is run parallel with the old. One might ask, however, to what extent such increases can still be thought of as being 'technologically induced'.

If the introduction of computer-aided technology leads to labor displacement, enterprises are usually concerned to employ displaced workers elsewhere in the enterprise, particularly if relatively highly qualified personnel are affected. Enterprises therefore do not seek only to fully utilize existing plant; they usually also attempt to fully utilize the existing work force, i.e. to avoid redundancies. This in turn provides an incentive to increase sales and/or to intensify efforts to exploit market openings and/or to expand production vertically in order to maintain staffing levels.

Even where the number of those directly operating modernized equipment remains constant, the increase in productive capacity resulting from the introduction of computer-aided technology often leads to positive employment effects in other departments. A typical example of this, and one we encountered frequently in the enterprises with which we conducted interviews involves the introduction of CNC machines in "parts' production". The greater productive capacity of CNC compared with conventional machines frequently involves a higher level of activity in other departments (e.g. storage, distribution, administration and, in particular assembly), which in some cases results in the hiring of additional labor. In most cases these indirect effects of the use of computer-aided technology in other departments are difficult to quantify and it is all but impossible to determine unambiguously whether or not they are technologically induced. Our analyses of the effects of the use of computer-aided technology in "parts' production" and "construction" (cf. section 4) clearly show that such effects (improved inputs for other departments, reduced need for additional post-production work) are relatively common.

It would seem an almost impossible task to estimate the indirect effects of the use of computer-aided technology resulting from the increased competitiveness arising from improvements in product quality or changes in product range. In order to define unambiguously the effects of such increases in competitiveness as technologically induced it would, strictly speaking, be necessary to know how the range and quality of products would have developed in the absence of the new technology. This leads to questions regarding the extent to which technologically induced changes in product range and quality have a positive impact on turnover and so on the overall development of the enterprise. The significance of the use of computer-aided technology on product quality can be seen from the fact that "improving product quality" is accorded the same importance as the reduction of production costs in terms of being the motive for the introduction of computer-aided technology (cf. figure 3.1). Our study has also shown that changes in product quality and range following the use of new technology are very far reaching and, in many cases, take effect during the initial, installation phase. The multifarious indirect effects of the use of computer-aided technology which we can scarcely begin to quantify with any degree of certainty cannot, however, simply be ignored when attempting to estimate the overall effects of its use. It is our belief that they are often of equal if not of greater importance for enterprise development than the direct labor-saving effects.

It is apparent that the net effect of the use of computer-aided technology on employment at enterprise level is closely linked to sales trends and the environment in which the new technology is employed. In the overwhelming majority of cases the absolute level of employment change is likely to be anything but dramatic. For this reason it is all but impossible to quantify exactly all those employment effects which go beyond the obvious labor-saving, rationalization effects. The fundamental impacts of the new technology will make themselves felt in the long term and will depend on the various ways in which the technology is applied, whereby it is debatable to what extent it is legitimate to put these long-term and indirect effects (such as changes in product quality and range) down to technology per se.

Even if it seems questionable whether it is possible to ascribe individual cases of displacement and hiring effects to the use of computer-aided technology, a clear pattern does emerge if the effects are differentiated by the qualifications and skills of the different groups of workers affected: the "ratio of hirings to displacements (exits from the enterprise) resulting from the use of computer-aided technology" is -20.5:1 (-14.0:1) for unskilled and semi-skilled workers (i.e. the exits outweigh the hirings considerably), The equivalent values for employees with a university degree are 6.8:1 (4.2:1) and for those with a polytechnic certificate 2.3:1 (1.5:1) are, on the other hand, positive. It is worth noting that skilled workers also appear to be negatively affected by the use of computer-aided technology. For this group the ratios are -2.8:1 (-1.9:1), i.e. they are also negative. In other words, on balance the enterprises interviewed by us have reduced the number of skilled workers slightly as a result of the use of computer-aided technology.

The Diffusion of New Technologies and their Effects in the Private Service Sector

Lisa Höflich-Häberlein and Hubertus Häbler

1. Data Sources and Analytical Methods

Infratest Sozialforschung's contribution to the Meta Study was conducted in the form of several building blocks or elements. In addition to the authors the work group included Matthias Thomae, responsible for the analyses with "Comtec", and Ulrich Pfeiffer, who collaborated in the project's conception and design. The study is based on macro data covering some 4 500 enterprises from the "Comtec" survey for the years 1982 to 1985. Some 3 000 data sets were obtained as panel data. The Comtec data provide an overview of the quantitative diffusion of information and communication technology (ICT) in the different branches of both manufacturing industry and the service sector.

In addition case studies were conducted in a total of 38 enterprises in order to analyze the principle determinants of the diffusion of technology within the enterprise and the quantitative, qualitative and organizational effects on employment of such diffusion.

On the basis of these case studies and the Comtec analyses, seven branches were selected which appeared particularly interesting in the context of their use of information and communication technology (table 1.1.). The selection criteria for the branches were average enterprise size, previous investment activity and the use of information and communication technology in the branch. A total of 238 users and 74 non-users were questioned either by telephone or in person, usually involving more than one representative of each enterprise. Some of the enterprises responded in writing.

In order to consolidate the relationships identified between the diffusion of technology, process and product innovation and their quantitative and qualitative employment effects, approximately 30 interviews were conducted with experts from the trade unions, employers' organizations and with academics.

Table 1.1: Respondents in the Survey by Branch

branches	case studies	enterprise survey users	non-users	experts
financial services	9	56	12	5
banks	6	93	3	6
wholesale/freight forwarders	9	29	5	5
travel agencies	2	25	2	5
pharmacists	1	9	10	2
engineering/architectural offices	3	16	10	2
doctors' practices	-	10	32	2
other	8	-	-	6
total	38	238	74	30

2. Motives for the Introduction of Micro-electronics and its Effects at Enterprise Level

2.1 Productivity and the Service Sector

The hypothesis that employment in the tertiary sector must continually increase because, in contrast to manufacturing industry, only a slow rate of productivity growth is supposedly possible has become relatively well established in the last two decades. This phenomenon, known in the trade as "Baumol's Disease", dates back to an analysis conducted by Baumol in 1967 (cf. Baumol 1967) in which two sectors are compared. In the first, productivity increases continuously over time. In the second output is measured as the volume of input. In such an economy productivity growth is zero, or very low, because the sector with constant productivity seeks to maintain a constant share of total output and so absorbs ever greater resources from the more productive sector.

Given that over time almost all resources become tied to the sector with low productivity, the overall growth rate of the economy tends asymptotically towards zero. A continual employment shift takes place towards the sector with stagnant productivity. Baumol's empirical definition of the sector with constant productivity is not clear. His analysis refers primarily to a state sector with no productivity growth. This, however, is precisely the view often taken of the private service sector. Empirically the analysis continues to have a certain relevance, as there remain sectors with stagnating productivity growth. The performance of a violin concerto does not allow for an increase in productivity, and the same applies to other personal services of a certain quality.

However, the introduction of information and communication technology (ICT) has rendered the classic distinction between the industrial sector, with a high rate of productivity growth, and the service sector, with stagnating growth rates, obsolete. ICT allows constant increases in rationalization, particularly in the field of information processing. In the light of these developments traditional views of employment trends in the tertiary sector must be reexamined.

The evidence gathered over the years shows that the affinity of different branches to the introduction of ICT varies considerably. The most suitable are those sectors concerned with the mass processing and transfer of information, e.g. remittances in the banking sector. At the same time, ICT has proved very adaptable, having found its way into all forms of administrative work in which large volumes of information and/or texts (e.g. correspondence) must be processed.

After the initial introductory phase ICTs have also moved into private households and human services. Homebanking is no longer a utopian vision. In doctors' practices patients' data is now increasingly processed using personal computers (PCs). Computers are also used in the care of patients, in the supervision of medicinal programs, the evaluation of analysis or the processing of their results etc. The use of educational computer programs in schools is still in its initial stages, while many areas of private households have not (yet) been affected.

It would be one-sided to consider the spread of ICT merely as a process of rationalization. One important development, for example, is the increase of computer networks and the growth in the capacity to process information is leading to the development of a new form of infrastructure, on the basis of which the modes of behavior of firms and private households are changing. The lines of demarcation between firms and branches are being redrawn.

These developments are also making it increasingly difficult to make prognoses about long-term employment trends in the tertiary sector. Rationalization due to process innovation may lead to job losses while product innovation is simultaneously creating employment. The more extensive use of the newly developed infrastructure is opening up new employment opportunities in other areas.

The concepts "process innovation" (or rationalization) and "product innovation" are unsuitable for coming to terms with the totality of these changes. New products come into being due to new demand within new communication systems and resulting from new exchange relations. In addition to process and product innovation new lifestyles develop and new demands are made on the quality of services, the density of the communication network and the currency and variety of information.

2.2 Motives for the Introduction of Micro-electronics

ICTs are introduced in a complex environment and for a variety of reasons. They may be the solution to overwork, provide freedom from tiresome routine, reduce costs, overcome deficits in quality, contribute to a more modern image, a better service or a better customer relationship. In many cases they are simply a necessary precondition for a firm to be able to take part in economic activity.

A distinction must be made between the situation in which ICT is introduced and the motives for its introduction. The following cases are typical of the situations in which such technology is introduced:

- The extent of the information which must be processed has become so great that the introduction of new technology and its constant modernization) is unavoidable. In many enterprises in the service sector the use of ICT is simply a precondition for survival;
- the complexity of economic and organizational relations within the enterprise is such that the introduction and constant updating of ICT seems advisable;
- an improvement in the quality of customer service and counseling makes the provision of computer-aided technology an indispensable tool in the competition with other firms;
- the need for flexible and immediate reactions to changed circumstances requires the introduction/adjustment of ICT.

As regards the areas in which ICT can be introduced, a distinction can be made between:

- "customer contact" areas, either for counseling, increasing the speed of services or the reliability of information (use of terminals for displaying information);
- in actual operating areas to enable information to be processed more precisely, more quickly and more rationally (e.g. statistics, text processing);
- in organizational/management areas to improve enterprise-level organization, controlling (e.g. data analysis etc.).

The motives for introduction can be grouped into:

- economic ("rational") motives, which may be formulated either actively or passively in the sense of enterprise-level decisions, based either on the enterprise's own calculations or on observations of competitor behavior;
- subjective motives, largely culturally determined, in the sense of wanting "to keep up", not to "miss the boat", or because of the desire for a new image.

In addition to the situation in which ICTs are introduced and the motives for their introduction, the extent of their use and their location in the enterprise are two factors which are of particular importance for evaluating employment effects.

The analysis of the employment effects of the use of ICT in the service sector thus involves the study of a complex network of causes and effects. The same applies to the evaluation of process and product innovations resulting from the use of such technology. Whereas the Ifo Institute restricts the definition of process innovations to investment, the nature of which is described as "innovative" by the firm itself, we consider each new aquisition of data or text processing or communication technology, as long as it is introduced with the aim of improving the production process in some way, as a process innovation. In view of the fundamental changes currently taking place in the service sector as a result of the use of ICT, this approach appears justifiable, at least for the present phase of development.

2.3 Rationalization through the Use of Micro-electronics

To the majority of service sector firms in our survey our questions on process and product innovation sounded very theoretical. Whereas in manufacturing industry computer-aided technology might be introduced in order to improve product quality, change batch sizes or the production process in general, in the service sector the introduction of ICT usually results from a complex of considerations and constraints, with the primary aim of improving the standard of service.

The most important field for process innovation in the service sector is the simplified automatic transfer of data in large integrated networks using compatible languages and programs, and the subsequent processing of this data. This process of rationalization has taken hold in all areas of the service sector in which large quantities of data need to be processed or which are dependent on the exchange of data. The input of data has become a routine part of office work, a trend which is reflected statistically in a fall in the number of data typists and booking personnel.

The second largest area is that of text or word processing, in which rationalization has been achieved through improvements in the equipment provided for employees responsible for compiling texts. The future will bring a more intensive integration of routine clerical work and text-composition, in particular with the aid of standardized text forms. With the spread of new technology and the increasing use of terminals and PCs the distinctions between clerical work, text formulation and text processing will become ever more blurred at ever higher levels in the enterprise hierarchy. In some German building societies, for example, there is no longer a typing service because text processing has been fully integrated into other forms of clerical work.

It is more difficult to give a general overview of product innovations in the service sector. Automatic cash dispensers which can be used outside opening hours are one example of a new product. The ability to book seats on flights in advance by telephone represents an innovation for customers and makes checking-in a less time-consuming process.

In the field of financial investments the financial institutions now offer a wide range of possibilities which it would be impossible to manage without the wide-ranging use of PCs. As a rule ICTs have improved product quality and increased product differentiation, especially in banking and insurance, to a considerable extent in travel agencies, but less so for doctors and architects.

2.4 Systemic Innovations through the Use of Micro-electronics

Even once the entire spectrum of product and process innovation has been identified, one major characteristic of the process of rationalization and innovation in the service sector remains unconsidered. Namely the networking between different firms (travel agencies and airlines, banks and their customers, transport firms and their customers), is one of the most prominent features of ICT. Rationalization is thus not restricted to certain jobs within certain enterprises or to the consequences of the more intensive use of ICT for such jobs.

Rationalizations can result from changes in the (cooperative) relationships resulting from information and communication networks. These "systemic" forms of rationalization (cf. Baethge/Oberbeck 1986) arise because suppliers induce or presuppose changes in customer behavior or changes in their organizational forms when they make changes to their internal system of information processing, or because relations of cooperation or supply are more closely linked and coordinated with one another using ICT systems. An example of systemic rationalization: Sales at the cash desk are automatically registered as a reduction in the inventory. Orders are placed automatically for goods once the level of stocks falls below a certain level. This order enters the information processing system of the supplier and it too is processed automatically or at least using computer-aided technology. The relationships between firms (bookings, commissions, deliveries) thus develop into an integrated system.

Such integrated systems enable forms of rationalization to be realized consisting of a number of separate but constituent developments. Such cases can be described as "systemic" rationalization as the different parts are related to one another and the major changes consist of modified relations of cooperation between different firms on the market. The ability to steer and anticipate the market increases. Links are established between customers, suppliers and other partners in a cooperative system.

Nevertheless, it was decided to complement the analytical concept of "systemic rationalization" with that of "systemic innovation". We use the term systemic innovation to describe a process which, over an extended period of time, leads to changes in a firm's external relationships and forms of cooperation and, derived from them, to internal effects. Systemic innovation thus means primarily a change in demand. This can lead to the opening of new markets and changes in the shares of existing markets. The following factors are of particular importance in this context:

- Technology is a constituent part of the relationship between firms and customers.
- Technology is a constituent part of a new ICT-based division of labor between different suppliers.
- Technology is a constituent part of the demands and requirements of customers *vis à vis* suppliers.
- Technology is a constituent part of enterprise performance.
- Technology is a constituent part of the "image transfer" between firms, customers and the market.

A further essential characteristic of systemic innovation is that it is not consciously planned by individual actors, but rather results from a series of adjustment strategies: changes in demand, for example, or in customer requirements induce reactions in firms which then feed back into customer reactions etc.

Systemic innovation basically means organizational adjustment on the part of firms to changing market structures. In retail trade, for example, this might affect the range of products on offer. The average grocery or general store of ten years ago bears no comparison to that of today. New technologies have transformed the entire way in which enterprises are organized and the range of goods on offer has, partially as a result of the use of new technology, been changed and extended. Meanwhile customers have adapted their behavior to the new supply of goods with changes in the frequency of visits to the shops, type of goods bought, integration of purchases of different products etc.

It is this interplay of changes in organization and supply of goods and the reorientation of customer behavior which makes it so difficult to separate the different effects of innovation and rationalization.

In some respects such processes resemble the introduction of the telephone. Here too an attempt could be made to determine the extent of the substitution of direct communication by indirect forms of information exchange. When the telephone was first introduced the fear was expressed that the extent of direct face-to-face contact would inevitably decrease. It became apparent, however, that people adjusted to the improved possibilities of communication such that contact between owners of telephones became more frequent and covered greater distances. This in turn had implications for the frequency and differentiation of face-to-face contacts. The question of the rationalization made possible by cutting down on traveling time is an idle one as the new network of communications has resulted in fundamental changes in lifestyles.

The concept of systemic innovation resolves the problem of the partial analysis of product innovations. It is particularly appropriate in the context of ICT and the service sector as its most prominent characteristic is the communication networks which have developed in recent years, and to which both suppliers and consumers have reacted accordingly. Because of these changes in behavior and in the form in which services are rendered and the systemic rationalization which, due to improved forms of cooperation between firms, reduces costs, systemic innovations create additional demand and open up job-creating opportunities. They also result in changes in skills and qualifications.

At the same time, the use of the concept "systemic innovation" does not imply that we have abandoned the analysis of the effects of product and process innovations. In many cases such analysis is a useful and satisfactory approach. The automatic cash dispenser outside banks, for example, is a a new product, or a new form of service. The use of terminals and PCs allows work to be accomplished more rapidly, efficiently and accurately. Nevertheless, the analysis of such tasks must be supplemented by the concept of systemic changes in which shifts in modes of behavior occur which make it more difficult to conduct a "before and after" analysis.

A consequence of the unconscious production of systemic innovations in individual firms is that such innovative processes are not usually recognized as such by the firms involved, or they are classified as an "improved service". This motive, so often stated as an aim of the introduction of micro-electronics, conceals a corporate adjustment strategy to changes in the market environment. This constellation will now be illustrated with reference to a number of branches.

2.5 The Motives for the Use of Micro-electronics in Different Departments within the Enterprise

In the enterprise the theoretical concepts of rationalization and innovation fall under the dual aims of cost reduction and the provision of new and/or improved services. Infratest's enterprise survey therefore contained questions on the use and the aims of new technology.

Table 2.1: The Uses and Aims of Technology by Department (in %)
- User Enterprises -

	technology users[1]	aim of[2] technology users	
		cost reduction	new services
internal administration			
bookkeeping/accounting	79	81	7
cost accounting	40	57	19
statistics	68	58	18
personnel	51	69	6
organization/coordination	38 (53)*	53	21
services and production			
procurement	14	64	15
documentation/data bank	34	64	15
storage	11	60	8
merchandise information system	11	56	15
text/correspondence	65	62	11
information exchange	34 (49)*	28	40
ordering/receiving orders	19	41	37
customer service	57 (77)*	18	56
counseling	54 (87)*	12	64
customer accounts	49 (72)*	55	18
logistics	10	46	29
marketing			
acquisition	42	10	64
customer service in the field	26	18	50
presentation	21	65	7
advertizing/media	16	19	35
sales promotion	31	18	58

( )* = ratings for banks (cf. table 3.4)

1 all user enterprises
2 % of users

Source: Enterprise survey 1987.

At present new technology is used primarily in administration, for book-keeping and other clerical tasks. The administrative sector thus formed the core of our analyses. 79% of the user-enterprises in our survey applied data processing technology in this department.

In administration technology is usually applied in order to reduce costs. Applications in cost accounting, the generation of statistics and for tasks associated with organization and coordination were reported by only one fifth of the enterprises in the survey. In marketing, on the other hand, the importance of the two major groups of motives was precisely the opposite. The major aim in this department is to provide new services with the help of computer technology, especially in areas of direct customer contact and in those involved with the internal and external exchange of information. Over half the service sector users of ICT in our survey applied micro-electronics in direct customer contact areas. While in these areas and the sales department micro-electronics were mainly used to develop new services, cost reduction was the dominant motive in all departments in which tasks are oriented to the internal needs of the enterprise (table 2.1).

It is interesting to note that those areas in which the use of new technology was aimed at the extension of the range of services were often mentioned by the enterprises in our survey as areas in which its introduction was especially important for competitiveness. This applies in particular to information exchange, customer counseling and services and marketing. Areas in which technology was introduced primarily to cut costs and which were considered especially important for competitiveness are cost accounting and text processing (all enterprises) and placing and taking orders, the merchandise information system (non-banks) and documentation and customer accounts (banks).

Firms which have not yet introduced new technology ("threshold users") see the areas bookkeeping and text processing as offering the best opportunities for its application. The difference between users and non-users regarding their respective motives for introducing computer-aided technology is clearly that the threshold users are more interested in cutting costs than the existing users. This tends to confirm the hypothesis that the initial users of micro-electronics were innovative and dynamic firms which built new technology into a strategy of expansion, whereas later users are forced to adapt to a changing market situation or to a shift in customer demand.

2.6 The Motives for the Use of Micro-electronics in Selected Branches

2.6.1 Banks

As far as the use of technology is concerned the banking sector is one of the most innovative in the West German service sector. New ICTs are employed in the following departments: internal bookkeeping and accounting (85%), the generation of statistics (78%) counseling (87%) and customer services (77%). The support of computer-aided technology for competitiveness is considered particularly important in the last two departments. 80% and 69% respectively of the banks in our survey indicated that the use of new technology in these areas was especially important for competitiveness.

While in administration the main aim of using ICT was to cut costs, in the departments concerned with customer service and counseling, product innovations and an improved counseling service were most often stated as the motive for the use of new technology. This also applies to acquisition.

Using modern technology, e.g. data processing, banks are in a position to evaluate investment opportunities more quickly and so offer advice which is more accurate and more closely tailored to customer needs (63%). Particularly important for banks is the time it takes to respond to a customer's request in the form of an offer to perform a given service. 85% of banks considered this to be a particularly significant product innovation.

In savings banks, especially those in large towns and cities, the use of computerized technology is very wide-spread. Each cash desk is equipped with a terminal and, at least in the cities, each branch has a self-service statement printer. Most of the staff responsible for customer counseling have terminals and printers at their disposal. The savings banks have invested so heavily in new technology in order to compete with other financial institutions and to reduce costs for routine tasks. They suffer in particular from the low (or even negative) profit margins associated with running salary accounts. For this reason they are the leaders in the diffusion of self-service technology.

In all financial institutions dealing with private customers the processing of salary accounts "eats" into the profits made from the interest differential between lending and borrowing. In other words, banks which can cut costs in the routine work of running salary accounts and advising customers can offer those same customers more attractive rates for borrowing and lending. For this reason savings banks, with their high proportion of salary accounts, are at a disadvantage compared with large banks or private institutions. They have reacted to this by inducing up to 90% of their private customers to use the self-service facilities.

A similar effect is to be seen with point-of-sale cash registers in which payment is made by creditcard and the sum is deducted automatically from the customer's account. Companies with an ever increasing level of financial transfers can now take advantage of cashless payment using cash-management systems. The major aim of freing labor from routine work appears to be to gain time for a more comprehensive customer advisory service, covering all forms of financial activity. This explains why "increasing customer contact" and "more time for individual counseling" are at the top of the banks' list of priorities when it comes to introducing new technology.

2.6.2 Retail and Wholesale Trade

In contrast to the extensive use of data processing technology in the administrative departments, as yet only a rather small proportion of enterprises make use of ICT in the actual "producing" departments of wholesale trade and freight forwarding. Although placing and receiving orders, the running of customer accounts the filling in of invoices and aspects of bookkeeping have been computerized in the majority of medium-sized enterprise (measured by turnover or number of employees) it is only in the larger enterprises (+100 employees) that the warehouse or vehicle fleet have been integrated.

However, it is in this area that the efforts of medium-sized firms are now concentrated. Their aim is to develop a system of merchandise information in which everything from the taking of orders to the calculation of bills and the debiting of customer accounts is computer-linked together and, furthermore, can be linked to the outside world (customers, suppliers) via data networks. The result is a management information system covering all aspects of the enterprise and greatly improving the efficiency of decision making and the productive process.

The use of ICT in marketing departments (acquisition, customer service, presentation, advertising and sales promotion) is still in its initial stages. Customer service and counseling in particular are to be improved with the support of computer technology. If the use of new ICTs in this direction proves feasible, so the view of the experts, such enterprises can be expected to offer qualitatively new services in future, a development which will be decisive for their competitive position as they battle for the customer's favor. The view of the enterprises themselves was also that, of the activities which are to enjoy support from ICTs, those oriented towards the "outside world" will be aimed at providing new services, whereas those used for administration and internal services are aimed primarily at cost reduction, and thus also at cutting staffing levels.

We studied pharmacists (drug stores) as being representative of retail trade because in this area the effects of the use of micro-electronics are not concealed or distorted by changes in sales' conditions. In pharmacies new ICTs are usually used as part of the sales procedure, indeed this is their major role. In most cases book-keeping and accounting have long been contracted out to a specialist service company, and so the firms in our survey did not usually consider them as a potential area for the use of new technology. Most pharmacists believe that a system of goods management will enable them to reduce their fixed costs. The most important costs in the short term are those for storage and inventory and, in the medium term, labor costs. The motto seems to be "the computer means a saving of half a worker". It should be noted that the question whether the use of new technology was aimed primarily at cost reduction or the provision of new services was often left blank in our survey, indicating that these aims are not considered alternatives but are being pursued in parallel.

2.6.3 Engineering and Architectural Offices

New technology is employed by engineering and architecture offices mainly to reduce costs. Orders are to be dealt with more efficiently and the documentation of services rendered is to be more thorough. It is usually the case that the computerization of routine tasks enables services to be offered and rendered, which were previously impractical. Project management can be optimized using network planning techniques. This enables more time to be spent considering alternatives which can then be presented to the customer. The customer can also play a more active role, for example by intervening in the rendering of a service via the exchange of data. This makes it possible for the customer to exert a guiding influence, increasing the dependence of the engineering or architecture firm on the client's demands. At the same time the involvement of the customer may increase the problem-solving capabilities (i.e. the major function) of such firms.

2.7 The Effects of the Use of Micro-electronics on Enterprise Performance

In the *ex post* judgement of the enterprises in our survey the main effect of the use of new technology was to improve the quality of customer service. Only a small number considered the most important effect to be the change in product range/services rendered or in staff reduction. Almost one third of the enterprises were undecided and placed equal value on all three effects (table 2.2).

Table 2.2: The Most Important Effect of the Use of Technology
- Banks and Non-Banks -*

	banks	non-banks
change in product range/ services rendered	3 %	4 %
staff reduction	7 %	12 %
improved customer service	58 %	51 %
all changes equally important	32 %	28 %
no details	-	5 %
	100 %	100 %

* The distinction was made because banks were overrepresented in the survey and because most of the enterprises form part of larger firms.

Source: Enterprise survey 1987.

Table 2.3: Anticipated Effects of the Use of New ICTs on the Enterprise (%)
- Threshold Users -

improved service	97 %
better control possibilities	83 %
better product quality	70 %
additional services possible	55 %
higher profits	45 %
reduced costs	40 %
increased turnover	27 %
total (more than one answer possible)	417 %

Source: Enterprise survey 1987.

Threshold users anticipate an improvement in service, better control possibilities, an improvement in quality and the opportunity to provide new services from the use of new ICT. Their expectations regarding economic effects are considerably more sober (table 2.3).

The majority of enterprises using ICT indicated in our survey that delivery times had been reduced, that product quality had increased (less faulty products) and that products were more in tune with customer requirements (table 2.4). The banks experienced a wider variety of changes than non-banks. The figures for threshold users reveal considerable reservation concerning expected changes in quality. It would seem that here the complexity of services rendered or the possibilities for applying new technology are lower than for users.

Table 2.4: Actual and Expected Effects of the Use of New Technology on the Quality of Products/Services
- Banks, Non-Banks, Threshold Users -

	actual effects		expected effects
	users		threshold
	banks	non-banks	users
effects			
cheaper products	19 %	11 %	14 %
shorter delivery times	85 %	69 %	33 %
smaller batches	13 %	9 %	14 %
greater product variety	40 %	23 %	19 %
responsive to customer requirements	63 %	46 %	33 %
delivery less seasonally dependent	33 %	15 %	17 %
more punctual delivery	39 %	33 %	28 %
products more standardized	25 %	18 %	31 %
reduction in faulty products	55 %	53 %	42 %
total (more than one answer possible)	372 %	277 %	231 %

Source: Enterprise survey 1987.

3. Inner-enterprise and Intra-enterprise Diffusion of Micro-electronics

3.1 The Determinants of the Diffusion Process in Different Branches

In the service sector the "early" users of computer-aided technology tend to be enterprises in which large volumes of information have to be generated, processed and transmitted. Such tasks tend to occur throughout the production process, are highly standardized and are relevant both for the workings of the enterprise itself (office automation) and for customer relations (networking). The most important branches in this context are financial institutions (banks, building societies), insurance agents and large retail and wholesale firms (cf. table 3.1).

An additional group of "early" users is made up of large firms in a variety of branches which do not require the mass processing of information in production but whose core activities are concentrated on distributive and consumer services. Above a certain enterprise size they are characterized by large-scale and early investment in data processing technology and other forms of office automation in administrative departments. Examples include transport firms, travel agencies, hotels publishers etc. After an initial period they also begin to apply these technologies to standardize and manage their intensive exchange of information with the outside world.

Early users of new information and communication technology gain an advantage over non-users by improving their capital stock, including the quality of their "human capital". They gain experience with new technologies, are able to test their effectiveness in different possible spheres of use, harmonize the use of hardware and software, and expand capacity utilization up to its limits. With a highly qualified labor force they are in a position to restructure their existing work force potential, optimalize it according to experience and use the changes in qualification structure as an accelerator for the development of additional services in competition with later users.

Certain classical firms where information processing plays a dominant role, such as brokers, freight forwarding companies and, to some extent, investment consultants seem to have opted for the use of data processing technology rather late. The common denominator here is that the use of computers only really becomes worthwhile when all, or nearly all the firms in a branch, (or, in the case of freight forwarding companies, for example, the majority of customers) have also installed appropriate equipment and it is then possible to construct a data network or data bank.

Table 3.1: The Use of ICT in the Enterprise
- by Branch, in % -

	Comtec-Panel*				Comtec-Retro*	
	1982		1985		1987	
	DV**	IuK-Tech.	DV**	IuK-Tech.	DV**	IuK-Tech.
total	5.6	8.1	8.2	17.1	10.0	22.4
manufacturing industry	9.1	14.0	9.9	20.0	10.2	24.3
construction	2.4	3.0	3.5	6.3	3.8	12.0
wholesale trade	12.6	17.6	21.4	39.0	16.4	38.2
retail trade	2.7	4.1	8.4	11.8	9.9	15.4
transport and communication	6.7	13.7	10.8	25.0	13.0	29.8
banking	13.3	13.7	62.3	65.3	38.2	44.8
insurances***	1.1	7.4	7.3	14.1		
private services (publishers, hostelries etc.)	1.4	2.0	2.5	4.5	3.0	12.9
professional services	6.7	8.9	11.8	18.9	5.8	26.0
health, science, education	3.5	4.6	5.5	13.2	6.4	18.8
public transport, charitable organizations, social insurance agencies	7.7	11.2	19.9	25.0	21.3	29.5

* Comtec-panel data are based on the same sample, Comtec-Retro is based on an own sample.

** Data processing as a whole or in special combinations with communication technology or the use of PCs.

*** Includes insurance brokers.

Source: Comtec

Clearly such major investment decisions are more difficult for small than for large firms. The wave of investment is not only contingent on the necessity for data networks, but also on the comparatively large capital sums required.

Nevertheless, the large investment is matched by the considerable potential for rationalization which it offers, a potential which was in fact realized by early users in some markets where this was facilitated by sufficiently large enterprise size, prominent market position and/or expansive market strategy (and subject to sufficient capital reserves and returns). A good example is provided by large broker firms.

There are of course numerous other branches in which a mass of information has to be processed. However, in contrast to those mentioned above, in the branches of this category - financial services, management consultants, doctors, lawyers, architects etc. - the generation and transmission of information is usually separate from the actual core activity. Most of the firms in such branches render personal services which are difficult to standardize, are taylored to meet individual requirements and demand qualified and specialized personnel.

In other branches the use of new technology in the internal, but, from the point of view of the central activity, peripheral administrative department is widespread and still growing. Due to the large difference between the "core activity" and "overhead activities" the decision to rationalize the internal administration is an easy one to make. However, in some of these branches, especially above a certain enterprise size, there is a tendency towards the increased use of ICT in the core areas. This results in the provision of additional services (product innovations) due to the creation of networks, decentralization and branch-specific software applications.

In the professional services and the banking sector the production of services is often supported by so-called expert systems. They are used for example to determine the consequences of particular forms of financial investment for income tax or the total return on certain portfolios in a given period of time. Certain forms of counseling have only become possible due to new technology (i.e. it takes the form of product innovations). The entire field of international capital investment, affected as it is by exchange-rate fluctuations, interest rate differentials, the state of the futures market etc. is, in its present form, a product of new technologies, as it is computer systems together with the appropriate software which allow the necessary optimalization calculations to be made for capital investment decisions.

For "intelligent and innovative" firms such as software developers, specialized engineering firms etc., the use of modern technology is usually an absolute necessity. However it is possible to make a distinction between two types of such firms. The early users in a given branch are normally those which work largely for other firms or who periodically take on work contracted out from other companies (e.g.

engineering offices and advertising agencies). They are then followed by firms which are not part of such satellite systems, or only to limited extent.

A third category of branches - private services - makes relatively little use of ICT in the departments of internal administration (here too, however, the influence of enterprise size is very strong). Examples include theaters, hairdressers, driving schools, launderettes, hotels, restaurants etc. In many of these branches the fact that ICT is not used is due to their small workforce (thus the exceptions are the larger enterprises), the small number of customers, and low output of each enterprise. Here branch-specific applications are still rather the exception as they usually require a certain enterprise size, professional marketing strategies and thus organizational demands.

In conclusion, the diffusion of new ICT in the offices of the private service sector differs from branch to branch. In the first category of branches, characterized by the need for mass information processing, firms have been making use of data processing since the 1970s and, in some cases, since the 1960s and of word processors and telecommunications since the 1980s. Diffusion in the second category (professional services with a considerable potential for the application of technology) and, after a time lag, in the third category (with the exception of large firms) did not begin until the early 1980s. In the latter categories electronic typewriters (with or without storage capacity) and telex machines were (and are still being) introduced before the decision for DP equipment, word processors and telecopiers was made. Bookkeeping and address filing are often contracted out to accountants or commercial data processing firms.

An important factor in deciding whether an enterprise can increase or rationalize its production using new technology is its size. Computer technology also offers specific advantages for firms consisting of more than one enterprise. Branches and subsidiary-enterprises of larger firms "inherit" hardware and software from the mother company and are increasingly linked to them by leased or switched line. Thus branches of a larger firm, even where they have a very small work force make up some of the most extensively equipped of all enterprises and thus enjoy competitive advantages.

In one-enterprise firms, on the other hand, new technology tends to be introduced only when a certain size has been reached, when it enables internal administration to be run more efficiently. The importance of enterprise size is reduced, however, where the core activity too can be wholly or partly standardized. This occurs when new technology directly promotes the rendering of services, where services can be diversified or new services offered using new technology (product innovations) or where it helps to rationalize production (process innovations). The fact remains though that the high level of expenditure and the increasing amount spent obtaining suitable software represents a considerable "investment threshold" for enterprises which, in other respects would be suitable for the application of such technology.

3.2 Diffusion within the Enterprise

Between 1982 and 1985 the average number of terminals installed in user enterprises in the service sector sank from 5.7 to 3.8. In manufacturing industry, on the other hand, the average increased from 6.5 to 9.3. This leads us to suppose that the trend in the service sector is towards the use of PCs, while in manufacturing industry the trend is to a more intensive use of DP technology. (In the Comtec data base PCs and terminals are counted separately.)

Table 3.2 shows the number of terminals per white-collar employee by branch and enterprise size in 1982 and 1985. The classification of enterprise size is according to the number of employees. The fall in the number of terminals per employee in small enterprises is to be explained by the fact that early users had invested particularly intensively in new technology. In the transport sector, especially in freight forwarding companies, the increased use of PCs is responsible for the drop in the figures for terminals per employee. An increase was registered for banks, professional services and manufacturing industry. Enterprise size clearly plays an important role in determining intensity of use. In small wholesale enterprises the figure of 0.7 terminals per white-collar employee represents saturation point, as, taking part-time and casual labor into account, almost every employee has a terminal at his or her disposal. In the banking and insurance sectors almost one third of work stations are equipped with a terminal, whereby it must also be taken into account that cash terminals can be used by several employees at once so that the number of users is probably considerably higher than the number of work stations.

The increase in the use of terminals, especially by banks, to a level of one third of all work stations is due to the introduction not merely of cash terminals but also of terminals, which can be used in direct contact with customers (counseling etc.).

In the case of professional services (doctors, lawyers, accountants, architects etc.) a trend can also be identified towards the provision of terminals at each white-collar work place. The fact that at present only one in 2.5 work places in small user enterprises are provided with terminals shows that there is a considerable potential for changes in the qualifications and the skill requirements of the employees in these branches.

Table 3.2: Terminals per Employee by Branch and Enterprise Size, 1982 and 1985
- User Enterprises -

	1-9 white-collar employees		10-99 white-collar employees		100 and more white-collar employees	
	1982	1985	1982	1985	1982	1985
total	0.90	0.41	0.15	0.16	0.07	0.13
manufacturing industry	0.19	0.31	0.10	0.17	0.06	0.15
construction	-	0.30	0.08	0.12	0.03	0.06
wholesale trade	1.60	0.73	0.20	0.19	0.07	0.14
retail trade	0.25	0.25	0.10	0.13	0.04	0.06
transport	0.33	0.28	0.21	0.15	0.02	0.03
banking/insurance	0.28	0.35	0.13	0.28	0.17	0.32
private services	0.50	0.43	0.22	0.24	0.11	0.15
professional services	0.27	0.42	0.20	0.29	0.00	0.10
other	-	0.34	0.29	0.09	0.07	0.06

Source: Comtec

Exceptions to the otherwise impressive trends towards greater use of terminals in user enterprises are retail trade, construction and the transport sector where there are considerable differences between large users (such as retail chain stores) and threshold users. In retail trade, and especially in newly founded firms, a terminal or even in some cases a PC are introduced to deal with ordering. With the spread of the wideband network run by the German Postal Service, there will be an accelerated diffusion of new technology within those firms which have survived.

Precise information on the absolute number of work stations and as a proportion of white-collar employment in both user and non-user enterprises is provided by table 3.3, which also shows the total number of white-collar workers in 1982 and 1985 and compares these values with the number of employees working with terminals, personal computers or word processing systems (total work stations). In 1982 a mere 2% of white-collar workers were employed at work stations; by 1985 the figure had climbed to almost 9%. The results confirm once more that banks and the wholesale trade are the "avant garde" in the use of information and communication technology.

Diffusion within the enterprise is dependent on the strength of enterprise organization, the employment profile during the period in which preparations were made for the introduction of DP technology. Working structure can be transposed into DP systems where the core activity of the enterprise is relatively formalized. Examples include insurance, travel agencies and wholesale trade.

3.3 The Use of New Technology in the Different Departments

The dominant area as regards the use of data processing technology is administration. The Comtec-Retro indicated the following chronological sequence of the application of data processing technology in the service sector:

1. Finances/bookkeeping
2. Other commercial areas
3. Aiding management control
4. Sales/distribution
5. Customer service
6. Computer centre
7. Warehouse/storage.

In the service sector new ICTs are usually first implemented in the internal administration. The process usually commences with tried and tested (and so riskless) applications in bookkeeping (including wages, financial, credit and debit bookkeeping). Where this department was previously contracted out, it is often reintegrated for cost reasons. The first tasks are usually the construction of data files on customers and goods and for correspondence (table 3.4).

The next steps are often to standardize the processing and administration of orders using DP as the first stage in the construction of an information system containing all the data relevant to the running of the enterprise. The information system enables statistics to be generated and production control to be organized for different product groups, branches and departments and even different workers. Finally it also offers the possibility of improving customer service.

Recent developments using PCs enable information to be processed and printed ready for presentation by laser printer (desk-top publishing). Such systems are already being used in some enterprises such as management consultancies and advertising agencies. In the last two to three years additional PCs are increasingly being used for special tasks such as cost accounting, statistics and presentation material. In customer service departments the use of portable computers is growing.

Table 3.3: Proportion of White-Collar Workers Using Data Processing and Word Processing Equipment
- by Branch, 1982 and 1985 -

	total w-c employees	employees working with terminals	PCs	text systems	total users
1982					
manufacturing industry	2,843,533	72,020	5,291	7,290	3.0 %
construction	420,071	1,695	823	-	0.6 %
wholesale trade	1,007,537	88,290	23,645	1,028	11.2 %
retail trade	1,630,819	6,876	1,366	987	0.6 %
transport	984,129	6,120	635	-	0.7 %
banking/insurance	884,759	23,136	2,280	1,188	3.0 %
private services	1,033,759	5,304	2,058	666	0.8 %
professional services	1,387,417	19,449	2,087	437	1.6 %
other	4,181,381	34,447	19,888	3,750	1.4 %
1985**					
manufacturing industry	3,051,054	305,011	32,182	28,132	12.0 %
construction	475,095	9,324	5,603	485	3.2 %
wholesale trade	1,114,034	138,173	42,271	10,670	17.2 %
retail trade	1,858,381	61,215	12,565	5,320	4.3 %
transport	1,005,717	28,685	3,267	1,955	3.4 %
banking/insurance	923,651	218,645	7,301	8,358	25.4 %
private services	1,175,681	16,951	3,938	706	1.8 %
professional services	1,590,843	102,308	40,830	29,872	10.9 %
other	4,429,667	149,518	97,052	14,967	5.9 %

* The number of white-collar employees was derived (in a similar way to the Comtec Survivor Analysis) using a weighting based on 1984. For this reason the absolute figures differ from the values indicated in the employment statistics. In order to obtain share values the number of users was set in relation to the number of white-collar employees in Comtec.

** Enterprises which survived.

Source: Comtec

Table 3.4: The Use of Technology by Department (in %)
- User Enterprises -

	all enterprises (n=238)	banks (93)	branch FS* (56)	WT* (29)	TA* (25)	E/A* (16)
internal administration						
bookkeeping/accounting	79	85	70	97	88	100
cost accounting	40	47	27	63	12	50
statistics	68	79	51	83	68	25
personnel	51	61	35	60	44	44
organization/coordination	38	53	33	17	32	13
services and production						
procurement	14	9	-	20	48	13
documentation/data bank	34	34	36	20	44	13
storage	11	7	4	30	4	-
merchandise information system	11	11	2	20	4	6
text/correspondence	65	65	84	37	60	94
information exchange	34	49	20	23	56	13
ordering/receiving orders	19	8	2	43	72	-
customer service	57	77	50	33	68	25
counseling	54	87	31	7	80	19
customer accounts	49	72	33	43	40	6
logistics	10	9	2	43	4	-
marketing						
acquisition	42	60	50	17	32	19
customer service in the field	26	36	31	23	8	13
presentation	21	24	31	10	12	25
advertizing/media	16	19	20	10	12	6
sales promotion	31	38	33	20	40	6

* FS = financial services
WT = wholesale trade/freight forwarding
TA = travel angencies
E/A = engineering/architectural consultants

Source: Enterprise survey 1987.

An increasing number of firms with several years' experience with ICT are beginning themselves to adapt their software to meet their own needs. This is generally done on the basis of existing programs which are then rewritten to improve user friendliness and debugging facilities.

A new quality in the use of ICT is to be seen in networking, both within the enterprise (in which word processing may be centralized) and that between branches, headquarters, suppliers and commercial and regular customers. Indeed it is this step which turns an instrument of automation into one of information and communication. Until this development became more wide-spread the use of technology remained in the form of "islands", in some cases of incompatible installations, e.g. internal and external communication were by traditional means (by telephone, face-to-face).

The following new developments are currently to be observed:

- the integration of data processing, word processing and information technology in multifunctional work stations,
- the development of information access and exchange between individual work stations,

Table 3.5: Direction of the Reorganization of User Enterprises (in %)
- Banks and Non-Banks -

direction of reorganization	banks (n=93)	non-banks (n=145)
provision of more (or all) office work places with new technology	70 %	45 %
improved efficiency	82 %	84 %
integration of word and data processing and communication	40 %	28 %
dialog between work stations	26 %	4 %
inner-enterprise networking	26 %	15 %
networking with suppliers	-	10 %
networking with headquarters/branches	33 %	9 %
link to data banks	26 %	13 %
total (more than one answer possible)	303 %	208 %

Source: Enterprise survey 1987.

- the decentralized access to the services of the central system via the technical linking of the systems with each other and/or of the systems with the central data bank, and
- the possibility of using computers decentrally at each work place with its specific requirements, even for non-formalized tasks aimed at individual service.

Approximately one quarter of the enterprises in our survey had not undertaken any notable reorganization of their technical equipment. Most enterprises had replaced their hardware or installed ICT for the first time. In the view of the enterprises themselves the primary aim was to improve efficiency. Banks provided office work places with access to new technology to a greater extent than non-banks. Approximately one third of enterprises were attempting to integrate different forms of ICT; again this was more pronounced among the banks (table 3.5).

3.4 Diffusion in Selected Branches

3.4.1 Banking

Practically all the banks in our survey were equipped with ICT. 63% used word processors, 93% terminals and 68% PCs. As for communication technology, 47% used BTX, 42% had Datex-L or Datex-P lines (rented from the Federal Post Office) and 62% leased lines. These figures show the banking sector to be one of the most innovative in the entire service sector.

The diffusion of real-time installations can be seen clearly with a look at the savings banks. Within 10 years (1973 to 1983) the use of such equipment increased from 5% to 97% of all enterprises. The number of on-line counter machines increased in the same period by a factor of eight, the number of terminals, on the other hand, did not increase noticeably until the end of the 1970s. Since 1983 the proportion of user enterprises has increased only slowly, although the number of terminals doubled once more (internal diffusion). The 1980s saw the introduction and the rapid diffusion of self-service equipment. Yet measured against the total number of branches (approx. 45 000), the number of cash dispensers (ca. 2 500 - 1986) and statement printers (6 300 - 1986) represents merely the beginning.

The process which led to the present system developed in several stages. As early as the 1960s central DP systems were introduced for current account credits and debits in order to cope with the rapid increase in payments. The motor behind this development was the trend towards paying wages and salaries direct into employees' accounts, which led to an explosion in the amount of data involved in running accounts and transferring money. According to all the experts we interviewed it would not have been possible to cope with such a volume of data reliably and in the required time without the use of central DP systems.

A further stage in the process of automation and the saving of expensive processing time was then achieved in the early 1980s with the introduction of self-service machines.

3.4.2 Wholesale Trade/Freight Forwarding

Data processing has been employed in the wholesale trade and in freight forwarding companies since the 1960s. Using magnetic ledger-card and punched-card computers large quantities of data for bookkeeping, making up invoices and later for the payroll accounts were processed. At this time DP users were primarily large enterprises with a substantial number of customers and regular orders and business practices.

In the 1970s the processing of orders was also integrated into (newly developed) DP systems. When orders were placed the data needed to fill in commissioning and suppliers' invoices, for making up bills and for statistical purposes were collected and could be read off from a terminal. When such DP systems were used for the first time, the lack of experience with the new technology usually meant that it was not possible to develop a fully integrated system for processing orders. The isolated application of the new technology in financial and payroll bookkeeping and in making up bills remained typical. An increasing number of wholesale and haulage firms performed their administrative work with support from central computer systems.

It was not until the beginning of the 1980s that the early users of DP systems (usually based on interactive data processing) had tried and tested software packages at their disposal, enabling them to integrate the whole system of processing orders and to take the first steps towards an integration of the warehouse and so to a merchandise information system. As a rule such firms or enterprises were large or medium-sized companies.

In these developments those firms which had been using DP systems since the 1960s and 1970s had the advantage of many years' experience and were able to restructure their production processes, complicated as they are given the large number of different products and customers and the rapid turnover of goods, without major dislocations. A further (and related) advantage was that they had a skilled labor force (e.g. with the ability to develop software) closely integrated into the firm. Many of the early users of DP technology have also had many years' experience with other forms of ICT such as word processors and PCs.

However, the majority of the current users of DP technology in these branches did not equip themselves with DP systems until 1983/84 and with PCs for special tasks until 1986/87. The efforts now being made to go beyond the computerization of specific and insular processes and areas are also a development which has only gained ground in recent years. A conscious aim is now the integration of all the instruments and areas within and linked to the enterprise using the whole spectrum of ICTs, especially the exchange of data with suppliers, customers and branches.

The results of the Comtec survey provide information on the extent of the use of technology in wholesale trade and freight forwarding enterprises in 1982 and 1985. They indicate that a considerable diffusion of ICT took place in these branches during this period. In 1985 39% of wholesale enterprises and 48% of freight forwarding companies (the figures here include travel agencies and tour operators) were equipped with at least one of these technologies.

Our enterprise survey provides the following picture of the users of ICT in the two branches: 83% were equipped with a PC or larger computer; the same number possessed terminals. 86% had electronic typewriters, 31% had word processors. As for telecommunications, 76% had Telex, 7% Teletex equipment while one third of enterprises were linked to Telefax and every fifth enterprise by BTX.

3.4.3 Travel Agencies and Tour Operators

The introduction of the new system for making bookings and reservations in the travel branch was primarily a question of having adequate capital and of the necessary cooperation of all the major tour operators.

By the end of 1987 the most important system for bookings and reservations, known as Start, was used in almost 2 500 travel agencies in the Federal Republic of Germany. Some 5 000 terminals were linked to the system via leased lines. In addition to Start there are a number of other systems, but these usually take the form of an in-house system for large operators linking their numerous branches.

Start was implemented in 1979 by Lufthansa, TUI (Germany's leading tour operator), the Federal Railway (Deutsche Bundesbahn) and the operator/travel agents ABR, Hapag-Lloyd and DER. In addition to reserving and booking the services offered by the initiators (and owners) of Start, the travel agents linked to the system also have access to the services provided by almost all other large tour operators, car rental firms, travel insurance agencies and hotel chains. Such services can be accessed, booked directly and, in most cases, confirmed on the spot. Furthermore, Start can also serve as an information system for management. The services offered here include: the collection of customer data or statistics on sales/ turnover in specific areas for specific periods of time.

Since 1984 BTX has also been used as an additional service, so that by the end of 1987 a further 1 200 registered travel agencies have access to Start via BTX. According to estimates by different experts, a BTX service (including that separate from Start) is now provided in 3 500 to 4 000 travel agencies.

Start is not only an information and reservation system; it can also be used for hard copy print outs, a function which is very intensively used. By 1986 the number of print-outs had risen to 37 million per year. This figure is made up of (approx) 13 mill. sales receipts, 6 mill. plane tickets, 15 mill. (non-air) tickets, 1.4 mill. confirmations and 1.3 mill. other forms of receipts and invoices.

The spread of Start facilities among travel agents has made it vital for tour operators and all firms offering services for tourists to be included in the Start reservation system. Despite the uncertainties of the cost-benefit calculation (the "entrance fee" at the end of 1987 was DM 200 000, and DM 10 was charged per booking) the permanent place which it secures on the tourism market is necessary for marketing reasons: those operators who have only recently joined the Start system will face a long pay-back period.

Nowadays a travel agency which is involved in flight booking simply cannot do without a computerized link to a central system. Even for agents and operators of train and bus/coach services the use of DP systems brings considerable advantages. It therefore comes as no surprise to note that in our enterprise survey almost all travel agents and tour operators had introduced Start terminals, whereas only a third of them used BTX. In addition, almost all of these enterprises/firms had electronic typewriters and telex machines at their disposal; 8% employed PCs and 13% a computer of a different type.

3.4.4 Engineering and Architectural Offices

Two major factors were responsible for the increasing use of micro computers and PCs in engineering and architectural offices: the fall in the price of hardware and the routine and formalized nature of many of the tasks in this field (planning and cost accounting, making and registering offers/tenders, monitoring offers, their execution, design, counseling, writing reports and making up invoices).

According to a survey conducted by the Federal Chamber of Architects (Cf. Deutsches Architektenblatt 1982/5, p. 596), in 1981 only 2-3% of the 20 000 architecture offices in West Germany were equipped with DP. Current estimates indicate that some 15% of enterprises in this field now use computers in their day-to-day work. In engineering offices diffusion is much more advanced. Experts estimate that about 90% are equipped with DP. Of those offices which undertake both kinds of work some 40 - 50% are estimated to use computer technology.

During the 1970 larger offices of both types began to introduce computers for tendering and billing. Special programs for design and statics were also used. Computer-aided design (CAD), on the other hand, was seldom used. It was not until hardware became much cheaper and much more compact, and software was improved (i.e. at the beginning of the 1980s) that the use of DP spread to medium-sized and to some small enterprises.

Interest is growing in cheap but powerful desk-top computers for word processing, cost control and project management. For many enterprises CAD technology, although it could be used, is simply too expensive. At the same time the pressure on engineering and architecture offices from their clients to use CAD for planning is increasing because, to quote one example, a blueprint stored on tape or disc is considerably more practical for those administering the building and for possible alterations etc. than traditional blueprints.

According to our enterprise survey about a third of the engineering and architecture offices using ICT have a medium-sized DP system while the rest use one or more PCs. Half of the enterprises use a word processing system while slightly more than half use telecommunication media such as Telex, Teletex or Telefax.

4. The Employment Effects of the Use of Micro-electronics

4.1 Department Analyses in Selected Branches

4.1.1 Department Analysis in the Banking Sector

Based on our case studies and interviews with experts we attempted to develop typical indicators for the changes in the demand for labor as a result of the specific use of technology.

According to expert opinion the introduction of terminals and PCs, particularly at the counter, means that between 20% and 30% less labor would be required at each post to deal with the same volume of business. This is due to the rationalization of data collection (booking and transferring data).

However, this does not mean that labor is necessarily cut from such positions. On the contrary, depending on the degree of "networking" within the branch the rationalization effect can affect a variety of positions. The enterprises we interviewed were unable to provide precise answers to this question:

- Rationalization effects occur due to the realization of specific and partial technical possibilities. For example, certain groups of account holders used to be assigned to specific employees. This rigid system has become more flexible with the use of VDUs, as they enable not only the balance but also a variety of other information including the holder's signature can be displayed. The additional information which can be accessed by VDU means that, in the case of transfers abroad for example, the customers home bank (where the customer's signature previously was held) need not participate in the transaction. The deployment of labor for counter work can be organized more flexibly, leading to rationalization effects which, however, could not be quantified. Indeed the effects are specific to each case, as is shown by a remark by one of our experts who claimed that the fact that it is no longer necessary to sort out receipts and invoices can save one to one-and-a-half working hours daily in a branch with seven to ten employees.
- Rationalization effects are more complex in composite processes such as granting loans. It was claimed that the simplification of the process of granting a real estate loan, for example, the amount of labor can be reduced by half.

However, such effects are derived from case studies of individual institutions and cannot be extrapolated to cover all other banks. Moreover, such figures for the partial rationalization effect cannot be taken to represent the overall effect. In political discussions on this theme one often encounters the notion that the labor saving effects of a new technology can be set against the additional costs which are incurred in a sort of profit and loss account. The view is that the overall rationalization effect of the new technology is transformed directly into job losses. In actual fact such effects frequently cannot be measured because they do not occur in the area/department in which the technology was introduced. Rather the rationalization effects normally diffuse through the entire network of employment relations in the bank (systemic rationalization). It is thus very difficult to estimate the overall effect, even for our experts, because a systematic analysis of the interrelationships within the banking sector has yet to be made. As yet precise figures are not even available for the number of employees working in the different departments mentioned above.

A further difficulty for the evaluation of rationalization effects is the constant reorganization of individual tasks within the enterprise. The extension of the counseling service, and the increasingly "tailor-made" nature of personal services often mean that the rationalization effects at enterprise level are fully compensated. The only visible change results from changes in qualifications in the "back office".

At the same time the numerous different and partial rationalization effects indicate that systemic rationalization effects may arise out of the linkage of these different elements and the close cooperation with customers. Even the simple fact that information on customers and accounts is always available and so the exchange of information between employees and customers is more comprehensive and rapid produces benefits in terms of the rationalization effects arising from the reduction in the time required to complete individual tasks.

If a direct link between branches and headquarters is established the resultant access to data banks and other services enables further savings to be made and services improved. Once such cooperation has been forged, further cooperation follows. Networking often leads to systemic rationalization and, as a result of a change in the division of labor between the different firms, to systemic innovations.

4.1.2 Department Analysis in Financial Service Enterprises

In contrast to the situation in the banking sector, contact between insurance or investment brokers and their clients occurs only at irregular intervals. A regular, computer-aided supply of information has proved a useful policy in order to induce customers not to "wander". A further benefit is to keep clients informed of additional services offered by the company. As far as quality of service is concerned the use of technology offers firms providing financial services the following advantages:

- they can serve their customers more quickly.
- they can offer a more extensive and comprehensive customer service.
- they can respond to the individual wishes of their customers.

In large enterprises such as large investment consultants, insurance brokers and the insurance companies themselves it was found that on-line access to data banks enabled employees involved in customer service to take on some of the work which had previously been performed by secretaries or in the "back office", without this resulting in a reduction in their skill requirements. This implies that the level of routine administrative work in the financial service sector will probably remain constant or even fall whereas the areas of customer counseling and the "intelligent" analysis of information will expand considerably.

This will result in changes in the skills required of the work force. Those offering advice to clients need no longer be expert dealers in securities with a specialist knowledge of a certain market. What will be required of investment consultants is a wider-ranging knowledge enabling him or her to provide reliable and dependable customer service.

As a rule consultants learn the necessary specialist and systemic knowledge in training on-the-job. The trend in software development is also towards greater user friendliness so relieving employees from the necessity to aquire specialized technical knowledge. For this reason the interview partners in our case studies were of the opinion that in future they will not suffer from the labor shortages in this field that they are currently experiencing.

Such changes in skill requirements will lead to the following changes in the hierarchical structure of the organization of work:

- Consultants with a comprehensive knowledge both of the markets and technological developments will become rarer, and thus more valuable for the enterprise, as it is only with their help that software development can be optimized.
- There follows a large group of consultants which, while they remain highly qualified, are in principle merely carrying out operative functions.

The trend in the expanding sector of financial services thus seems to be towards a polarization of skill requirements.

Part-time agents (which currently number approx. 250 000 in the Federal Republic) are likely to lose ground to full-time agents if they are not equipped with terminals. Given that it can hardly be expected that they will achieve the same terminal-density as their full-time colleagues, a process of differentiation, a new division of labor, will set in within the branch. While part-time brokers will concentrate to a still greater extent on pure acquisition and contact mediation, full-time brokers may well start to employ pert-time agents as partners. Already there is a boom of so-called "forward sale companies" which employ part-time agents purely for establishing contacts by telephone. The full-time agent then conducts business once the contact has been arranged.

In such a rapidly changing market the quantitative employment effects resulting from the use of ICT can only be derived indirectly. The trend is clearly towards the transfer of work previously performed by secretaries and routine "back office" clerical work onto the shoulders of the brokers/consultants themselves. The result is a simplification of organizational structures and of the identification of the sources of costs and revenues. The consequences for the overall level of employment, however, remain open, because as yet it is far from clear whether the transfer of responsibilities mentioned above will lead to quantitative employment effects. In the medium term the improvement in the control over work and the degree of utilization will result, ceteris paribus, in an increase in productivity, so that a negative employment effect seems more likely than a positive trend.

4.1.3 Department Analysis in Wholesale Trade and Freight Forwarding Companies

Compared with other branches the use of technology in wholesale trade and freight forwarding companies has had a much less significant effect on the service offered to customers and suppliers. A quarter of the enterprises in our survey detected no improvement at all in this regard, and from the long list of possible improvements in most cases only one or two aspects were reported to have occurred. It is only in punctuality in meeting delivery dates that more enterprises from these branches registered a greater improvement than other branches in the service sector. The counseling service, for example, has been improved in only a very small number of enterprises in these branches (table 4.1).

Table 4.1: The Effects of the Use of Technology on Service
- Wholesale Trade and Freight Forwarding Companies -

improvement in	
speed of correspondence	33 %
regularity of information/customer contact	23 %
layout of correspondence/letters	20 %
punctuality/meeting deadlines	37 %
counseling	7 %
time spent on induvidual customer contact	10 %
service could not be improved	27 %

Source: Enterprise survey 1987.

Communication with customers and carriers plays an important role in procuring transport and (re-)loading facilities. The use of extensive telephone systems and links between them and DP systems has greatly simplified such communication. The same applies to maintaining storage facilities (warehousing) collecting goods etc. For example it is now possible to organize the passage of a consignment through customs via a PC-linked computer connection with the central customs authorities within half a day. By the traditional method of written correspondence, two weeks were normally required. This has resulted in an equivalent reduction in the time transported goods spend in storage. The logistical problems which such time lags used to cause customers has also been reduced. Cooperation between carriers and freight forwarding firms has also been greatly simplified by the use of information networks, resulting in a considerable saving in the costs of communication (Telex/Telefax at various regional levels). An information network offers the added advantage that freight brokers do not have to contact each freight forwarding company individually in order to ascertain the most favorable route for a given product: this information is provided automatically through the network.

This market constellation induced the more efficient freight forwarding companies to adopt the necessary technology very rapidly in order to be able to hook into the information networks and dispatch points. However, despite the strong incentives, the diffusion of communication technologies in this branch proceeded relatively slowly. For example, 18 months after the introduction of the possibility of settling customs formalities via remote data transmission (by PC) only eight international freight forwarding companies (of which five were medium-sized) were actually hooked up to the system. Nevertheless, it is expected that international freight forwarding companies will link up to the system within the next five years. In the wholesale trade the share of total product value accounted for by product innovations is hard to judge, because it is not the character of the products themselves which has changed but rather distribution, inventory management, distribution and product range.

Here the most direct link is probably that between product innovation and product range. The logistics developed in firms with a high level of technology enable a range of goods to be held, the breadth and depth of which is far in excess of that normally available. A greater number of more differentiated products can be stored and sold, while software optimizes warehouse use and procurement and so reduces the amount of tied-up capital and enabling products to be delivered at shorter notice. This fulfills three major conditions for product innovation. Our survey confirms that such product innovations, introduced as part of the overall logistical program in wholesale trade, are making the branch more independent of both customers and suppliers. As one of our employers put it: "Logistics has become my *raison d'etre*".

A large proportion of the work in wholesale and freight forwarding companies consists of routine and standardized procedures. These include accepting and placing orders, filling in invoices and bills, cost accounting and bookkeeping. Software for processing orders identifies the order and quantity with a customer and article number and processes it immediately, so that the order can be executed at the appropriate time without the need for further data preparation. This enables labor to be saved in certain order-processing departments.

In integrated merchandise information systems, i.e. when suppliers or customers are involved the processing of their orders through telecommunication or data networks these labor-saving effects may encompass further tasks. Those affected by such developments are secretaries and all grades of clerical workers. Redundancies or a policy of "natural wastage" usually affect the less qualified clerical workers. Frequently there is an exchange of workers, in the course of which employees with additional commercial skills are hired. They are to take on the tasks of counseling and customer service which, due to the computerization of order processing, are becoming ever more important.

A further source of staff reductions may come from remote data transmission. The equipping of pharmacists and bookshops with terminals for ordering goods, which began in the 1970s, led to job losses in the wholesale trade among staff receiving incoming orders. Similar effects were registered among wholesalers in the food industry following the recent introduction of mobile data collection appliances in the retail trade.

New workers, usually highly skilled, are being hired by many enterprises in the wholesale and forwarding branches. Furthermore, in medium-sized and large enterprises new applications have arisen for ICT, which have led to new jobs for computer personnel with responsibility for running and maintaining hardware and software.

According to the majority of enterprises in our survey the use of ICT in these branches has led to changes in the skill structure of their respective work forces. Knowledge of DP techniques has become more important right across the board, and is now an important criterion when hiring new workers. For half of the enterprises retraining measures have become necessary. A loss of low-skill jobs was reported by only a small number of enterprises. Where it did occur it affected routine administrative work and unskilled warehouse jobs, which made it possible to redeploy labor within the enterprise.

Rationalization of the processing of orders and the use of merchandise information systems have had labor-saving effects. Workers previously active in these areas are, however, frequently redeployed in procurement and sales. In our case studies a substantial fall in employment was registered primarily in storages and in the truck/lorry park in those enterprises in which it proved possible to integrate them into the central working structures. The introduction of logistics systems has reduced volume of goods held in storage and warehouses and the average time goods spend there. The inventory is now normally run by a white-collar employee with the help of an archive file system. In the warehouse itself only blue-collar workers are now employed, responsible for loading and marshaling vehicles. The overall impression gained from the case studies is that in both wholesale trade and freight forwarding companies the effects of product innovation are greater than that of process innovation.

However, labor-saving rationalization effects do occur in enterprises both upstream and downstream of these branches: for example, in firms purchasing services from freight forwarding companies through the reduction of time spent by their goods in storage and consequently in the logistical efforts required, in carriers due to the reduction in volume of orders, and finally in retail trade due to the reduced necessity of data transmission.

4.2 The Employment Effects of Product and Process Innovations

In most cases the use of ICT results in employment effects of some kind. The intensity of the quantitative and qualitative effects on the work force is dependent on the extent of the qualitative improvement in productivity, organization and operating income which can be achieved through the use of new technology. Further criteria determining the employment effects are the quality and degree of specialization of the software used and the level at which it is introduced.

Quantitative employment effects can occur:

- due to the more rapid, efficient and flexible accomplishment of required tasks/work at a work station (improved control, improved access to information, reduction in on-the-job overheads),
- because functions between the next work station upstream and downstream are no longer necessary (e.g. on-line access by workers from different departments to a data archive, through the implementation of software programs which make quality-control stages redundant etc.),
- through reduced administration and coordination between the different work stations, (e.g. monitoring costs directly on the word processor enabling external cost monitoring to be dispensed with).

As is evident from this 3-way division of employment effects, not all employment effects are directly visible. Rationalization effects which occur in those work stations in which new technology is applied are easily identified. However, the rationalization effects experienced by the workers employed there are frequently compensated by redeployment to other areas. When a secretarial office is converted from electronic typewriters to word processors, for example, there is an immediate rationalization effect (process innovation), as the possibilities for data storage are greater and floppy disks are more flexible than traditional magnetic storage media so that errors are more easily corrected. At the same time word processing software offers additional possibilities in the form of word wrap and various types of advanced formatting (secondary rationalization effects).

On the other hand, these developments are leading to an increase in the quality expected of letters and manuscripts. The new technical opportunities are creating a more widely differentiated demand for well laid-out texts. Still, the demands made on clerical workers and secretaries to improve product quality can hardly be considered a real product innovation. On the contrary, the change in demand lessens the rationalization effect of the process innovation.

The conversion to the use of PCs can be accompanied by additional applications - such as supporting time scheduling, the internal monitoring of costs and the documentation of correspondence. If the possibilities offered by PCs are fully exploited in this way the work of secretaries often acquires aspects of tasks performed by clerical workers. The theoretical rationalization effect which would have occurred under ceteris paribus conditions (i.e. if output remains constant), is usually at least partially compensated (indeed sometimes more than compensated) by a series of inner-enterprise product and service innovations, additional skill requirements and changes in the existing redivision of labor. Together these effects shed a new light on the employment effect in the user work stations themselves.

In our case studies firms' representatives did not usually associate employment effects which arise in other than the user work stations (as a result of, for example, a change in the plant-level communication network) with the introduction of technology. Where such effects were mentioned the rationalization effect was usually very pronounced, leading to improvements at the organizational level. Following the introduction of on-line equipment in building societies, the labor required for processing and transporting files and for maintaining archives fell to almost zero. In this case it was possible to ascribe the rationalization effect of the enterprise network to these specific areas.

It is far more difficult to measure and to ascribe the effect of the fact that, for example, on-line access enables two employees simultaneously to fetch information from the same file. There will be less use of the telephone within the enterprise and reduced delays due to files being unavailable; in short the enterprise embarks upon a process of systemic rationalization. This reduction in required labor is, however, to be contrasted with the increase in telephone contacts with customers resulting from the (technologically induced) flexibility of the product range (more flexible choice of building society loans). Thus the effect of the product innovation (new products) complicates the measurement of the process innovation.

Data processing generally allows the tasks at a work station to be more efficiently organized due to a higher degree of standardization. DP makes it easier to check, control and evaluate work. This may permit a reduction in the amount of administrative work necessary for coordination within the enterprise. The evaluation of our case studies showed, however, that the increase in the volume of information and the improvement in the possibilities for controlling work bring about possibilities for improved supervision (systems of data collection on individual workers and the enterprise as a whole). At the same time it was only in exceptional cases that such data collection systems were used to their full effect, e.g. in relation to collective or individual wage negotiations.

Given the complexity of the rationalization effects of process innovations it cannot be claimed for sure that the use of new technology in the service sector has an overall potential for rationalization. Rather this potential is heavily dependent on the installation level (the number and type of installations), which of the possible applications are in fact used, the current status of the process of implementation and whether the enterprise has already achieved an optimal organizational structure which corresponds to the use of DP technology.

The majority of the enterprises studied by Infratest registered an overall (i.e. despite job losses in some departments) increase in employment, with constant or falling employment rather less common. As the use of technology frequently induced an expansion of the services rendered by the enterprise and consequently of the enterprise as a whole, it was often only possible to identify a hypothetical employment effect. Job losses (assuming a constant level of output) occurred mainly in secretarial work (data input, card processing, correspondence), bookkeeping and routine clerical work and in some cases in sales jobs where the emphasis is on customer counseling. In these areas the use of ICT induced considerable productivity effects. The more innovative an enterprise - particularly early users with an offensive market strategy - the more likely are the workers in the areas mentioned above to benefit from compensation effects, as they are normally redeployed within the enterprise. Our survey also showed that in one fifth of the enterprises the introduction of new technology led to new hirings, whereby skilled white-collar workers were usually the beneficiaries (12%).

In some cases negative employment effects resulting from investment in technology and the consequent expansion of the enterprises involved may occur in upstream and downstream markets. Our case studies provide examples of enterprises where the reintegration of correspondence or bookkeeping work led to a loss of orders for the firms to which such work had previously been contracted out. In the publishing industry the use of word processors has made the work of the setters in printing firms redundant. However, it proved difficult to quantify these effects. Job losses occurred in slightly less than 20% of enterprises. In the vast majority of cases the process of natural wastage was used. Some evidence was found (e.g. from our expert interviews) that in manufacturing industry certain types of work are being contracted out in the course of enterprise reorganization. Examples include packing, design, data and word processing, maintenance, marketing etc. This may enable enterprises in the service sector, including those affected by rationalization, to fill their order books.

The rationalization effects of process innovation are often concealed by the use of natural wastage for staff reduction. At the same time hirings often consist of workers with DP skills, so that the overall effect is to increase the average skill level of the work force. In some cases a higher overall employment level is necessary during the implementation phase as DP experts are taken on and the existing work force is maintained. Existing workers are (provisionally) kept on so that they may pass on their knowledge and experience of the workings of the enterprise in order to ensure the optimal adjustment of the new systems to the existing production process. This temporary increase in employment is then reduced at the end of the investment cycle; during this period part of the work force is being retrained. Some workers will be trained in simple skills for standardized, repetitive tasks (data input, correspondence), while others are trained in the sophisticated skills required for complex, directing activities (planning, programming, controlling). In some of the enterprises in our case study some workers, particularly older women workers did not want to work with VDUs and thus, more or less voluntarily, resigned.

No evidence of direct redundancies as a result of rationalization investment was found in our case studies. Where redundancies did occur they were without exception attributed to a loss of sales/turnover due to the overall business cycle or to economic factors at the micro level.

Only very seldom do measurable employment effects stem directly from the use of new ICT alone, but more frequently in association with improvements in organizational structures, new controlling and steering processes and the withdrawal from unprofitable profit centers. Exceptions to this come primarily in the form of job creation in DP departments (newly hired computer specialists) and job losses due to the application of ICT in the office environment (e.g. in inventory and dispatch departments merchandise information systems), the dissolution of unprofitable branches (enterprise data collection systems) and the use of DP in bookkeeping.

To summarize: the view of micro-electronics as a "job killer", widely held in the FRG, is not confirmed by the different studies carried out by Infratest. The intensity of the use of micro-electronics is not significantly correlated to differences in the employment trend in different enterprises. There is, on the other hand, a very close relation between increasing turnover and the use of micro-electronics. It was particularly evident from our case studies that an increase in turnover, and thus an expansion or stabilization of market position is in most cases all but impossible without the use of micro-electronics.

This is also confirmed by the fact that more than half of the enterprises in our survey reported that they would require on average 17% more staff without micro-electronics. Such an increase in staffing levels would, however, be accompanied by a rapid increase in administrative costs and thus a considerable deterioration in their competitive position. According to our survey it had not proved possible to diversify production to any great extent.

4.3 The Effects of the Use of Micro-electronics on the Skills and Qualifications of the Work Force

Considerable changes in the employment structure are usually seen to occur in the course of the introduction of ICT, and this general trend was confirmed by the views of most of the enterprises in our survey. The majority reported an overall increase in the level of skills and qualifications following the introduction of new technology: often internal or external (re)training was necessary; DP skills are becoming ever more important. Only a minority were of the opinion that flexibility is more important than a moderate level of skills and qualifications: specialist skills and a high level of general education remain of great importance. In a small number of cases specialist skills have been replaced by the use of DP while jobs with a low skill level have become redundant (table 4.2).

Table 4.2: The Effects of the Use of New Technology on the Skills and Qualifications of the Work Force
- Banks and Non-Banks -

	banks (n=93)	non-banks (n=145)
increased qualifications	72 %	52 %
retraining necessary	84 %	49 %
DP-skills more important	67 %	59 %
flexibility more important	34 %	25 %
reduction in jobs with low skill level	10 %	6 %

Source: Enterprise survey 1987.

The most important trend effects on the structure of employment can be summarized as follows:

- "The customer is king": the reintroduction of the comprehensive customer service is once again *à la mode*, at least in the branches in which new technology leads to the rationalization of standardized tasks and decision-making.
- The trend is towards increasing organization and differentiation of work, as the use of DP forces the enterprise to rationalize its organizational structures.
- The overall volume of standardized tasks is decreasing, often accompanied by a simultaneous expansion of the services offered (better presentation, greater variety).
- Product innovations often lead to greater flexibility and autonomy at the place of work; monotonous, repetitive work is on the decrease.
- At all skill levels the classical skill-profile requirements are giving way to demands for experience in the use of DP (or the willingness to come to terms with it).

The fact that ICTs require a broadly based increase in skill levels can be seen clearly from the payroll statistics and expenditure on training and education measures by the firms in our survey. There is a shift in the volume of work from simple information processing (correspondence, data input) to (semi-) skilled commercial work (granting loans, negotiating contracts, advising customers). To this extent the overall average level of skills and qualifications in the service sector has increased.

This does not represent an adequate description of the changes involved, however, as the use of VDUs and PCs and the extensive supply of context-specific, supporting software have increased the responsibilities of the individual worker. This extension is sometimes achieved at the cost of deskilling because some of the operations are performed automatically by the new technology. Such workers in financial service companies, for example, can now present a customer with effective interest rates quickly and without effort while explaining to customers the effects of the various conditions attached to a given loan. By entering certain important data, such as income, monthly repayments, value of collateral, he or she can do the groundwork for the granting of a loan; appropriate programs guide him or her through difficult procedures. This is one reason for the claim which is often heard that the computerization of such work in the service sector leads to deskilling. At the same time responsibility for different types of work increases because workers with the same skills are able to deal with a wider range of tasks. In this way traditional knowledge of the specific task (how to calculate graduated interest, the provisions which must be taken into account when clearing goods through customs) is being replaced by operational or systemic knowledge. Employees in these areas must know how to use the different hardware and software and how to apply them to specific cases. They must be able to respond correctly to information provided by the terminal. Last but not least, even when many processes are carried out automatically, staff must be able to explain them to clients, i.e. they must retain at least a passive knowledge of the tasks involved.

In the banking sector too the widespread use of ICT has brought considerable changes in the employment structure. The number of employees entering data or performing booking functions (i.e. simple clerical work) has decreased. An increasing number of routine tasks are being carried out by computer, especially for remittances. The support of computers in this area is also leading to a new integration of tasks and skills. In many banks the specialization between the savings, current account and loans counters has been (partially) done away with. The experts interviewed by us were agreed that the average skill groups, i.e. those with a training in the banking sector of approx. three years, will in future make up a larger proportion of the total work force, while the share of the lowest skilled will decrease still further. During the implementation phase of micro-electronics the financial institutions increased their expenditure on employee training accordingly. The annual report (for 1986) of a large German bank, for example, states that a total of 20 800 employees had taken part in training schemes in that year. This is equivalent to c. 50% of the work force. The same institute spent DM 140 million on retraining in 1986, over 6% of the total wage and salary bill.

It is generally agreed that after the initial implementation phase, the intensity of retraining is likely to tail off again. This is also due to the fact that the requalification of the work force is a result, not only of the retraining programs, but also of the exit of less qualified employees and the hiring of better qualified workers. Despite the positive overall trend towards higher qualifications, a certain amount of deskilling is also likely to occur, especially in financial institutions in the field of customer service. This is due to the changing status of factual knowledge which can now be stored and accessed by computer.

Nowadays it is often impossible for an employee in a freight forwarding company to complete a customs declaration or application for clearance without DP support. The program is necessary as it provides the specimen pattern for the declaration. It is often the case now that at least some bank personnel are no longer able to complete a savings book by hand.

In our conversation with experts they often conjured up the horror vision of general computer failure in which case no-one would be able to perform complicated business transactions. This is a reflection of the fact that our experts have had many years of experience in the trade and had been used to solving problems at a completely different level of technical development. Nevertheless, such criticism, with its touch of nostalgia, gives some idea of the extent of the changes which have occurred in the nature of work in these areas.

As a rule employees are also required to have a greater volume of information at their command. It remains an open question whether the increase in operating or systemic knowledge will lead to an overall increase in skill levels. However this may be, it is clear that, in addition to traditional skills, the new systemic knowledge will become more and more indispensable.

The "traditional" type of knowledge can be divided into

- fundamental knowledge, which is important for understanding the way in which the enterprise/firm is organized, and
- knowledge which is simply no longer required (e.g. calculating interest rates).

The relative importance of each type of knowledge would have to be studied in each context before an overall trend towards reprofessionalization can be claimed. Whatever the result of such an analysis, it is clear that there is a dramatic process of skill and qualification restructuring taking place at present, which is leading to considerable expenditure on retraining measures, both within the enterprise and externally. This process of restructuring will bring about lasting changes in the existing pattern of skills in the Federal Republic of Germany.

To take retail trade as an example, the skills of department managers, sales personnel and purchasers/management clerks are being dramatically affected by the organizational changes brought about by the introduction of ICT. The centralization induced by the use of merchandise information systems relieves department managers of administrative work, reduces his or her responsibility for procurement while increasing that for personnel work and labor deployment. For sales personnel and purchasers centralized controlling means changes in their time budgets (deployment by turnover/customer frequency) and performance control (with its disciplinary and performance-stimulating effects). The net result for these groups is an intensification of work and increased stress.

As the analysis of our case studies shows, the qualitative employment effects which were emphasized by the enterprises in our survey, and which they normally considered to be in the direction of higher qualifications, are primarily based on the scarcity of (technically) qualified skilled labor. However, this will apply only in the short term.

In the course of the innovation process (in the case of the conversion to word processors this lasted between four and five years according to our experts) a corresponding development of the skills of the work force takes place. Whereas good secretaries with systemic knowledge were rather the exception four years ago, the market is now more or less in equilibrium. When new secretaries are hired a great emphasis is placed on experience with word processing, with the result that in the near future there will probably be very few without the necessary qualifications.

The support of new technology is also changing the role played by employees in the enterprise environment. In future the terminal will be the "colleague" with which the end-user has greatest contact. Workers' autonomy will increase to the extent that "holistic" working structures are reintegrated into the enterprise. The effects of this new autonomy and independence together with an increase in external control, not by immediate superiors, but rather (particularly in large enterprises) by controlling groups not directly involved in the labor process, will force firms to consider how leadership structures within the enterprise and the role of white-collar workers are to be changed.

A distinction must be made between two skill levels:

- skills based on knowledge (input skills) defined as analytical, systemic and contextual knowledge, and
- operating skills (output skills) which are determined by the performance of the employee at his or her place of work.

While the second form of qualification can be controlled in the context of the use of ICT, input qualifications, together with a third component of qualifications ("social skills") are more difficult to verify. Whereas input qualifications and social skills were considered to be of great importance when they were evaluated by immediate superiors, there is the danger that their significance may decrease in the future.

Changes in Enterprise Size and Employment Levels in the Branches of the Federal Republic of Germany 1980 to 1986

A Longitudinal Analysis of Employment in Enterprises Active between 1980 and 1986, "New" and "Non-Surviving" Enterprises

Andreas König and Gernot Weißhuhn in collaboration with Jürgen Seetzen

1. Introduction

In the current discussion on the role of enterprise size in changes in employment levels the empirical results which have been obtained have been a subject of controversial debate. (For a summary of these results cf. Fritsch/Hull 1987a; for a critical analysis of German, American and British studies cf. Fritsch/Hull 1987 b, Eckart/v. Einem/Stahl 1987, Storey/Johnson 1987). The main reason for this is the difficulty of empirically testing the hypotheses made regarding the importance for employment of firms and enterprises of different sizes. The greatest problem here is the availability of statistical data, in particular of time series data on enterprise-size-specific developments, as it is longitudinal data on firms and enterprises rather than cross-sectional information in which we are interested.

The research presented here aims at a longitudinal analysis based on recent data of the enterprise-size-specific changes in employment in the various branches in the Federal Republic of Germany. This study is one of the results obtained by the enterprise-size-specific analysis of employment changes conducted within module 3 (development of new statistics frameworks) as part of the Meta Study. The *leitmotif* of our study was the view that the deployment of labor ought to be studied at enterprise level in order to create a bridge between technical progress, which proceeds at very different rates in different branches and enterprises in different size categories, and employment and its structure.

The results which we have obtained are of a descriptive nature, indicating changes in the number of enterprises in the different branches and change in employment between 1980 and 1986 (June 30 of each year). In view of the massive amount of data which had to be processed - the study of changes in some 1.3 million enterprises in the branches analyzed - no attempt has been made in this study at an empirical test of theoretical hypotheses.

While there can be no doubt that in times of high unemployment it is important to analyze the employment developments in firms and enterprises of different sizes, it must not be forgotten that this indicator is in itself not sufficient to analyze the role played by small medium and large firms for economic growth (Bade 1987). Such a task would require the integration of both supply-side conditions for growth with demand-side aspects in a size-specific analysis of enterprise growth.

As far as employment theory is concerned the question remains unanswered as to the causes behind these developments. This would require an econometric analysis on the basis of short and long-term employment functions in order to explain the optimal demand for labor. A number of econometric studies are indeed available (Jäger 1980, Schulte zur Surlage 1985, Deutsch 1988), but were all conducted at the macroeconomic or branch level rather than being differentiated by enterprise size. This analytical deficit is primarily due to the lack of suitable data, such as firm or enterprise-specific factor costs.

In addition to the theoretical aspects of the role of enterprises of different sizes there is also the question of the "level" of study. This is not primarily a problem of classifying enterprise size groups (e.g. as measured by number of employees, turnover, value-added) as levels of aggregation can be selected at will provided the data base allows a sufficient degree of differentiation. The much more important distinction is that between "firms" and "enterprises". This study had to be based on the "enterprise" in the sense of "place of production", as the data base used did not permit a classification in terms of the "firm". Thus changes at firm level (multi-enterprise firms) as a legal entity and the grouping of firms in concerns and trusts were ignored in our analysis. On the other hand, the use of the "enterprise" as the analytical basis enables questions of the regional economy to be addressed in a much more satisfactory way.

Based on the same data (employment statistics for the period 1977 to 1985) a longitudinal study of enterprise-size-specific changes in employment has recently been conducted for the FRG (Cramer 1987 and for the service sector Tengler/Dahrenmöller/Cramer 1987). The major difference between this study and our analysis is that while Cramer et al also differentiate between enterprises of different sizes (according to the number of employees) no differentiation was attempted between "new" and "non-surviving" enterprises by branch. Furthermore, no mention is made of movements of the "surviving" enterprises between the different enterprise-size groups.

Our study consisted of an integrated analysis at branch level (excluding agriculture and forestry, transport, post, the public sector, nonprofit organizations and private households). Specifically we attempted to answer the following questions for each branch and enterprise-size category (cf. tables 1.1 and 1.2):

a) How many "survivor" enterprises were there over the period 1980 to 1986?
b) How many "new" enterprises were formed in that time?
c) How many enterprises did not survive?

The answers to these questions serve as a basis for an analysis of the employment gains and losses in cases a) to c), differentiated by enterprise-size. This allows us to take account of the employment losses due to some enterprise closing, rather than dealing exclusively with "survivor" enterprises, as this would lead to a distortion of the figures on the changes in employment levels (on this "survivor-bias" cf. Fritsch/Hull 1987b). The exclusion of "new" enterprises also produces a distortion, as it leads to an underestimation of the number of jobs created (Fritsch/Hull 1987b).

An integrated study also enables us to analyze changes in enterprise-size structure from two perspectives: into which size categories have enterprises shifted, in which categories have new enterprises started up business, and to what structural shifts has this given rise (target year perspective); from which size groups in the base year (1980) have the enterprises developed, in which groups have enterprises failed to survive and what have been the employment effects (base year perspective).

At this point it is necessary to point to two limitations of the analysis which must be borne in mind when interpreting the results. The longitudinal analysis was based on the enterprise numbers, whereby this is subject to certain problems (for details see section 2). Furthermore, due to data protection legislation a number of points on the transfer matrices 1980/86 (tables 1.1 and 1.2) were not filled in order to prevent direct reference to individual enterprises.

Table 1.1: Sector 08 Iron, Steel, Nonferrous Production

size category in 1986:	0	1	2	3	4	5	6	7	8	9	total	total	total
		number of enterprises											
size cat. in 1980	(abandoned)	1-4	5-9	10-19	20-49	50-99	100-199	200-499	500-999	>999	Sur-vivors 1980	"new" enter-prises	entrprs 1980
0 "new"	0	73	22	20	15	9	8	*	*	*		147	
1 1- 4	78	72	17	8	*	0	*	0	0	0	97		175
2 5- 9	22	12	30	17	*	*	0	0	0	0	59		81
3 10- 19	18	*	16	44	12	*	*	0	0	0	72		90
4 20- 49	11	*	*	9	53	18	0	0	0	0	80		91
5 50- 99	7	*	*	0	7	28	9	*	0	0	44		51
6 100- 199	*	0	0	0	*	14	39	*	0	0	53		53
7 200- 499	*	*	0	0	*	0	*	44	*	*	44		44
8 500- 999	*	0	0	0	*	0	0	5	28	*	33		33
9 > 999	*	*	0	0	0	0	0	0	10	52	62		62
Σ survivors 1986		84	63	78	72	60	48	49	38	52	544		
Σ "abandoned"	136										enterprises 1980: 680		
Σ all entrprs. 1986		157	85	98	87	69	56	49	38	52	enterprises 1986: 691		

Cells representing less than five enterprises are marked with * and are excluded in totals.

gains/losses of employees (absolute figures)

size category in 1986:	0	1	2	3	4	5	6	7	8	9	total	total	total
size cat. in 1980	(abandoned)	1-4	5-9	10-19	20-49	50-99	100-199	200-499	500-999	>999	Sur- vivors 1980	"new" enter- prises	entrprs 1980
0 "new"	0	133	135	293	445	582	1.220	*	*	*		8.655	
1 1- 4	-135	-13	38	90	*	0	*	0	0	0	281		146
2 5- 9	-136	-39	-7	94	*	*	0	0	0	0	141		5
3 10- 19	-242	*	-74	23	155	*	*	0	0	0	343		101
4 20- 49	-305	*	*	-103	109	367	0	0	0	0	193		-112
5 50- 99	-512	*	*	0	-117	103	283	*	0	0	206		-306
6 100- 199	*	0	0	0	*	-527	130	*	0	0	-515		*
7 200- 499	*	*	0	0	*	0	*	-459	*	*	673		*
8 500- 999	*	0	0	0	*	0	0	-738	-818	*	-1.767		-3.337
9 > 999	*	*	0	0	0	0	0	0	*	-52.342	-68.862		-73.812
Σ survivors 1986		-3.017	-294	104	-1.252	117	417	-834	-14.979	-49.569	-69.307		
Σ "abandoned"	-8.799												
Σ all entrprs. 1986		-2.884	-159	397	-807	699	1.637	-222	*	*		total gains	-69.451

Cells containing figures that could disclose informations about single enterprises are marked with *. The totals however contain these figures.

Source:own computations basing on the "Beschäftigtenstatistik 1980 and 1986 (June 30)"

Table 1.2: Sector 37 Private Law and Business Services, Architecture, Engineering

						number of enterprises							
size category in 1986:	0	1	2	3	4	5	6	7	8	9	total	total	total
size cat. in 1980	(abandoned)	1-4	5-9	10-19	20-49	50-99	100-199	200-499	500-999	>999	Sur-vivors 1980	"new" enter-prises	entrprs 1980
0 "new"	0	27.962	3.706	1.175	378	68	23	10	*	*		33.322	
1 1- 4	15.590	24.259	4.405	513	65	7	*	0	0	0	29.249		44.839
2 5- 9	2.138	3.747	6.780	1.804	124	5	0	0	0	0	12.460		14.598
3 10- 19	705	438	1.358	3.151	532	18	*	0	0	0	5.497		6.202
4 20- 49	248	67	92	405	1.059	133	15	*	0	0	1.771		2.019
5 50- 99	48	9	5	10	84	181	33	*	0	0	322		370
6 100- 199	16	5	*	0	*	22	85	23	0	0	135		151
7 200- 499	*	*	*	0	*	*	9	39	5	0	53		53
8 500- 999	*	*	0	0	0	0	*	*	8	0	8		8
9 > 999	0	0	0	0	0	0	0	*	*	9	9		9
Σ survivors 1986		28.525	12.640	5.883	1.864	366	142	62	13	9	49.504		
Σ "abandoned"	18.745										enterprises 1980: 68.249		
Σ all entrprs. 1986		56.487	16.346	7.058	2.242	434	165	72	13	9	enterprises 1986: 82.826		

Cells representing less than five enterprises are marked with * and are excluded in totals.

gains/losses of employees (absolute figures)

size category in 1986: / size cat. in 1980	0 (abandoned)	1 1-4	2 5-9	3 10-19	4 20-49	5 50-99	6 100-199	7 200-499	8 500-999	9 >999	total Survivors 1980	total "new" enterprises	total entrprs 1980
0 "new"	0	47.319	23.508	15.270	10.757	4.621	3.069	2.816	*	*		109.245	
1 1- 4	-25.911	547	13.812	4.942	1.666	*	*	0	0	0	21.635		-4.276
2 5- 9	-13.730	-11.690	894	8.198	2.354	341	0	0	0	0	97		-13.633
3 10- 19	-9.249	-4.302	-6.143	1.402	5.280	*	*	0	0	0	-2.478		-11.727
4 20- 49	-7.247	-1.731	-1.795	-3.666	*	3.540	1.363	*	0	0	-860		-8.107
5 50- 99	-3.217	*	-269	-450	-1.901	396	1.635	*	0	0	-863		-4.080
6 100- 199	-2.158	-638	*	0	*	-846	220	2.084	0	0	45		-2.113
7 200- 499	*	*	*	0	*	*	-860	388	1.463	0	*		-1.611
8 500- 999	*	*	0	0	0	0	*	*	*	0	*		-1.799
9 > 999	0	0	0	0	0	0	0	*	*	2.873	2.493		2.493
Σ survivors 1986		-19.572	5.447	10.426	7.816	4.095	2.373	3.220	1.008	2.873	17.686		
Σ "abandoned"	-62.539												
Σ all entrprs. 1986		27.747	28.955	25.696	18.573	8.716	5.442	6.036	*	*		total gains	64.392

Cells containing figures that could disclose informations about single enterprises are marked with * The totals however contain these figures.

Source: own computations basing on the "Beschäftigtenstatistik 1980 and 1986 (June, 30)"

2. Data Base and Method

2.1 Employment Statistics

The study is based on an evaluation of the data provided by the employment statistics for the 30th June 1980 and 1986. The statistics available provide information on the following attributes of each employee subject to social insurance contributions:

- enterprise number
- district
- sex
- branch
- occupation
- position within firm
- skill level
- age

The enterprise number was used for the aggregation of the data; the number of employees for each enterprise number was counted for each of the two reference dates. Using the enterprise number the two data bases were then collated into a longitudinal data base, such that the changes in the number of employees could be studied for each enterprise number.

However, this approach entails a not insignificant problem. Enterprise numbers are assigned by the Federal Employment Agencies and this is not standardized. This means that certain enterprises may be assigned more than one number or the number may be changed due to a change in ownership or legal status, although from an economic point of view the enterprise as such continues to exist much as before. Equally, a number of enterprises may be assigned to just one number if they belong together in administrative terms and are registered in the same region and branch. In some cases the employer may even be allowed to keep "his" number following the closure of an enterprise, and this same number may then appear once more if he decides to found another enterprise.

A final point is that workers not subject to social insurance contributions (the self-employed, family workers, those working fewer than the minimum hours for compulsory insurance contributions and civil servants) are not included in the employment statistics. However, the exclusion of civil servants poses no problem in that state and quasi-state enterprises did not form part of our study (see below).

The above factors lead to distortions in all three types of enterprise considered, the "survivors", "non-survivors" and the "new" enterprises. For this reason the analyses of enterprise size will only lead to accurate results if the distortions resulting from the problem of enterprise-number assignment are relatively evenly distributed.

For the purposes of the study the attribute "branch" was transformed from a 3-digit classification by the Federal Labor Office into a systematic of 42 branch classification. Considerable attention was paid in the construction of the systematic that all state enterprises (local and municipal administration, social insurance) and enterprises owned by non profit-making organizations (trade unions, political parties etc.) were separated from the rest and then excluded from the analysis. The high level of differentiation was maintained not only for manufacturing industry but also for the service sector. The transport and communications sectors were also excluded because the Federal Railway and Postal Services - both public concerns - are dominant in the two sectors respectively. Similarly the agriculture and forestry sector was not studied due to the high proportion of self-employed workers, so that many enterprises do not employ any workers which are subject to social insurance, and which thus do not appear in the statistics.

As was mentioned at the start of this contribution, the remaining attributes listed in the employment statistics were not evaluated. In this study a merely descriptive "picture" of the developments in the different enterprise-size categories is evaluated. The effects of other explanatory variables such as regional structure (the proportion of large enterprises, regional unemployment etc.) or skill structures and their changes over time are, however, to be included in further studies.

Trainees and apprentices were counted simply as employees. Differentiation on this point would indeed have been desirable, because as yet no suitable studies have been conducted which provide information on the number and nature of enterprises providing places for trainees. However, the sheer volume of the data meant that this question too must be postponed for further study.

2.2 The Consistency of Data in the Case of a Change of Branch

The aim of our study was to describe the structures of enterprise size and their changes *within* individual branches. In order to obtain a consistent picture it is necessary to make allowances for those enterprises which were assigned to a different branch in 1980 compared to 1986.

In this context it was necessary first to consider a small group of enterprises whose employees were assigned in one year to more than one branch. These result from coding errors, which were corrected as a first step in improving the data on the principle of relative majority: each enterprise number whose employees were registered under different branch codes were ascribed to the branch most often registered. In the case of a "tie" the first branch mentioned was chosen. In 1980 this correction applied to 19 468 employees in 4 792 enterprises. In total these enterprises employed 170 625 workers, i.e. not more than 11 %. The picture in 1986 was similar: 10 985 employees in 2 926 enterprises were corrected for; total employment in these enterprises was 117 941.

The adjustment proved more difficult for those enterprises which were assigned differently in the two years studied. As the branch given in the employment statistics represents the main activity of a firm, it cannot simply be assumed that those whose branch ascription has changed represent errors in classification. The slightest shift in production may result in a change in classification. Thus our analysis required a preliminary examination of the data in order to determine from which and into which branches enterprises had moved.

This was followed by a correction of the data. All enterprises registered in both 1980 and 1986 (survivors) were assigned to the branch ascribed to them by the employment statistics for 1986. This approach was based on the view that any changes in production must have been slight and that the enterprise must have been at least partially active in the relevant sector at the beginning of the period studied. Furthermore it can be assumed that there is a time lag between the change of branch and this being registered in the statistics, so that the error resulting from our approach ought to be negligible.

According to this method approximately 15 000 enterprises with slightly less than 280 000 employees (1980) changed their classification during the period studied. At a higher level of differentiation the figures increase correspondingly (e.g. using the 3-digit systematic of the Federal Labor Office a change in classification would apply to 20 000 enterprises with some 460 000 employees).

The most important "losers" were retail and wholesale trade with a combined net loss of 1 287 enterprises. The automobile industry was the most prominent "winner", but light metals, construction, transport, hostelry and "administration of land, housing and property" also made net gains.

Taking all branches together, enterprises which changed their branch classification had a total of 7 042 less employees in 1986 compared with 1980. If the development (growth) of these enterprises is differentiated by branch the picture is almost completely uniform, neither particularly rapid growth or decline. The one exception to this being enterprises which have changed between wholesale/retail trade and manufacturing industry. Those changing from trade to industry increased employment by 8.6% (+2 704) while those moving in the other direction suffered an employment loss of 40% (-5 538).

The figures presented so far do not provide great insights into the question which enterprises have left and entered which branches. The matrix of all 42 branches which documents these changes is, however, much too extensive and contains far too many insignificant results to be presented here. Table 2.1 is a summary of the most notable results of the full matrix, showing the movements in both directions:

Table 2.1: Selected Changes of Economic Sectors

		Change of Sector Number of Enterprises (in parenthesis number of employees in 1986)	
Sector "A"	Sector "B"	Change "A" --> "B"	Change "B" --> "A"
wholesale trade	retail trade	692 (9,890)	619 (6,435)
retail trade	motor industries	428 (6,622)	101 (1,275)
wholesale trade	motor industries	362 (6,024)	64 (429)
retail trade	building, constr.	257 (2,217)	159 (973)
motor industries	machinery	31 (6,386)	50 (1,211)
wholesale & retailing	processing ind.	2,071 (34,086)	1,158 (13,833)
wholesale & retailing	private services	824 (7,055)	455 (3,782)
processing ind.	private services	483 (8,153)	480 (15,134)

2.3 Evaluation

Within the framework of the integrated analysis the changes in employment in each branch were analyzed for the period 1980 to 1986, i.e. in a longitudinal approach. Based on the data from the employment statistics for the two years in question (see above) movements in the following variables can be detected:
- survivor enterprises (enterprise number available for both 1980 and 1986)
- non-survivor enterprises (enterprise number only registered for 1980)
- new enterprises (enterprise number only recorded for 1986)

Based on this differentiation it is possible to examine the sectoral changes in the number of enterprises and employment in the survivor enterprises, together with changes in enterprise-size category between 1980 and 1986. For the non-survivors the employment loss can be differentiated by branch and enterprise-size as can the employment gain in the new enterprises. Figure 2.1 provides a schematic model of these findings:

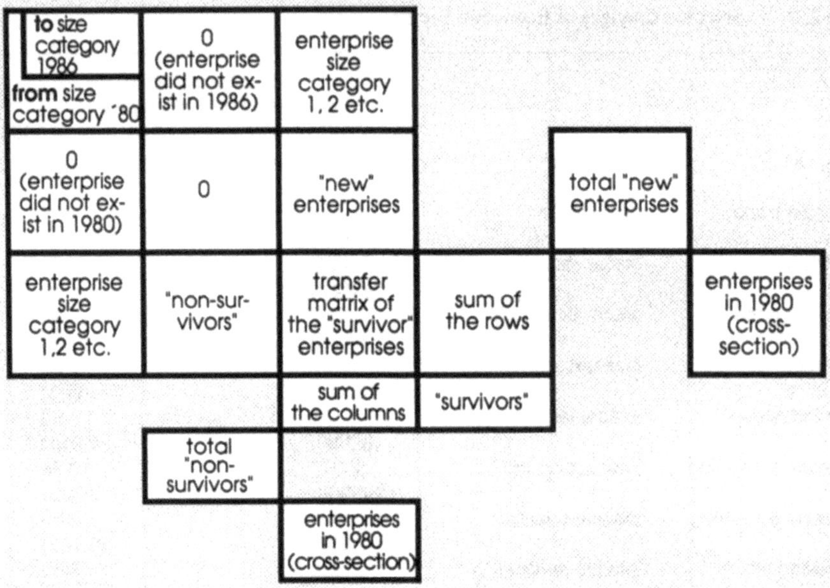

Figure 2.1: Number of Enterprises and Employment Gains and Losses (absolute)

3. The Major Empirical Results

3.1 The Extent of the Enterprise-size-specific Change in Employment

This section will first consider the extent of changes in employment for the economy as a whole. The results are summarized in table 3.1. The overall employment loss (of workers subject to social insurance contributions) amounted to approx. 223 000 persons between 1980 and 1986. This change consisted of two major components (the sectors "agriculture" and "transport" can be excluded due to their negligible changes in employment). In the public sector and among the nonprofit organizations employment increased by 328 000 whereas in the private sector branches studied here employment fell by 554 000. In section 3.2 the changes in employment in the private sector will be analyzed at a higher level of disaggregation on the basis of the integrated methodology described above.

Table 3.1: Sectoral Development of Employment in the Federal Republic of Germany 1980 - 1986

employees	30.6.1980 in 1,000	30.6.1986 in 1,000	gains/losses 1980-1986 in 1,000
total (obliged to social security system), June 30	20,953.5	20,730.1	-223.4
sectors of occupation:			
agriculture	219.2	231.3	12.1
traffic and transportation, postal services	998.6	994.7	-3.9
public sector, nonprofit organizations	3,066.5	3,394.0	327.5
without sectoral classification	6.4	1.0	-5.4
analysed private sectors: energy, industries, construction, banking, insurances, services	16,662.8	16,109.1	-553.7

source: own computations of the "Beschäftigtenstatistik"

3.2 Sectoral Patterns of Enterprise-size-specific Changes in Employment

For the longitudinal analysis of the loss of employment in the private sector branches included in our study were examined separately as the differences between developments in each sector were thought likely to be considerable. This was complemented by an enterprise-size-specific analysis of the survivor, non-surviving and new enterprises. Table 3.2 shows the employment trends in the different sectors differentiated by enterprise-size categories and by the three different types of enterprise.

Table 3.2: Development of Employment 1980 to 1986 by Economic Sectors and Categories of Size of Enterprises in "Survivor" Enterprises, in "New" and in "Non-Survivor" Enterprises (absolute figures)

		gains and losses in employment					
		surv.	surv.	surv.	surv.		total
			size of enterprise 1980			total	empl.
		1-19	20-49	50-499	>499	surv.	1980
02	energy, water supply mining	3,173	3,226	4,783	-12,618	-1,436	482,487
03	chemical ind., mineral oil	1,578	1,075	838	-8,241	-4,750	586,160
04	plastics processing ind.	5,465	2,115	2,981	-3,365	7,196	243,321
05	rubber and asbestos	365	67	484	-8,810	-78,94	114,253
06	stone and clay	-2,439	-6,136	-15,908	-5,624	-30,107	220,518
07	ceramics and glass	393	-462	-5,251	-14,132	-19,452	156,209
08	iron, steel, nonferrous prd.	765	193	364	-70,629	-69,307	333,799
09	foundries, rolling, drawing	11,660	1,694	-256	-17,055	-3,957	399,531
10	light metal, fitting,rep.	5,102	-1,810	-8,910	-8,524	-14,142	352,783
11	machinery	14,056	7,283	4,131	-49,349	-23,879	1,033,783
12	motor industries	1,945	-9,610	-4,306	24,553	12,582	973,279
13	ship industries	-8	-48	-146	-11,955	-12,157	60,030
14	electrical,electron. ind.	11,498	6,963	12,467	-18,891	12,037	1,117,415
15	precision mechanics, optical goods	2,915	-1,526	1,160	-6,402	-3,853	212,985
16	iron and metal products	4,211	1,230	-3,782	-5,357	-3,698	394,853
17	wood products	6,395	-4,432	-22,928	-10,374	-31,339	454,948
18	paper processing ind.	519	-532	-4,180	-2,737	-6,930	171,368
19	printing industries	3,676	-1,390	-4,354	-5,950	-8,018	224,990
20	leather processing ind.	491	-1,177	-6,969	-2,084	-9,739	103,210
21	textile industries	1,052	-1,213	-20,750	-17,185	-380,96	327,208
22	clothing industries	537	-3,362	-20,117	-2,737	-25,679	283,505
23	food industries	30,396	1,524	-13,896	-16,794	1,230	702,577
24	tobacco industries	-20	-54	-1,017	-1,346	-2,437	20,171
25	remaining industries	220	-934	-3,690	-2,953	-7,357	162,729
26	building, construction	7,332	-36,353	-86,216	-17,674	-132,911	1,698,509
27	wholesale trade	23,826	-12,897	-39,365	-9,466	-37,902	1,173,869
28	Retail trade	7,233	-19,818	-56,591	-41,172	-110,348	1,718,692
30	banking	14,932	10,884	29,123	5,710	60,649	538,204
31	insurance	3,610	-603	-3,119	-1,170	-1,282	224,019
32	private hotels, restaurants	16,220	2,073	1,141	-87	19,347	425,906
33	private movies, TV, radio	246	-153	-545	2,407	1,955	44,693
34	newspaper/books publ.	3,350	228	141	-3,509	210	120,918
35	private educ., science, culture,sport	4,519	1,197	719	132	6,567	73,462
36	prv. health, veterinary,	7,385	1,739	7,925	943	17,992	373,961
37	prv. law & business services/architects	19,254	-860	-1,916	1,208	17,686	415,695
38	private property & housing management	1,014	-1,165	-5,036	-1,182	-6,369	127,236
39	private advertising	2,690	-489	-550	-559	1,092	37,339
40	remaining private services	19,165	4,722	2,837	-903	25,821	558,163

Table 3.2: Development of Employment 1980 to 1986 by Economic Sectors and Categories of Size of Enterprises in "Survivor" Enterprises, in "New" and in "Non-Survivor" Enterprises (absolute figures)

gains in employment					losses in employment				
new	new	new	new		non-surv.	non-surv.	non-surv.	non-surv.	
	size of enterprise 1986			new		size of enterprise 1980			non-surv.
1-19	20-49	50-499	>499	total	1-19	20-49	50-499	>499	total
2,154	1,585	4,289	2,228	10,256	-1,797	-1,198	-4,049	-12,794	-19,838
2,767	2,068	7,695	5,544	18,074	-3,011	-2,115	-7,977	-1,217	-14,320
7,549	4,562	8,746	1,718	22,575	-6,456	-5,713	-11,764	-618	-24,551
723	472	1,633	3,012	5,840	-773	-790	-2,019	-1,350	-4,932
7,682	2,675	3,131	1,333	14,821	-11,036	-6,849	-7,867	0	-25,752
1,847	867	4,844	1,382	8,940	-1,683	-1,016	-7,626	-1,609	-11,934
561	445	2,414	5,235	8,655	-513	-305	-1,461	-6,520	-8,799
15,257	5,310	8,042	2,432	31,041	-13,805	-6,391	-16,794	-2,784	-39,774
21,722	8,304	10,731	1,109	41,866	-20,272	-11,269	-19,137	-1,524	-52,202
16,392	9,648	28,353	14,960	69,353	-12,984	-11,604	-40,773	-9,313	-74,674
30,920	7,404	7,585	1,419	47,328	-23,335	-9,116	-10,910	-3,433	-46,794
517	144	1,251	1,090	3,002	-475	-449	-1,894	-2,834	-5,652
18,928	9,700	22,757	23,411	74,796	-12,788	-7,635	-28,074	-24,906	-73,403
14,797	3,766	5,063	1,767	25,393	-7,292	-3,390	-5,677	-2,558	-18,917
7,708	4,161	9,415	2,600	23,884	-7,509	-6,991	-17,779	-5,130	-37,409
23,180	5,826	9,876	0	38,882	-26,850	-13,247	-34,041	-3,370	-77,508
2,023	1,528	5,857	0	94,08	-2,454	-2,317	-8,365	-536	-13,672
10,854	3,673	4,719	785	20,031	-8,857	-4,154	-7,757	-1,403	-22,171
2,933	1,021	4,069	0	8,023	-4,518	-4,081	-9,187	0	-17,786
3,698	2,372	6,511	1,174	13,755	-6,195	-7,188	-29,321	-5,783	-48,487
8,765	7,306	9,437	787	26,295	-14,466	-15,480	-40,552	-1,588	-72,086
30,513	7,614	15,865	1,819	55,811	-44,540	-11,610	-22,484	-3,229	-81,863
24	0	242	585	851	-167	-523	-1,276	0	-1,966
4,836	1,018	2,531	548	8,933	-5,254	-1,960	-4,568	0	-11,782
128,188	40,592	33,324	1,650	203,754	-153,520	-79,589	-92,202	-2,745	-328,056
102,662	27,022	31,722	1,921	163,327	-109,412	-40,019	-40,248	-581	-190,260
210,646	45,392	40,387	4,078	300,503	-196,121	-35,904	-40,267	-2,262	-274,554
8,683	2,060	2,477	1,722	14,942	-10,227	-2,354	-4,754	0	-17,335
10,835	1,093	2,096	0	14,024	-6,552	-2,130	-1,617	0	-10,299
125,359	22,949	19,132	0	167,440	-109,080	-15,224	-8,109	0	-132,413
4,590	830	408	0	5,828	-2,622	-515	-328	0	-3,465
5,037	1,615	3,697	0	10,349	-3,396	-1,419	-2,170	-528	-7,513
18,874	2,695	4,332	1,472	27,373	-10,695	-1,253	-895	0	-12,843
101,422	3,291	5,561	748	111,022	-50,169	-2,175	-3,719	0	-56,063
86,097	10,757	10,506	1,885	109,245	-48,890	-7,247	-5,888	-514	-62,539
26,668	3,432	2,823	960	33,883	-17,826	-2,145	-3,307	0	-23,278
9,853	1,265	543	0	11,661	-5,826	-1,033	-542	0	-7,401
81,770	22,492	31,825	3,635	139,722	-64,064	-12,252	-16,751	-524	-93,591

The employment trend in each sector is composed of these different elements. However, while it is obviously of importance to determine the absolute total figures, the study was complemented by the following relative values (for each sector and enterprise-size category):

a) the overall sectoral increase in employment in the survivor enterprises 1986/80 (absolute employment change between 1980 and 1986 with respect to employment in 1980)
b) enterprise-size-specific growth of employment in the survivor enterprises (enterprise-size-specific employment gain/loss 1986/80 with respect to employment in each sector in 1980)
c) distribution of the employment gains in new enterprises by enterprise-size category (share values in %)
d) distribution of the employment losses in non-surviving enterprises by enterprise-size category (share values in %).

These relative values are shown in table 3.3.

The pattern of employment trends and their component parts by enterprise-size category differs widely in the various sectors, so that an aggregation to the level of large sectors (e.g. manufacturing industry) would not appear to provide valuable results: this would merely obscure contradictory trends in different sectors. Given the plethora of individual results only a selection can be presented and discussed in this contribution. The absolute figures for the number of enterprises and employment, for example, are not documented here; the interested reader is referred to Weißhuhn/König/Sakkas/Seetzen 1988):

- Energy, water supply and mining:
  The number of enterprises in this sector did not increase noticeably; at the same time employment fell by about 11 000. In survivor enterprises as a whole the loss of employment was low; losses were particularly high in enterprises which in 1980 employed more than 500 workers. The net loss of employment is due to enterprise closures among medium and large enterprises, which were not compensated by the creation of new enterprises (which were relatively evenly distributed among the size categories).
- Manufacturing industry:
  Pharmaceuticals: The number of enterprises decreased slightly, but the overall loss of employment was very slight (c. 1 000). The gains from new enterprises were greater than the losses from non-survivors (particularly in the medium-sized enterprises). One noticeable trend were the particularly high losses in enterprises which employed more than 500 workers in 1980.
  Iron, steel, non-ferrous metal production: The number of enterprises increased slightly whereas these industries experienced job losses of 69 000. The net effect

of new and non-survivor enterprises was insignificant (slight losses balanced by slight gains). The large losses were thus due almost entirely to the survivor enterprises. Although firms which in 1980 were in the small and medium-sized categories increased employment slightly, the losses in enterprises in which more than 500 workers were registered in 1980 were very large.

Steel and light metal construction, railway engineering, assembly, ventilation repair etc.: The number of enterprises increased by about 700, while overall employment dropped by 24 000. Survivor enterprises with upwards of 50 employees (1980) contributed more than proportionately to job losses. The largest share of both new and non-surviving enterprises were in enterprises with less than 500 workers.

Engineering: The number of enterprises increased by some 600, with a total employment loss of 29 000, mostly due to job losses in survivor enterprises (24 000), in particular those with more than 500 employees. Gains and losses from new and non-survivor enterprises respectively were fairly evenly distributed among the size categories, with some concentration among medium-sized firms (50-499 employees).

Automobile industry: The number of enterprises increased considerably by 3 000 with corresponding employment gains of 13 000. The net effect of new and non-surviving enterprises is close to zero - whereby both were concentrated in the smallest size category (1-19 employees). Medium-sized survivor enterprises suffered employment losses, whereas the largest firms (+500) expanded employment considerably.

EDP-equipment, office machinery, electrical engineering: The number of enterprises increased by approx. 1 700, overall employment by 13 000. The employment gains of new enterprises and losses of non-survivors were roughly balanced and spread evenly over all size groups. Although survivor enterprises increased employment by 12 000, those with over 500 workers in 1980 experienced losses.

Textiles: Between 1980 and 1986 the number of enterprises in this branch was reduced by 750, while employment fell by 73 000. Survivors lost approx. 38 000 workers, primarily in large and medium-sized plants. These size categories also experienced the majority of plant closures. New enterprises did create jobs in the classes 1-499.

clothing: the number of enterprises fell by 1 900; employment by 71 000. The greatest job losses among survivor enterprises were from those in the medium-sized categories in 1980. The net employment effect of new and failed enterprises is clearly negative, with particularly large losses among the medium-sized enterprises (50-499). The gains from newly founded enterprises are concentrated among small and medium plants.

Table 3.3: Relative Indices of the Development of Employment 1980 - 1986 by Economic Sectors and Categories of Size of Enterprises

	index: gains (losses) by "survivor" enterprises in each size category devided by total employment 1980 in % (size of enterprise by number of employees in 1980)					gains by "new" enterprises in each size category devided by total gains in "new" enterprises (size of enterprise by number of employees in 1986)				losses by "non-survivor" enterprises in each size category devided by total losses in "non-survivor" enterprises (size of enterprise by number of employees in 1980)			
	1-19	20-49	50-499	>499	total	1-19	20-49	50-499	>499	1-19	20-49	50-499	>499
02 energy, water supply, mining	0.66	0.67	0.99	-2.62	-0.30	21.0	15.5	41.8	21.7	9.1	6.0	20.4	64.5
03 chem.ind., crude oil	0.27	0.18	0.14	-1.41	-0.81	15.3	11.4	42.6	30.7	21.0	14.8	55.7	8.5
04 plastics proc. ind.	2.25	0.87	1.23	-1.38	2.96	33.4	20.2	38.7	7.6	26.3	23.3	47.9	2.5
05 rubber and asbestos	0.32	0.06	0.42	-7.71	-6.91	12.4	8.1	28.0	51.6	15.7	16.0	40.9	27.4
06 stone and clay	-1.11	-2.78	-7.21	-2.55	-13.65	51.8	18.0	21.1	9.0	42.9	26.6	30.5	0.0
07 ceramics and glass	0.25	-0.30	-3.36	-9.05	-12.45	20.7	9.7	54.2	15.5	14.1	8.5	63.9	13.5
08 iron, steel, non-ferrous production	0.23	0.06	0.11	-21.16	-20.76	6.5	5.1	27.9	60.5	5.8	3.5	16.6	74.1
09 foundries, rolling, drawing	2.92	0.42	-0.06	-4.27	-0.99	49.2	17.1	25.9	7.8	34.7	16.1	42.2	7.0
10 light metal, fitting, rep.	1.45	-0.51	-2.53	-2.42	-4.01	51.9	19.8	25.6	2.6	38.8	21.6	36.7	2.9
11 machinery	1.36	0.70	0.40	-4.77	-2.31	23.6	13.9	40.9	21.6	17.4	15.5	54.6	12.5
12 motor industries	0.20	-0.99	-0.44	2.52	1.29	65.3	15.6	16.0	3.0	49.9	19.5	23.3	7.3
13 ship industries	-0.01	-0.08	-0.24	-19.92	-20.25	17.2	4.8	41.7	36.3	8.4	7.9	33.5	50.1
14 electrical/electronic industry	1.03	0.62	1.12	-1.69	1.08	25.3	13.0	30.4	31.3	17.4	10.4	38.2	33.9
15 prec. mechanics, optical goods	1.37	-0.72	0.54	-3.01	-1.81	58.3	14.8	19.9	7.0	38.5	17.9	30.0	13.5
16 iron, metal products	1.07	0.31	-0.96	-1.36	-0.94	32.3	17.4	39.4	10.9	20.1	18.7	47.5	13.7
17 wood products	1.41	-0.97	-5.04	-2.28	-6.89	59.6	15.0	25.4	0.0	34.6	17.1	43.9	4.3
18 paper processing ind.	0.30	-0.31	-2.44	-1.60	-4.04	21.5	16.2	62.3	0.0	17.9	16.9	61.2	3.9
19 printing industries	1.63	-0.62	-1.94	-2.64	-3.56	54.2	18.3	23.6	3.9	39.9	18.7	35.0	6.3
20 leather proc. ind.	0.48	-1.14	-6.75	-2.02	-9.44	36.6	12.7	50.7	0.0	25.4	22.9	51.7	0.0

continuation of table 3.3

	index: gains (losses) by "survivor" enterprises in each size category devided by total employment 1980 in % (size of enterprise by number of employees in 1980)					gains by "new" enterprises in each size category devided by total gains in "new" enterprises (size of enterprise by number of employees in 1986)				losses by "non-survivor" enterprises in each size category devided by total losses in "non-survivor" enterprises (size of enterprise by number of employees in 1980)			
	1-19	20-49	50-499	>499	total	1-19	20-49	50-499	>499	1-19	20-49	50-499	>499
21 textile industries	0.32	-0.37	-6.34	-5.25	-11.64	26.9	17.2	47.3	8.5	12.8	14.8	60.5	11.9
22 clothing industries	0.19	-1.19	-7.10	-0.97	-9.06	33.3	27.8	35.9	3.0	20.1	21.5	56.3	2.2
23 food industries	4.33	0.22	-1.98	-2.39	0.18	54.7	13.6	28.4	3.3	54.4	14.2	27.5	3.9
24 tobacco industries	-0.10	-0.27	-5.04	-6.67	-12.08	2.8	0.0	28.4	68.7	8.5	26.6	64.9	0.0
25 remaining industries	0.14	-0.57	-2.27	-1.81	-4.52	54.1	11.4	28.3	6.1	44.6	16.6	38.8	0.0
26 building, construction	0.43	-2.14	-5.08	-1.04	-7.83	62.9	19.9	16.4	0.8	46.8	24.3	28.1	0.8
27 wholesale trade	2.03	-1.10	-3.35	-0.81	-3.23	62.9	16.5	19.4	1.2	57.5	21.0	21.2	0.3
28 retail trade	0.42	-1.15	-3.29	-2.40	-6.42	70.1	15.1	13.4	1.4	71.4	13.1	14.7	0.8
30 banking	2.77	2.02	5.41	1.06	11.27	58.1	13.8	16.6	11.5	59.0	13.6	27.4	0.0
31 insurance	1.61	-0.27	-1.39	-0.52	-0.57	77.3	7.8	14.9	0.0	63.6	20.7	15.7	0.0
32 private hotels, restaurants	3.81	0.49	0.27	-0.02	4.54	74.9	13.7	11.4	0.0	82.4	11.5	6.1	0.0
33 private movies, TV, radio	0.55	-0.34	-1.22	5.39	4.37	78.8	14.2	7.0	0.0	75.7	14.9	9.5	0.0
34 newspaper/books publishing	2.77	0.19	0.12	-2.90	0.17	48.7	15.6	35.7	0.0	45.2	18.9	28.9	7.0
35 prv. educ., science, culture, sport	6.15	1.63	0.98	0.18	8.94	69.0	9.8	15.8	5.4	83.3	9.8	7.0	0.0
36 prv. health, veterinary	1.97	0.47	2.12	0.25	4.81	91.4	3.0	5.0	0.7	89.5	3.9	6.6	0.0
37 prv. law & business services, architects	4.63	-0.21	-0.46	0.29	4.25	78.8	9.8	9.6	1.7	78.2	11.6	9.4	0.8
38 prv. property & housing management	0.80	-0.92	-3.96	-0.93	-5.01	78.7	10.1	8.3	2.8	76.6	9.2	14.2	0.0
39 prv. advertising	7.20	-1.31	-1.47	-1.50	2.92	84.5	10.8	4.7	0.0	78.7	14.0	7.3	0.0
40 remaining prv. serv.	3.43	0.85	0.51	-0.16	4.63	58.5	16.1	22.8	2.6	68.5	13.1	17.9	0.6

- Construction:
  Overall the number of enterprises fell by approx. 4 200; employment by 257 000, of which survivors contributed 133 000 losses, despite an increase among the very smallest survivors (some 29 000); the losses among the large enterprises were less than proportional. Job losses due to enterprise closure were concentrated among small and medium-sized firms, while gains (170 000) were registered in small firms; hardly any new large enterprises were founded.
- Retail trade:
  This sector registered an increase in the number of enterprises of about 2 600, while employment fell by 84 000. This figure is the net result of considerable movements between the three types of enterprise: The survivors lost some 110 000 jobs, despite employment gains in the small (1980) survivor enterprises. Enterprise failures with considerable job losses are concentrated among small and medium-sized plants: larger enterprises were scarcely affected by closures. In the categories 1-499 employees, and particularly in the smallest enterprises, new enterprises were founded creating a large number of jobs.
- Service sector:
  Financial institutions: There was a slight reduction in the number of enterprises in this branch (400), while employment increased by 58 000. This was due primarily to medium-sized survivor enterprises (20-499). New and failed enterprises were mostly to be found in the smallest size category.
  Hostelries etc.: The number of enterprises rose by about 5 000 while employment expanded by 54 000, of which 19 000 was due to employment growth in survivor firms. Differentiated by size categories, almost all groups of survivors registered employment gains. The balance of new and failed enterprises is positive, resulting from a high level of job losses (some 132 000) more than compensated by the gains in new enterprises (167 000). These gains and losses were to be found largely among small firms (1-49).
  Education, science, culture, sport (private enterprises): Here the number of enterprises grew by 3 600 and employment by approx. 21 000. In survivor enterprises employment increased by about 6 500, concentrated among the smaller firms (1-49). New enterprises had a positive effect on employment in small and medium-sized enterprises, whereas losses due to firm closure were prevalent in small firms and particularly in the smallest category (1-19).
  Health and veterinary services (private firms): The total number of production units increased by 13 000, employment by about 73 000, of which 18 000 jobs were created in survivor enterprises, primarily in two categories - 1-19 and 50-499. A comparison of new and failed enterprises indicates the source of most of

the jobs (55 000) created in this branch. New enterprises mostly employ between 1 and 19 workers; closures affected those with 1-49 employees.

Legal and management consultancy, architectural and engineering offices: In this segment of the private service sector there were approx 14 600 more enterprises and 64 000 more jobs in 1986 than 1980. Some 18 000 jobs were created within survivor enterprises. Differentiating by enterprise-size, it is apparent that these gains were largely made in the very smallest enterprises. There was a considerable "surplus" of jobs created in new enterprises over those lost in non-survivors (some 46 000). Both new and failed enterprises were mainly to be found among small and medium-sized enterprises, particularly the smallest category (1-19).

Other service enterprises (personal services, cleaning, fotographic trade, hygienic and similar facilities, auctioneers etc.): The number of enterprises in this heterogeneous residual part of the service sector increased by 5 300; employment rising by about 72 000, of which 26 000 were created in survivor enterprises, particularly in the small and medium-sized categories. Some 94 000 jobs were lost through plant closure, this also occurring mainly outside the large firms. New enterprises employed an additional 140 000 jobs, again primarily in small and medium-sized enterprises.

4. Summary and Outlook

A look at the overall results for the period 1980 to 1986 confirms a well-established picture: employment fell in the sectors "energy, water supply, mining", "manufacturing industry" (with the exception of plastic processing, automobile, computer hardware production, electrical engineering, precision engineering and optics) "construction" and "wholesale and retail trade", whereas the number of those employed in "finance and insurance", in the private service sector increased. The study described in this paper entailed the differentiation of these overall results within the individual sectors on the basis of an integrated, longitudinal analysis.

Our analysis revealed that the different components of the employment trends were extremely highly differentiated within each sector, so that it was not possible to identify typical "patterns" for aggregated sectors (e.g. energy, industry, services). This is especially true of the question as to the role of enterprises of different sizes in job creation which has gained in importance in recent discussions on employment. From our study it is clear that any attempt to answer such a question will have to be based on a sector-specific analysis combined with an integrated, longitudinal study of the different components of employment. At the same time and with the appropriate caution it would seem possible to posit the existence of the following overall employment trends and their components:

- The public and nonprofit sectors have made a considerable contribution to job creation.
- In the "goods" producing sectors (energy, water supply, mining, manufacturing industry, construction) the employment trend among survivor enterprises was almost uniformly negative.
- These negative employment effects of survivor enterprises were concentrated in the size categories 50-499 and 500 and greater. The employment gains in new enterprises were, on the other hand, concentrated in the smallest size groups and those between 50 and 499, although in some sectors there is a high proportion of new large enterprises. The picture in the case of employment losses due to enterprise failure is very heterogeneous. However, large enterprises do appear to play a less important role.
- In the service sector (finance and insurance, private services) survivor enterprises have a positive employment trend in almost all branches, with the smallest enterprises (1.19) playing a particularly important role, although isolated incidences of positive effects in large enterprises remain. Similarly, employment gains through new enterprises are mainly to be found in small but also in medium-sized plants. However, these positive effects are to a great extent compensated by the negative effects of closures in these size categories.

The analysis presented here is subject to certain limitations, as mentioned early in this paper, due to the data base used and the descriptive methodology of our approach. However, this study is to be supplemented by analyses of the empirical findings on the basis of theoretical hypotheses. Furthermore, it is intended to broaden the analysis by further differentiation among the labor force, for example, by the level of qualification, skilled workers, part-time workers etc.

PART II

INDUSTRIES

Innovation, Growth and Employment

Innovative Activity at Plant, Sectoral and Intersectoral Level and its Effects on the West German Economy in the 1980s

Lothar Scholz, Horst Penzkofer and Heinz Schmalholz, in collaboration with Jörg Beutel

1. The Theoretical Framework

Seen from the perspective of the economist, technical progress is a complex phenomenon which is subject to a variety of causes and which can induce a whole range of economic and social consequences. The pioneering work of Robert M. Solow (1957) in the 1950s provided the economics profession with a decisive impetus to deal with this phenomenon from a new angle. The importance of technical progress for economic growth was demonstrated empirically on the basis of modern production theory and with the aid of the residual method. However this method measured technical progress merely as a residual component, and thus represented a "measure of ignorance" (Abramovitz 1956, p. 11). Since then a series of hypotheses have been put forward, theoretical approaches developed and empirical experiments conducted on the causes and effects of the "inexplicable phenomenon" (d'Alcantara 1986). Yet the questions remain unanswered (Cyert/Mowery 1987), questions which, however, *must* be answered if policies on growth (Oppenländer 1988) and research and technology (Nelson/Winter 1977) are not to lack a firm basis in economic theory (Gahlen 1972).

The variety of the manifestations of technical progress was shown by Edward F. Denison in the early 1970s (Denison 1974), while a decade earlier Murry Brown (Brown/De Cani 1963; Brown 1966) had pointed out that technical progress amounted to a "basket of components" which in the course of time can constitute themselves in very different ways. This results in "technological epoches" which are characterized by a number of specific constellations and impacts, i.e. they lead to different production functions. Brown also detected an "abstract technology" concealed behind the parameters of the (estimated) production function (Brown 1966 p. 12). However, due to methodological problems and the lack of empirical data, the concrete phenomena which characterized this abstract technology could not be determined.

Economic theories of technical progress can be distinguished according to the assumptions on which they rest and the degree of complexity of the approach. On the basis of these theories it has proved possible systematically to structure the various forms of technical progress and their effects, but it still remains unclear precisely which constellations with which effects lead to different phases of economic and social development. Research into innovation has produced a large number of empirical results using a range of different partial analytical approaches, but these results are often contradictory and so do not permit generalized hypotheses to be made. There has thus been little fundamental change in the state of research since the time of E.M. Rogers (Nelson 1981).

Growth theoreticians have dealt intensively with the conditions - derived from growth theories - for "steady growth". The fact that they are now analyzing the phenomenon of growth cycles - as opposed to the short-term business cycle - must be seen as a step forward. In such a context the rate and direction of technical progress is seen as a decisive explanatory factor. However, where attempts are made to furnish this factor with explanatory value, the most remarkable thing is the creativity with which scientists seek to find a name for this measure of ignorance. In formalized approaches using models to illustrate such growth cycles this cause of the fluctuations in growth rates is termed "entrepreneurial activity" or the "level of economic activity" (Krelle 1986, p. 92). This represents a further retreat from an attempt to establish a relationship between abstract technology and the concrete manifestations of technical change due to modern technologies, and to analyze its effects.

2. The Methodological Approach

2.1 Problems of Measuring "Technical" Progress

In economic theory "technical" progress can occur in an economy even if no technical changes are to be registered in that economy. Instead progress may be merely due to, for example, quantitative structural changes in the product mix (structural effects) or the neglect or uneven addition of external effects (e.g. expenditure on R&D or education, environmental effects). These "errors" in the measurement of the rate and direction of technical progress could only be avoided if sufficient light were shed on these relationships and the relevant data were available.

The Meta Study (phase II) bears the title "The Labor Market Effects of New Technologies". In addition to technical progress general effects are exerted on the labor market by economic policy and more specific impacts result from the processes of collective bargaining. Such effects can only be depicted on the basis of macroeconomic model analyses. However, this incurs the problem that a past value for technical progress, calculated using econometric techniques, is distorted by the sum of ex-post effects. Only on the assumption that the direction and intensity of such influences will continue unchanged in the future is it legitimate to base prognoses or simulation calculations on a rate of technical progress calculated in the (ex post) way. Yet econometric calculations have shown that the rate of technical progress is far from remaining constant in the long term (Krelle 1986). This leads to the problem of developing an "activity index", which assesses changes in the level of technology in the literal sense of the word, and of obtaining the requisite data.

Within the framework of the theory of technical progress the point of departure has always been a definition based on qualitative changes in output or in the factors of production (Ott 1959). However, when it comes to econometric measurement and the use of the available data it has not proved possible precisely to meet this restrictive definition. It was thus also impossible to test the theoretical approaches "cleanly" (i.e. econometrically) for their account of the way in which qualitative changes in output (product innovation) and/or in the factors of production labor and capital (process innovation) influence the productive efficiency of the economy. If, on the other hand, the definition of technical progress which presupposes qualitative changes in the form of product and/or process innovation is retained, then another measuring concept has to be developed.

2.2 Approaches to the Measurement of Innovative Activity

The measurement concept for product and process innovation developed by the Ifo-Institute takes account of the experience gained in the entire range of empirical studies in the field of innovation research dating back to the 1960s (Oppenländer 1971). One important point which has emerged from these case studies is that while product innovations can occur without process innovations and vice versa, it is often the effect of complementary product and process innovations which significantly influences the efficiency of production. With this in mind an approach was developed with which innovations could be structured (Scholz 1974) and which also contained the so-called technical variables ("Technik-Variablen", Scholz 1977). This implied the attempt to build a bridge between the various engineering disciplines and economic theory in order to enable the "abstract technology" of the production function to be transposed onto the actual technological changes in output and the production process.

While it was possible on the basis of case studies which analyzed individual processes of innovation or "bundles of innovation" (Rogers/Shoemaker 1971, p. 171) to isolate the nature and effects of technological changes in the relevant departments of innovative firms, the analysis of causal relationships at firm level, and all the more so at market and branch levels faced insuperable difficulties due to the twin problems of isolation and aggregation. As informative as the results of these studies were, they provided at best illustrative examples for certain developments, the causes of which, however, could not be attributed to specific factors.

A precondition for product and process innovation in the form of technical changes is new technical know-how, or at least a new combination of existing knowledge; in other words, research and development. According to statistics from the "Stifterverband für die deutsche Wissenschaft" public and private expenditure on R&D shows relatively constant trends over time. The point in time at which this R&D-based knowledge led to successful projects can be seen for some sectors of the economy from statistical analysis of patent applications (Faust 1987). However, this still leaves the question unanswered if and when this new know-how is actually used economically, and thus contributes to technical progress. Moreover, knowledge which is not or cannot be protected by a patent (e.g. in design) may also induce changes in the rate of technical progress.

For theses reasons the measure of technical progress developed by the Ifo-Institute, the "degree of economic activity", ("ökonomischer Aktivitätsgrad", Krelle 1986) is dependent of the indicator "innovative activity" (Schmalholz/Scholz 1985). The activity index covers the "producers of technology", i.e. the firms of manufacturing industry. On this view technical changes in products and production are necessary but not sufficient conditions for technical progress. This is because the answer to the question whether a product innovation contributes to technical progress depends not only on the supply of that innovation but also the demand for it (acceptance and purchasing power). Moreover, the efficient use of product and process innovations often requires software problems to be solved, presupposes organizational changes and often causes skill and training problems in user firms. Such factors may reduce the increase in efficiency or may cause delays before the full potential of the new technology is realized. This implies that an innovation indicator must be developed which measures the structure and intensity of the supply of innovation, while at the same time the output from these innovations must also be measured.

2.3 The Intersectoral Flow of Innovation

The aim of the Ifo Institute's project within the framework of the Meta Study was to analyze the relationship between innovation, growth and employment (and NOT the macroeconomic effects on the labor market) a) on the basis of data from individual enterprises, b) at the branch-level of aggregation and c) with respect to the final-demand sectors. The approach could therefore be described as a multi-layered technology-assessment study. In addition to the isolation and aggregation problems mentioned above, partial analytical and sector-specific technology-assessment studies frequently also have the disadvantage that they only consider the innovative activity which can be registered in the enterprises or branches studied. Innovation-exchange effects which arise from the fact that the innovative sector can obtain innovative inputs (intersectoral exchange of product innovations) or processes (intersectoral exchange of process innovations) are in many cases of great significance for the impacts of technology within the sector studied. An extreme case is represented by a sector whose level of innovative activity, measured directly, is very low, but which, due to effects of indirect innovation, exhibits economic and social consequences which cannot be directly attributed to the use of modern technology.

For this reason one of the focal points of Ifo's research in the Meta Study lay in the development and analysis of innovation-flow matrices both for manufacturing industry and the entire economy. Our work in this field was founded on the input-output projection procedure developed by the Ifo-Institute. The necessity of developing work in this field in the direction described has been recognized by other economists (Scherer 1982). The "Technology-Economy-Labor Market Model" developed by the Ifo Institute in 1978/79 for the so-called "Kabinettstudie" (Ifo/ISI/Infratest 1980) was so designed that such intersectoral innovation effects could be formalized: at that time, however, a suitable empirical data base was lacking.

3. Data Base

3.1 Surveys Conducted by the Ifo Institute

On the basis of a variety of survey techniques, the Ifo Institute has developed a panel of approximately 5 000 industrial enterprises (functionally differentiated by line of production) and 3 000 industrial firms (institutionally differentiated by branch) which provides partial data on turnover, employment and investment. By the mid-1970s the lack of suitable statistical data had set limitations on the institute's work on innovation and on the theory and measurement of technical progress. A way round these restrictions was sought leading in 1978/79 to the development of the concept of the "Ifo Innovation Test" (Reinhard/Scholz 1979). Since 1979 the data from this annual survey based on our panel of industrial firms and enterprises has provided the empirical foundations for our analyses of innovative activity in West German industry. In the first three years the results were used to test and modify the functioning of this new instrument (Schmalholz/Scholz 1985). Within the framework of phase II of the Meta Study the results of different Ifo surveys, which until then had been kept separately, were adjusted to ensure consistency and were then collated (table 3.1). We therefore have at our disposal a data base for the period 1979 to 1986 which enables us to conduct theoretically-based, empirical analyses of innovative activity in West German industry. The data from the panel of firms and enterprises was prepared - using weighting and extrapolation techniques developed especially for the purpose - so that it could also be used for analyses at branch level. The data from the R&D and investment statistics served partly as a means of testing these values.

Table 3.1: Number of Firms/Enterprises in Ifo Tests 1979 to 1986

	1979	1980	1981	1982	1983	1984	1985	1986
KT/IT	1538	782	1633	1643	1690	1680	1627	1527
KT/INT	402	526	690	1262	1386	1403	1208	1472
KT/INT/IT	226	293	377	750	845	866	754	874

explanation: KT = business-cycle test
IT = investment test
INT = innovation test

Source: Ifo Institute

Scientific research into innovation has shown that there are considerable time lags between invention, innovation and the diffusion of technical advances, i.e. before inventions contribute to technical progress in the economic sense of the word (fig. 3.1). Such lags can amount to years if not decades (Scholz 1974), so clearly such processes cannot be adequately captured using a data base of just 8 years. An attempt was therefore made to reconstruct an innovation indicator - expenditure on innovation in industry - for the years 1962 to 1978, initially just for manufacturing industry as a whole. This too was based on the R&D statistics and the surveys conducted by Ifo as part of the Investment Test (table 3.2).

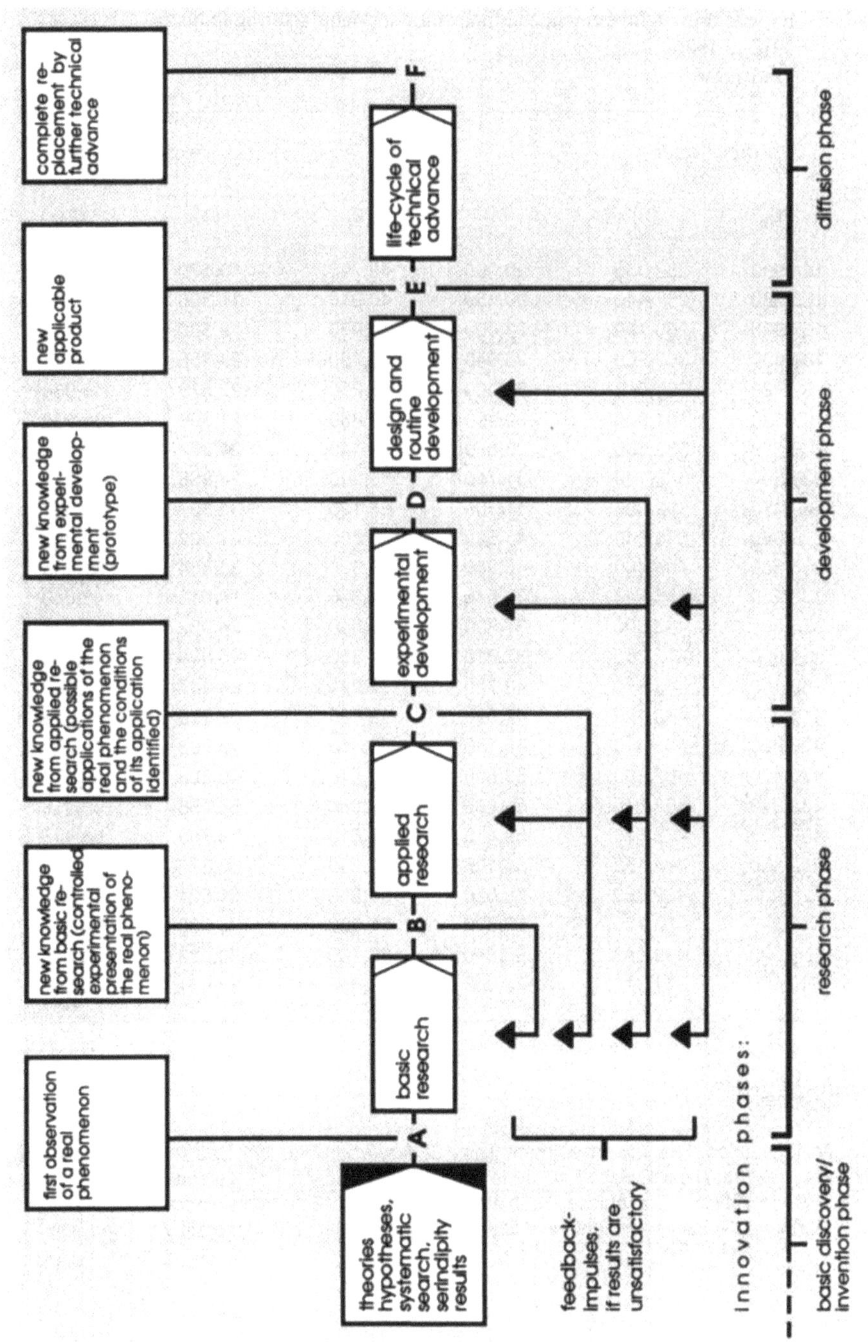

Figure 3.1: The Application of Technical Knowledge and the Innovation Process

Table 3.2: Expenditure on Investment and Innovation in Manufacturing Industry 1962 to 1986
- mill. DM -

	gross value added		gross investment		expenditure on innovation	
	(1)	(2)	(1)	(2)	(1)	(2)
1962	144,990	252,330	20,950	42,790	18,179	37,096
1963	150,580	257,440	20,650	40,940	17,398	34,493
1964	167,520	280,120	23,000	44,080	19,520	37,411
1965	184,890	302,260	26,040	48,150	21,856	40,413
1966	192,920	307,910	26,960	48,250	22,679	40,588
1967	193,580	301,900	22,950	40,850	21,398	38,088
1968	200,660	333,580	23,670	42,440	21,377	38,329
1969	229,590	373,140	31,740	55,200	24,858	43,231
1970	259,450	392,580	41,100	66,120	30,357	48,837
1971	278,040	396,540	43,530	65,800	31,322	47,346
1972	296,560	409,460	40,060	58,900	34,864	51,235
1973	333,250	435,540	38,540	54,370	34,053	48,040
1974	355,620	432,500	37,290	48,540	34,982	45,536
1975	354,060	412,350	37,010	44,800	35,823	43,363
1976	389,700	443,970	40,720	47,240	36,447	42,283
1977	414,420	453,120	44,090	49,800	37,815	42,712
1978	437,920	461,400	46,150	50,360	45,093	49,207
1979	470,420	482,800	52,600	55,510	49,628	52,374
1980	482,840	482,840	59,680	59,680	56,458	56,458
1981	489,730	475,520	59,410	56,610	59,763	56,946
1982	502,850	464,150	56,190	51,250	62,876	57,348
1983	524,930	469,360	58,520	51,960	66,521	59,064
1984	547,240	482,170	58,870	51,290	73,700	64,135
1985	584,020	499,060	68,640	58,160	78,628	66,646
1986	642,070	511,310	74,968	62,848	84,631	70,949

(1): current prices
(2): in 1980 prices

Source: Ifo Innovation Test; Ifo Investment Test; Stifterverband für die Deutsche Wissenschaft; Federal Statistical Office (Fachserie 18 - Volkswirtschaftliche Gesamtrechnung, Reihe S, 8, Revidierte Ergebnisse 1960-1984; Fachserie 18 - Volkswirtschaftliche Gesamtrechnung, Reihe 1) Accounts and Standard Tables 1986; calculations by the Ifo Institute

3.2 Innovation-Flow Matrices

In order to be able to analyze intersectoral exchanges of innovation it was necessary to create a data base specifically for this purpose. For the required input-output calculations we were forced to resort to the input-output tables of the Federal Statistical Office for 1980. An estimation approach was developed which, in its basic conception, is similar to the RAS or MODOP procedures, but which is based on the "activity analysis approach" and includes all the quadrants of the I-O table in the process of iteration (fig. 3.2).

The estimation of the I-O table for 1981 to 1986 was based on readily available official data. This method was chosen because the methods used previously to estimate input-output tables had indicated the existence of various shortcomings, such as "implausible" changes in coefficient. Without being able to give a final opinion on the reliability of the estimation procedure used, the closeness of the match with actual trends and its suitability for prognoses, this approach was selected with the aim of achieving a methodological step forward in the analysis of intersectoral innovation exchange.

Due to the shortness of the time-series data base developed and prepared by the Ifo Institute it is only possible to draw empirical conclusions for the first half of the 1980s. In order to be able to conduct these innovation analyses in the macroeconomic context it was necessary to estimate innovative activity for sectors which are not part of manufacturing industry. As far as the service sector is concerned we drew on the data from Infratest (Module 1 of the Meta Study). Regarding the innovative content of imports it was assumed that on average it was at the same level as in the corresponding domestic sectors.

The above points illustrate the fact that the Ifo research team faced difficulties with assumptions and estimation requirements comparable to those faced by Robert M. Solow in the 1950s with his data base, which earned him the title of "bushman", unswervingly cutting a path through the impenetrable forest. At the same time one of the aims of the Meta Study was to provide the basis for novel methodological approaches, or at least to open up new perspectives in this direction.

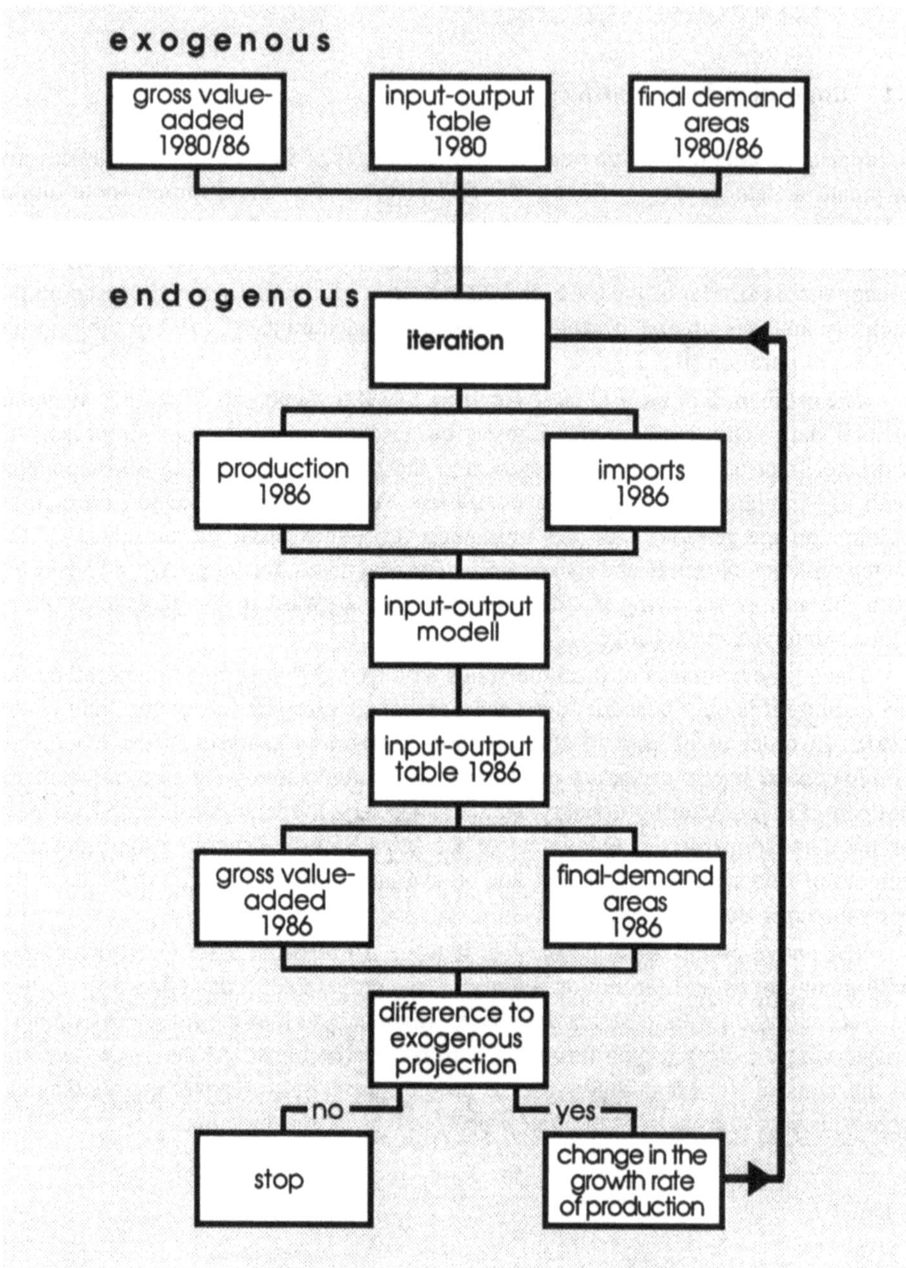

Figure 3.2: Projection of the Input-Output Table 1986

4. Empirical Results

4.1 The Conditions for Innovation

According to the methodological concept developed by Krelle (1986) to formalize the relationship between technical progress and economic growth, our approach has three exogenous variables: the supply of labor, the propensity to save and the propensity to invest in research. The two propensities are seen as "...being determined primarily by socio-psychological and political variables", which in turn determine "entrepreneurial activity" or the "level of economic activity" (ibid. p.92). On this view it is primarily the state of the economy as a whole which exerts an influence on innovative activity on the supply side.

Table 4.1: The Aims of Product Innovators in Manufacturing Industry 1982 - 1986
- in % -

aims of innovation: to	1982	1983	1984	1985	1986
create products to succeed discontinued products	57.4	62.4	62.0	63.0	55.7
widen product range					
- within the existing field of operation	81.2	75.4	79.1	78.3	80.9
- outside the existing field of operation	10.5	14.6	12.8	13.1	12.7
maintain market share	70.1	74.7	72.5	68.7	70.1
open up new markets					
- with respect to foreign markets	47.5	46.4	46.8	44.5	50.9
- with respect to new customer target groups	46.5	40.5	43.5	45.3	41.8

more than one answer possible

Source: Ifo Innovation Test

In the Ifo Innovation Test the macroeconomic conditions impinging on innovation are accounted for qualitatively in the form of the aims of, and the impulses and barriers to innovation. Overall it is apparent that the conditions influencing innovation have not changed dramatically for manufacturing industry during the 1980s.

Ignoring the differences between sectors and enterprises of different sizes, for the moment the following empirical results were obtained:

- The aims of innovation: New or radically improved products introduced to the market (product innovations) are either intended to broaden the product range ("rounding off" the production program) or are designed to replace those about to be discontinued (product substitution). The main aim appeared to be the defense of market shares (table 4.1). In the case of process innovators the dominant aims were to reduce labor costs as a share of total costs (factor substitution) and to increase the flexibility of production (equalization of capacity fluctuations) (table 4.2).

Table 4.2: The Aims of Process Innovators in Manufacturing Industry 1982 - 1986
- in % -

aims of innovation: to	1982	1983	1984	1985	1986
increase the flexibility of production	72.4	72.3	70.2	77.7	75.7
reduce production costs by reducing					
- share of labor costs	72.6	80.9	83.3	80.8	73.9
- material input	53.0	40.6	41.4	47.3	41.0
- engergy consumption	35.6	28.2	28.3	29.3	28.3
- waste	54.3	38.5	49.0	50.5	52.3
improve working conditions	51.7	38.3	40.1	46.0	42.2
reduce environmental pollution	38.2	28.6	29.1	33.6	31.0

more than one answer possible

Source: Ifo Innovation Test

- Impulses to innovate: Small and medium-sized firms in particular were induced by customer-oriented market demands to develop new products. Larger firms, on the other hand, tend to *give* R&D-based technology impulses to the market. The "stimuli" resulting from external conditions set by the state (e.g. public education and research, the patent system, legislation and research and technology policy) on the other hand, seem to be comparatively weak (table 4.3).

Table 4.3: The Source of Ideas for Innovation in Manufacturing Industry 1982 - 1985
- in % -

source	1982	1983	1984	1985
internal				
- research and development	62.7	57.3	63.8	63.2
- production. materials management	31.5	34.3	34.9	35.4
- marketing	71.8	59.2	71.4	68.3
- enterprise-level system for evaluating suggestions	9.6	8.4	11.2	8.1
- firm's management	54.4	48.8	49.2	50.1
external				
- partner firms	17.4	21.0	21.7	18.2
- competition	42.7	45.4	47.9	47.5
- suppliers	14.8	20.7	15.7	15.0
- customers	53.1	58.3	62.1	60.7
- specialist literature	13.5	14.3	12.5	15.4
- scientiic/academic field	9.4	10.2	9.2	9.4
- printed patent specifications	7.4	5.9	5.8	7.1
- trade fairs, congresses	23.0	24.5	24.3	26.4
- legislation	8.3	12.4	14.3	11.1
- state-financed R&D support programs	10.8	12.0	14.7	13.1

more than one answer possible

Source: Ifo Innovation Test

- Barriers to innovation: Non-innovators are characterized above all by a low equity capital base and by the expectation that product innovations will not be sufficiently profitable (table 4.4). To a much lesser extent this is also true of innovators, but they tend to point also to the difficulties of recruiting suitable R&D personnel (table 4.5).

From these findings it is possible to draw conclusions not only for research and technology policy but also for economic and financial policy regarding ways in which to improve the conditions for innovative activity. They do not, however, permit quantitative analyses of the relationship between the "degree of economic activity" oriented towards innovation and its effects of growth and employment. They merely provide latent variables which could be incorporated into a macroeconomic model as indicators of innovation.

Table 4.4: Barriers to Innovation Reported by Non-Innovators 1982 - 1986
- in % -

barriers	1982	1983	1984	1985	1986
lack of equity capital	50.4	33.4	50.6	57.4	44.1
lack of loan capital	17.0	8.0	18.4	21.4	18.0
insufficient expected margins on product innovation because					
- required investment too high	34.0	23.7	26.1	35.4	27.7
- pay-off period too long	13.0	9.5	15.9	3.0	10.7
- market trends too uncertain	52.0	44.7	59.3	53.3	29.8
insufficient willingness to innovate on the part of					
- the employees	6.7	4.2	7.1	4.6	1.0
- the work council	1.9	1.1	3.3	0.5	0.2
- management	9.0	9.0	12.0	7.5	1.9
organizational problems	7.4	10.9	7.0	12.9	8.9
agreements protecting workers from the effects of rationalization	4.9	0.5	0.5	-	0.4
personnel problems due to difficulties of recruiting suitable workers for					
- R&D	8.4	7.4	4.9	11.1	2.8
- production	4.1	5.3	10.3	7.5	6.1
- sales and marketing	3.6	6.3	9.8	5.3	6.4
lack of cooperation from					
- firms in the same branch	12.3	3.2	2.0	4.8	5.2
- suppliers or customers	5.8	3.2	2.9	8.2	11.5
further innovation not possible because of the advanced level of existing technology	7.4	14.1	16.1	12.5	14.7
lack of information on externally available know-how	9.1	11.7	6.9	5.4	4.1
lack of transforming technical knowledge into marketable products	13.4	14.7	8.2	7.9	4.4

more than one answer possible

Source: Ifo Innovation Test

Table 4.5: Barriers to Innovation Reported by Innovators 1982 - 1986
- in % -

barriers	1982	1983	1984	1985	1986
lack of equity capital	30.4	21.3	23.7	19.9	20.6
lack of loan capital	5.4	4.3	4.2	2.6	2.9
insufficient expected margins on product innovation because					
- required investment too high	36.7	31.2	37.6	33.6	39.8
- pay-off period too long	34.4	26.8	30.8	29.8	36.7
- market trends too uncertain	31.5	53.1	51.7	50.0	46.9
insufficient willingness to innovate on the part of					
- the employees	5.9	3.5	3.6	5.1	3.8
- the works council	2.9	3.0	4.9	3.5	2.2
- management	4.7	3.5	4.7	6.1	4.3
organizational problems	14.4	11.9	12.7	15.2	14.5
agreements protecting workers from the effects of rationalization	4.0	3.4	2.3	1.1	0.8
personnel problems due to difficulties of recruiting suitable workers for					
- R&D	19.6	27.5	31.7	36.3	39.5
- production	6.6	6.1	9.3	13.4	14.0
- sales and marketing	6.9	8.2	6.4	9.2	11.3
lack of cooperation from					
- firms of the same branch	6.8	5.0	5.1	4.1	6.6
- suppliers or customers	3.8	4.0	5.8	5.0	6.5
further innovation not possible because of the advanced level of existing technology	11.2	7.1	6.0	8.2	6.8
lack of information on externally available know-how	8.5	3.8	3.8	4.5	8.8
lack of transforming technical knowledge into marketable products	16.0	10.7	15.2	14.2	15.6

more than one answer possible

Source: Ifo Innovation Test

4.2 The Input and Output of Innovative Activity

The aim of Ifo's research concept was to reveal in as quantitative a way as possible the relationships between innovation, growth and employment, and so to structure our concept of innovation - which initially had been defined purely qualitatively - more stringently. As the figures for expenditure on R&D do not on their own provide any way of judging if and when the R&D-based accumulation of technical knowledge is actually used in economic terms, our point of departure for the "input measurement" of innovative activity was the total expenditure of innovative firms on innovation. Thus, in addition to R&D expenditure, our approach also took into account "innovative expenditure" on design, patents and licenses, production and market preparation and process innovations (table 4.6). This figure for total innovation expenditure was then differentiated between product and process innovations (table 4.7). As far as the "output measurement" of innovative activity is concerned, one point of departure was the turnover of products which are still in the phase of "market introduction" (i.e. product innovations) (table 4.8), another was the overall trend of output and employment. This then is the data base on which our multi-layered analysis was based.

4.3 The Innovation-Output Ratio

With the help of data on innovation expenditure in manufacturing industry for the period 1962 to 1986 (table 3.2; the data for the period 1962 to 1978 were reconstructed), regression analyses were conducted on the "incremental innovation-output ratio", an analogous concept to the "incremental capital-output ratio". This resulted in a statistically significant relation between expenditure on innovation and gross value-added. However, attempts to determine short and long-term effects and the time lags between innovation and growth using the same data base were not successful. This may be due to methodological difficulties or technical problems with the data, but may also result from fundamental changes between innovation-input and economic growth, i.e. the rate and effects of technical progress over the period 1962 to 1986. Due to the lack of a precise empirical data base for West German industry in the years between 1962 and 1978, it is not possible to test this further. However, the results obtained by Krelle (1986) would seem to confirm the latter view. This would imply that great caution must be exercised with prognoses made on the basis of average values for the relationship between innovation and output resulting from ex-post calculations.

Table 4.6: Expenditure on Innovation in Manufacturing Industry by Branch 1979 - 1986
- million DM at current prices -

branch	1979	1980	1981	1982	1983	1984	1985	1986
stone and clay	702	847	972	719	858	1.009	961	768
non-ferrous metals	202	331	343	323	325	466	380	308
crude oil processing	426	547	599	659	682	779	861	930
wood working	318	337	343	153	315	370	205	135
paper production	119	145	150	181	228	222	291	218
rubber processing	161	181	299	362	241	310	322	479
steel and light metal products	272	474	391	426	338	434	350	276
machinery	4,867	5,043	5,367	5,580	5,741	6,304	6,483	6,179
automatic data processing ind.	1,116	1,162	1,331	1,525	2,343	2,245	2,849	2,610
road vehicles	9,118	11,094	12,375	13,443	13,604	15,731	16,168	19,961
ship building	220	218	179	147	213	77	68	103
electrical goods	9,169	10,229	11,063	11,278	12,780	13,947	16,491	17,436
precision engineering, optics	423	571	754	908	801	1,072	1,354	1,380
steel construction	141	172	212	210	307	424	484	562
iron, sheet metal and metal goods	928	1,123	1,137	1,426	1,285	1,693	1,370	1,719
fine ceramics	150	112	98	89	106	125	149	123
glass production and processing	268	387	259	431	381	448	554	647
wood processing	350	542	572	600	556	629	658	570
toys production	332	323	319	356	316	401	441	429
paper processing	281	262	200	278	210	277	268	253
printing	491	455	354	308	516	529	640	707
plastics production	763	793	732	624	726	824	1,128	1,064
leather industry	150	154	139	89	76	101	102	99
textile industry	982	919	1,010	1,037	1,120	1,336	1,528	1,345
clothing	128	218	294	156	141	225	355	209
food, drinks, tobaccos	1,254	1,947	2,159	2,540	2,642	3,196	3,228	3,096
manufacturing industry (branches covered by survey)	33,331	38,586	41,651	43,848	46,852	53,174	57,688	61,606

Source: Ifo Innovation Test

Table 4.7: Expenditure on Product Innovation in Manufacturing Industry by Branch 1979 - 1986
- in % -

branch	1979	1980	1981	1982	1983	1984	1985	1986
stone and clay	50.3	13.0	31.4	20.3	44.0	25.4	27.6	30.1
non-ferrous metals	70.9	32.0	53.0	10.0	16.0	13.2	20.7	29.3
crude oil processing	5.0	67.0	5.0	9.7	2.0	2.0	25.0	41.0
wood working	49.0	45.0	4.7	64.6	42.0	25.9	32.6	63.7
paper production	78.7	57.0	16.3	71.0	51.0	30.5	28.0	10.9
rubber processing	72.7	53.0	48.5	27.8	73.5	18.6	37.9	32.9
steel and light metal products	62.0	76.0	44.2	45.0	18.9	17.0	17.1	17.3
machinery	71.3	64.0	72.5	70.8	70.6	66.4	64.6	66.0
automatic data processing ind.	100.0	55.0	54.1	12.0	35.6	100.0	68.8	71.4
road vehicles	91.0	70.0	57.0	45.0	44.2	39.7	49.1	54.8
ship building	95.0	95.0	75.3	49.0	57.5	57.6	65.9	81.5
electrical goods	73.7	65.0	66.6	64.0	63.4	55.4	59.6	57.6
precision engineering, optics	73.1	68.0	75.0	65.0	62.3	64.9	68.7	58.3
steel construction	22.5	30.0	47.0	30.0	31.1	17.3	9.8	25.7
iron, sheet metal and metal goods	49.4	49.0	49.8	56.7	47.7	50.5	55.7	53.9
fine ceramics	55.0	55.0	38.7	55.4	42.7	50.4	52.8	62.8
glass production and processing	19.6	56.0	71.7	23.6	26.6	32.7	18.7	35.4
wood processing	76.0	61.0	66.0	64.5	66.8	62.8	69.9	61.2
toys production	71.0	72.0	59.8	57.8	50.7	59.1	60.4	38.0
paper processing	38.0	56.0	31.4	83.0	41.0	57.8	53.2	50.3
printing	7.8	12.0	15.7	8.3	3.5	11.8	4.5	11.1
plastics production	49.5	51.0	46.0	48.3	50.0	41.6	42.3	49.9
leather industry	79.3	56.0	70.0	77.0	37.0	64.0	51.0	69.6
textile industry	52.0	41.0	53.6	48.3	54.4	53.8	57.6	44.5
clothing	40.0	50.0	68.0	43.0	19.7	48.9	8.7	35.8
food, drinks, tobaccos	66.6	36.0	54.0	38.3	47.3	58.0	49.4	42.6
manufacturing industry (branches covered by survey)	67.5	56.0	61.0	56.0	49.8	48.9	49.6	49.0

Source: Ifo Innovation Test

Table 4.8: Expenditure of Product Innovation in 1980 and Turnover Share of Products in the "Market Introduction" Phase in 1986
- in % -

	expenditure on innovation*	expenditure on product innovation*	share of turnover in the "market introduction" phase
chemical products	12.10	1.48	8.50
crude oil products	0.62	0.06	9.70
plastic and rubber products	2.64	1.35	9.52
stone and clay, fine ceramics and glass	3.11	0.85	5.28
iron and steel, non-ferrous metals etc.	1.30	0.41	7.65
steel and light metal products	2.20	1.67	5.60
machinery products	4.11	2.63	11.70
office machines, automatic data proc.	16.00	8.80	17.80
automobile industry	8.74	6.12	9.90
electrical engineering products	9.07	5.90	14.50
precision engineering, iron and sheet metal goods, musical instruments	3.47	2.04	11.81
wood working and processing	2.18	1.22	11.27
paper, printing	1.79	0.60	3.68
leather	1.76	0.99	8.70
textiles	2.78	1.14	11.40
clothing	1.05	0.52	8.80
food, drinks, tobacco	1.38	0.55	10.09

* as a percentage of the value of production

Source: Ifo Business Cycle Test, Ifo Innovation Test, Ifo Investment Test, Federal Statistical Office, "Stifterverband für die Deutsche Wissenschaft", calculations by the Ifo Institute.

4.4 Growth and Employment Trends of Different Types of Innovators at Sectoral Level

Reliable data on innovative activity in West German industry are only available for the period 1979 to 1986. Our analyses conducted on the basis of these figures produced the following findings:

- The results of innovation processes at enterprise level are predominantly discontinuous in nature, although certain enterprise-size-specific trends are apparent (table 4.9).
- Innovative firms took on more workers during upswings and made fewer redundant during downswings than the average for industry as a whole, whereas non-innovators, despite an increase in turnover, were not able to maintain their level of employment; indeed employment fell in each year of the period analyzed (table 4.10).

Table 4.9: Innovative Behavior over Time - in % -

sectors enterprise-size categories	distribution of enterprises with continuous		discontinuous innovative activity in the period		no	
	1979/86	1982/86	1979/86	1979/86	1972/86	1972/86
manufacturing industry*	23.8	30.3	70.7	58.1	5.5	11.6
basic and intermediary goods	4.3	14.4	83.3	60.6	12.4	25.0
investment goods (producing sector)	32.1	36.9	64.9	54.8	3.0	8.3
consumer goods (producing sector)	12.7	23.6	79.7	61.8	7.6	14.6
food industry	13.2	16.9	75.2	69.9	11.6	13.2
enterprises with						
20-49 employees	2.6	12.4	73.9	54.6	23.5	33.0
50-199 employees	13.4	19.3	83.4	62.6	3.2	18.1
200-499 employees	18.9	26.8	71.8	65.4	9.3	7.8
500-999 employees	22.7	30.3	72.8	63.1	4.5	6.6
1000 a.m. employees	41.0	45.2	59.0	49.1	-	5.7

* manufacturing industry excluding production and processing of fossile material, iron industry (foundries etc.), chemical industry, aerospace industry

Source: Ifo Business-Cycle Test, Ifo Investment Test

Table 4.10: Employment and Turnover between 1979 and 1986, Innovative and Non-Innovative Enterprises

	percentage change in													
	1979/80		1980/81		1981/82		1982/83		1983/84		1984/85		1985/86	
	concerning													
	E	T	E	T	E	T	E	T	E	T	E	T	E	T
official statistics*	0.7	9.4	-2.5	4.5	-3.7	2.5	-4.2	2.1	-0.8	5.7	1.6	7.4	1.9	-0.9
all enterprises*	1.3	8.2	-1.9	3.2	-3.4	2.0	-3.1	2.9	0.0	6.4	1.4	7.4	1.1	0.2
innovators	1.7	8.9	-1.6	3.9	-3.0	1.9	-2.9	3.9	0.9	7.5	2.0	8.7	1.7	0.8
of which:														
- product innovators	1.1	8.4	-1.1	4.7	-3.1	2.3	-2.8	2.6	-1.1	5.5	1.2	6.2	2.0	2.5
- process innovators	0.6	10.4	-2.3	5.5	-2.0	1.0	-3.0	3.6	-0.2	4.6	-0.2	5.9	-0.3	-3.8
- combined product and process innovators	2.6	8.6	-1.5	3.0	-3.3	2.0	-2.8	4.7	2.1	9.0	2.7	10.3	2.1	1.2
enterprises with projects in the planning stage	1.5	6.6	-2.5	3.2	-4.2	1.4	-4.0	1.7	-1.1	4.0	0.9	5.1	0.0	-0.9
non-innovators	-0.3	7.9	-3.3	-1.8	-3.9	2.8	-3.1	0.1	-2.8	4.6	-0.6	3.4	-0.9	-1.4

E = employment
T = turnover

* manufacturing industry excluding production and processing of fossile material, iron industry (foundries etc.), chemical industry, aerospace industry

Source: Ifo Business Cycle Test, Ifo Investment Test, Federal Statistical Office (series 4, row 4.2.2.), calculations by the Ifo Institute.

- At branch level the sectors characterized by dynamic innovators exhibit more pronounced employment and turnover effects than the average, and have thus increased their share of output over this period of structural and sectoral change (table 4.11).
- Firms with a combination of both product and process innovations tend to have enjoyed more favorable changes in employment and turnover than those involved with only one of these forms of innovation (table 4.10).

Table 4.11: Expenditure on Innovation and Economic Growth
- in % -

	real growth rates 1980/86		expenditure on innovation as a % of output in 1980	
selected sectors	output*	employment	direct	cumulative**
chemical products	2.3	-0.2	12.1	17.4
crude oil products	0.0	-3.2	0.6	3.0
plastics production	3.8	1.5	3.1	7.6
iron and steel	-2.2	-4.9	1.3	3.0
engineering	1.8	-0.5	4.1	6.5
office machinery	10.2	3.9	16.0	19.9
automobile industry	2.5	0.1	8.7	12.4
aerospace industry	4.3	2.0	22.2	28.4
electrical engineering	3.4	-0.2	9.1	12.0
textiles	0.0	-4.6	2.8	5.8

* in 1980 prices
** incl. intermediary, investment and imported goods

Source: Ifo Business-Cycle Test, Ifo Innovation Test, Ifo Investment Test, Federal Statistical Office, "Stifterverband für die Deutsche Wissenschaft", calculations by the Ifo Institute.

The fact that it is not possible using this data base to achieve clearer differences between innovative activity, growth and employment is due to time-lag effects, the analysis of which would require a longer time series from the Innovation Test surveys. This preliminary approach only takes account of the direct effects of innovative activity at enterprise and branch-level. Based on the analyses of innovation exchange, which enable both the direct and indirect effects on innovation at sectoral level to be incorporated, cross-sectional analyses showed a good fit for the relationship between production and innovation. The relation between expenditure on innovation and employment was analyzed using 4 types of flow-matrices, (figures 4.1 and 4.2). While sectors with above-average direct and indirect expenditure on innovation (type 1) showed a more favorable employment trend of up to two percentage points compared with the industrial average for the period 1980 to 1986, branches whose innovation expenditure was below average experienced a reduction in employment (table 4.12).

Table 4.12: Changes in Employment and Gross Value-added in Different Types of Innovation Flow (1980 - 1986)
- in % -

sector belonged in 1980 to type ...	all branches		of which: manufacturing industry	
	E	V	E	V
I	-0.4	16.6	-0.6	20.7
II	8.2	25.5	-9.4	-4.3
III	-3.8	5.1	-12.1	9.2
IV	-3.5	5.2	-12.1	-2.7
total	-2.2	9.3	-8.1	5.6

explanation: E = change in employment
V = change in gross value-added at 1980 prices

Source: Ifo Business-Cycle Test, Ifo Innovation Test, Ifo Investment Test, Federal Statistical Office, "Stifterverband für die Deutsche Wissenschaft", calculations by the Ifo Institute.

As can be seen from figures 4.1 and 4.2 the branches with the highest level of direct and indirect innovative activity include:
- chemical industry
- office machinery (data processing equipment)
- automobile industry
- electrical engineering

Within the framework of the intersectoral division of labor, these leading innovative sectors combine direct and indirect effects in such a way as to achieve positive effects on growth and employment. As they are furthermore all export-oriented industries facing stiff competition on world markets from other technologically advanced countries this innovation network represents the cornerstone of the international competitiveness of West German industry.

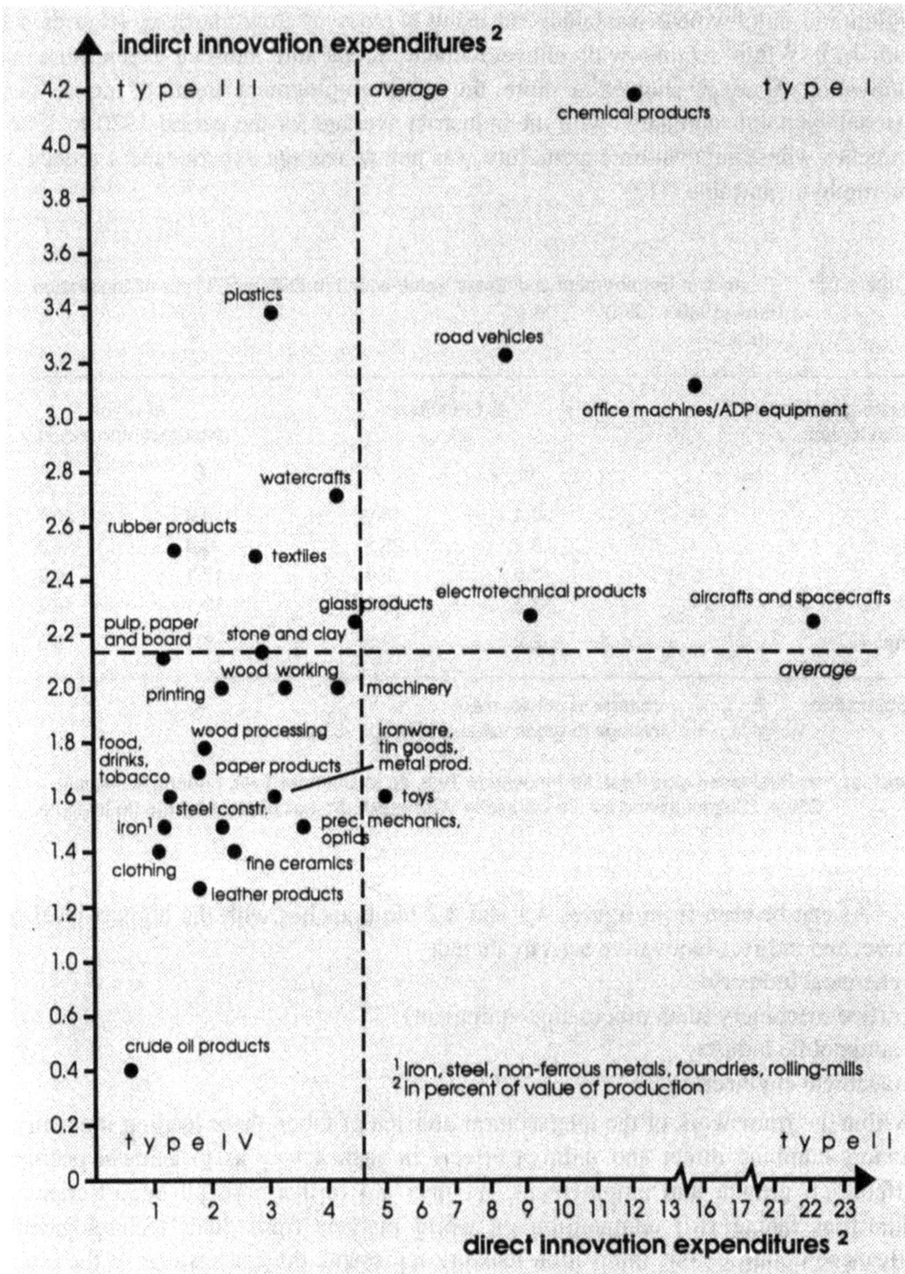

Figure 4.1: Intersectoral Innovation-Bundle of the German Industry (1980)

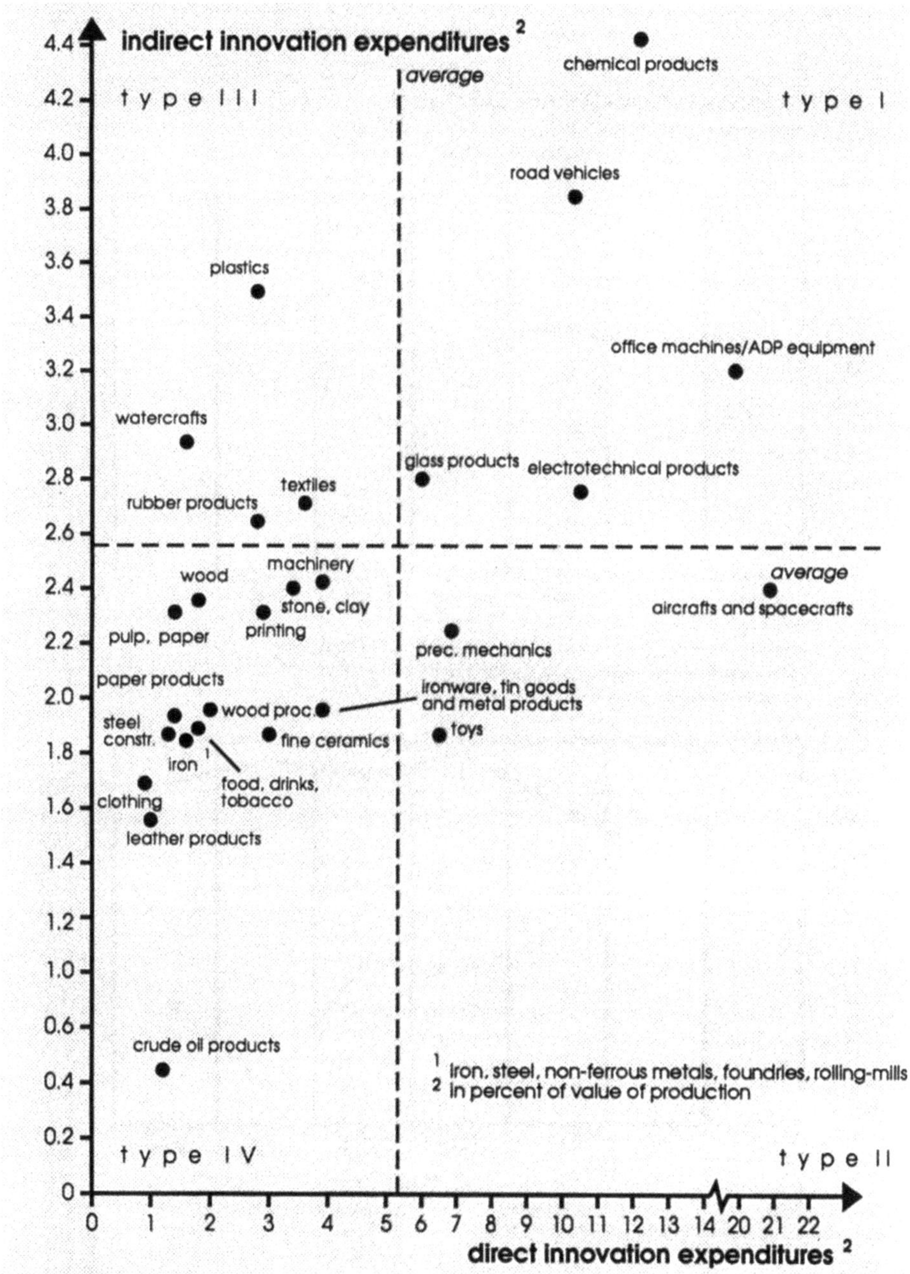

Figure 4.2: Intersectoral Innovation-Bundle of the German Industry (1986)

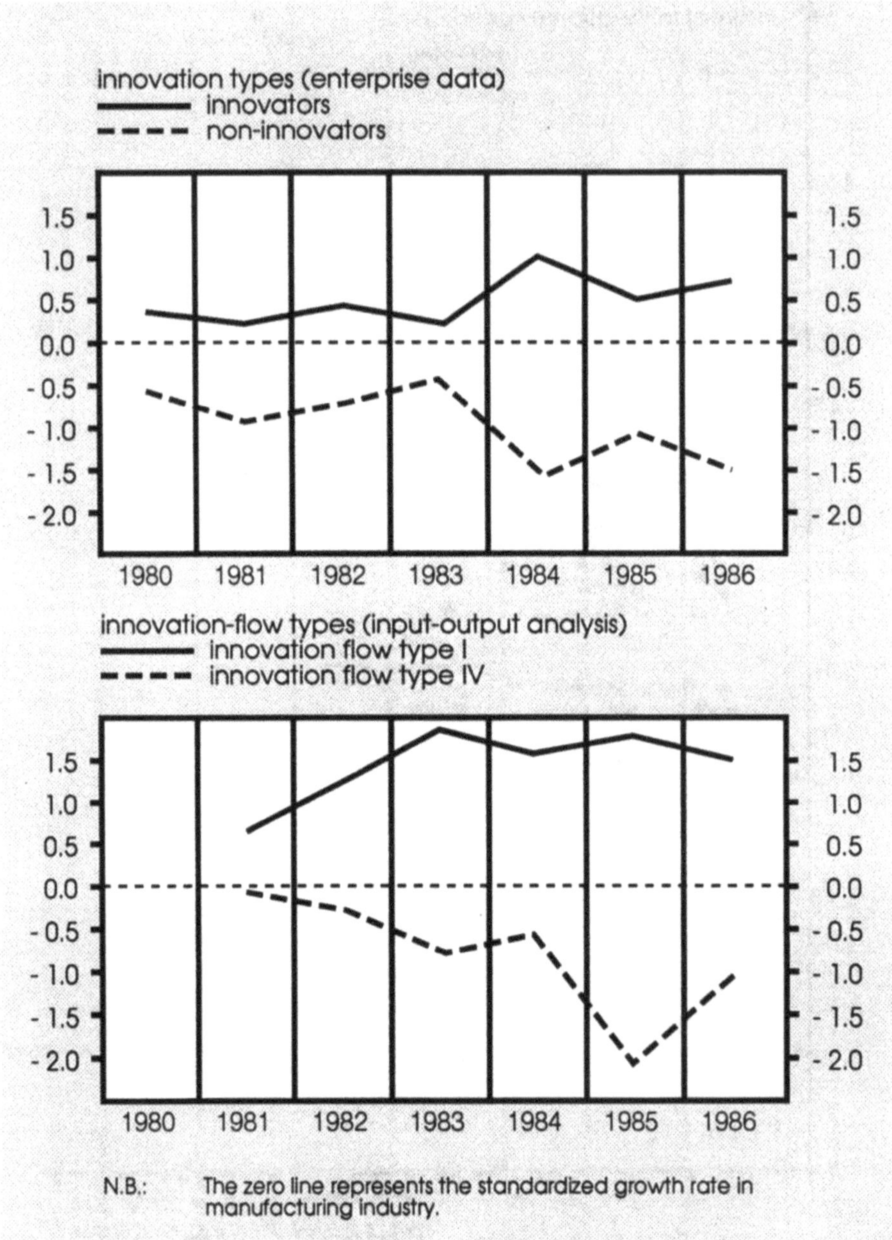

Figure 4.3: Employment Trends in Selected Innovation and Exchange Types

On the other hand, industrial sectors which neither invest little in direct innovative activity nor receive more than minimal innovative impulses from other sectors tend to suffer from a lack of competitiveness and are under increased pressure to adjust to structural change (Gerstenberger 1988). The fact that the engineering industry, the heart of the manufacturing sector in the FRG belongs to the latter group must give cause for concern, all the more so since its position worsened during the period studied due to a fall in direct expenditure on innovation (cf. figs. 4.1 and 4.2). However, these trends refer to the sector as a whole and may conceal the fact that the sectoral product mix consists of highly innovative products and those with a low level of innovation.

It should also be noted that while overall and at branch level the basic trend in manufacturing industry is for a positive relationship between innovation and growth and employment for the period studied (fig. 4.3), it is possible that at a higher level of disaggregation (e.g. individual markets) there may be compensating effects if innovative firms grow at the cost of less innovative ones. Moreover, innovative activity in manufacturing industry may increase the demand for market and production-oriented services. The rather diffuse sector "other market-oriented services" registered a particularly favorable employment trend. In order to analyze such problems it would be necessary to conduct intersectoral studies beyond the limits of manufacturing, as displacement effects in industry as a result of innovative activity may lead, via changes in the division of labor and follow-on effects of technical progress, to compensation effects in the service sector (Freeman/Soete 1987, p.250).

4.5 The Innovative Content of Final Demand

As one of the primary ways of responding to the pressures of competition and structural adjustment is through innovation, it is advisable from a macroeconomic perspective to pursue the statistics for expenditure on innovation to the components of final demand (table 4.13). Private consumption and exports each account for approximately one third of overall expenditure on innovation (table 4.14); the remaining third consists of public consumption, investment and changes in stocks. Over the period studied the relative shares shifted slightly in favor of exports. As far as the absolute level of expenditure in the industrial sectors is concerned, private consumption is stimulated by innovation to a particularly great extent in the following branches: the automobile, chemical, electrical engineering and food (non-drink) industries. The first two sectors account for approximately 50%, the remaining branches for a further 20% of total expenditure on innovation aimed towards private consumption.

Table 4.13: The Innovation Budget of the FRG 1980 to 1986
- million DM at 1980 prices -

year	total innovation budget	total GDP	share of budget in GDP	manufacturing industry abs.	manufacturing industry % of total	investment good industry abs.	investment good industry % of total
1980	98,062	1,371,460	7.2	58,760	59.9	31,423	32.0
1981	99,951	1,374,139	7.3	58,616	58.6	32,574	32.6
1982	100,952	1,367,431	7.4	58,837	58.3	32,979	32.7
1983	105,008	1,388,004	7.6	61,617	58.7	34,608	33.0
1984	112,901	1,429,409	7.9	66,695	59.1	37,746	33.4
1985	117,583	1,461,792	8.0	68,835	58.5	39,236	33.4
1986	127,200	1,500,249	8.5	74,701	58.7	42,332	33.3

Source: Ifo Business-Cycle Test, Ifo Innovation Test, Ifo Investment Test, Federal Statistical Office, "Stifterverband für die Deutsche Wissenschaft", calculations by the Ifo Institute.

Table 4.14: Innovative Content of Final Demand
- bill. DM -

components of final demand	total 1980	total 1986*	expenditure on innovation 1980	expenditure on innovation 1986*	expenditure per unit 1980	expenditure per unit 1986*
private consumption	681	736	35	44	5.1	6.0
public consumption	298	323	12	15	4.0	4.6
investment (equipment)	102	107	9	11	8.8	10.3
investment (buildings)	193	176	7	7	3.6	4.0
changes in stock	19	12	2	1	10.5	8.3
exports	356	460	31	46	8.7	10.0

* in 1980 prices

Source: Ifo Business-Cycle Test, Ifo Innovation Test, Ifo Investment Test, Federal Statistical Office, "Stifterverband für die Deutsche Wissenschaft", calculations by the Ifo Institute.

Three quarters of the export-oriented innovation budget is spent producing products in the chemical, automobile, electrical and engineering sectors. The last three sectors also account for about 75% of the innovation expenditure which ends up in capital investment.

However, the comparison of these absolute relations does not reveal that figures for the innovative content of each unit of output (tables 4.14 and 4.15) indicate a clear orientation of innovative activity in West German industry towards export and investment. Expenditure on innovation per unit of final demand in 1980 (1986) was 9% (10%) in investment and exports but only 5% (6%) in the consumption sectors. However, it cannot be concluded from this fact alone that this is a reflection of a relatively lower innovative growth effect with respect to private demand. Nevertheless, it is a noteworthy fact that West German industry invests more in innovation on the investment than on the consumption side. It may be assumed that this results in an increase in international competition via the export of innovative investment goods. The consequences which this has for the German economy as a whole can only be determined by macroeconomic models which simulate foreign trade in a sufficiently rigorous way.

Table 4.15: Innovative Content (product innovation) of Final Demand

components of final demand	1980 in bill. DM	1980 in %	1980 expenditure per unit	1986 in bill. DM*	1986 in %	1986 expenditure per unit
private consumption	9.7	30.0	1.4	11.4	29.1	1.6
public consumption	2.1	6.5	0.7	2.6	6.6	0.8
investment (equipment)	5.0	15.5	4.9	5.6	14.3	5.2
investment (buildings)	2.0	6.2	1.1	1.9	4.8	1.1
changes in stock	1.1	3.4	5.9	0.7	1.8	6.3
exports	12.4	38.4	3.5	17.0	43.4	3.7
total	32.3	100.0	2.0	39.2	100.0	2.2

* in 1980 prices

Source: Ifo Business-Cycle Test, Ifo Innovation Test, Ifo Investment Test, Federal Statistical Office, "Stifterverband für die Deutsche Wissenschaft", calculations by the Ifo Institute.

5. Innovation Indicators

From a macroeconomic perspective the conclusion can be drawn that productivity increases based on product innovations are an important prerequisite for income growth. Whether this increase in purchasing power is used to increase purchases of goods which are already available on the market, or to buy innovative products depends on both consumer preferences and the supply of innovative products. Regression analyses have established a close correlation between the innovative content of investment goods (process innovation) and the trend of labor productivity (gross value-added per employee), and between the expenditure on product innovation and the share of total turnover accounted for by products in the introductory phase (product innovations). It would thus be possible, once a suitable macroeconomic model has been developed, to attempt to formulate "optimal" innovation strategies (with respect to overall growth rates) taking into account the effects of innovation on the terms of trade and foreign trade as a whole. This would simultaneously reveal the rate and direction of technical progress in a form which would depict the relationship between technological innovation and its economic and social effects more concretely than economists have been able to until now. However, this would require both an improved data basis - including statistics for other countries - and further methodological advances. All the same, on the basis of the Ifo Innovation Test and the empirical and methodological progress made by the Meta Study, we are already in a position to draw some conclusions about the future direction and intensity of innovative activity in the medium term, based on the "leading edge" of current innovative activity. The foundations for such an analysis are provided by the system of innovation indicators developed within the framework of the Meta Study (table 5.1 and 5.2).

In accordance with the aims of its contribution to the Meta Study, the Ifo Institute developed sectoral and macroeconomic (quantitative) innovation indicators which measure the direct and indirect innovation input on the basis of the expenditure on innovation and the innovation output with the help of the product innovations recently introduced to the market. These represent quantitative innovation indicators describing the dynamics of innovation between 1980 and 1986. However, the time series is too short to conduct econometric analyses of the relationship between innovation input and output. As the Ifo surveys proceed, however, the basis for such calculations is gradually being created.

Table 5.1: Innovation Indicators for the Federal Republic of Germany

indicators	1980 in Mrd. DM	1980 in % of GDP	1986 in Mrd. DM*	1986 in % of GDP
direct expenditure on innovation	98.1	7.2	127.2	8.5
- research & development, design	38.6	2.8	50.2	3.3
- investment, market preparation	59.5	4.3	77.0	5.1
- product innovation	32.7	2.4	39.8	2.7
- process innovation	65.4	4.8	87.4	5.8
indirect expenditure on innovation	64.6	4.7	82.0	5.5
- domestic intermediary goods	43.2	3.2	55.6	3.7
- investment goods	10.4	0.8	12.5	0.8
- imported intermediary goods	11.0	0.8	13.9	0.9
direct and indirect expenditure on innovation	162.7	11.9	209.2	13.9

* in 1980 prices

Source: Ifo Business-Cycle Test, Ifo Innovation Test, Ifo Investment Test, Federal Statistical Office "Stifterverband für die Deutsche Wissenschaft", calculations by the Ifo Institute.

The attempt to develop our methodological concept for modeling the sectoral and macroeconomic relationships between innovation, growth and employment within the framework of the Meta Study has proved successful. However, the fact that we were able to shed light on the quantitative relationships should not blind us to the fact that innovation is always a qualitative phenomenon. With this in mind the significance of the qualitative innovation indicators developed within the Ifo Innovation Test becomes apparent. One of the most important features of this is the figures on the "technology areas" of product and process innovation (cf. table 5.3 and 5.4), as they provide information on the technological structure of innovative activity. So long as this structure remains constant, it is possible to make medium-term prognoses on the basis of quantitative, ex-post data. If, however, the structural pattern of innovation changes over time this may imply the existence of "technological epoches". The survey results of the Ifo Innovation Test for 1986 indicate that such a change is indeed apparent (tables 5.3 and 5.4).

Table 5.2: Innovation Indicators for Selected Sectors (1986)

	chemical products		road vehicles		electrical engineering	
indicators	mill. DM*	%	mill. DM*	%	mill. DM*	%
direct expenditure on innovation	18,167	12.1	15,374	10.3	13,609	11.0
- research, development, design	6,864	4.6	5,074	3.4	7,519	6.1
- investment, market preparation	11,364	7.5	10,301	6.9	6,090	4.9
- product innovation	2,703	1.8	8,413	5.6	7,835	6.4
- process innovation	15,465	10.3	6,961	4.6	5,774	4.7
indirect expenditure on innovation	6,635	4.4	5,714	3.8	3,385	2.7
- domestic intermediary goods	6,162	4.1	4,955	3.3	2,757	2.2
- investment goods	473	0.3	759	0.5	628	0.5
total indirect expenditure on innovation	8,546	5.7	6,615	4.4	4,364	3.5
- domestic intermed. goods/investment goods	6,635	4.4	5,714	3.8	3,385	2.7
- imported intermediary goods	1,911	1.3	901	0.6	979	0.8
direct and indirect expend. on innovation	26,713	17.8	21,989	14.7	17,973	14.6
cumulative expend. on innovation1980-1986	159,423	106.2	126,188	84.3	99,698	80.8
gross fixed capital	129,550	86.3	117,510	78.5	83,940	68.0
gross fixed capital formation	8,046	5.4	12,458	8.3	10,301	8.3
wage and salary earners**	613	0.4	975	0.7	1,108	0.9
gross output	150,143	100.0	149,701	100.0	123,379	100.0
- domestic intermediary goods	82,739	55.1	79,588	53.2	47,654	38.6
- imported intermediary goods	25,437	16.9	13,238	8.8	12,824	10.4
- gross value-added	41,967	28.0	56,875	38.0	62,901	51.0
output of recently introduced goods	12,762	8.5	14,820	9.9	17,890	14.5
input of recently introduced goods	6,294	4.2	7,650	5.1	4,784	3.9

* in 1980 prices

** 1000 persons per mill. DM output

Source: Ifo Business Cycle Test, Ifo Innovation Test, Ifo Investment Test, Federal Statistical Office, "Stifterverband für die Deutsche Wissenschaft", calculations by the Ifo Institute.

Table 5.3: The Technological Areas of Current Innovative Activity (1986)

in ... % of firms the following technological areas were significant for	product innovation				process innovation (production)			process innovation (administration)		
branches	new materials	new inputs	new functional solutions	fundamentally new products	new production technology	further automation	new techn.-organ. structures	data processing	word processing	cmmunication technology
stone and clay	39.2	21.5	50.1	22.5	80.7	44.3	4.3	95.1	31.5	11.6
non-ferrous metals (incl. foundries)	62.2	31.1	37.6	25.2	85.2	77.2	22.8	100.0	75.2	71.0
crude oil processing	.	.	.	.	.	.	.	.	.	.
special chemical products	38.9	37.9	56.5	67.0	99.1	93.5	61.6	97.9	80.9	78.9
rubber processing	.	.	.	.	.	.	.	.	.	.
wood working	18.1	2.1	64.0	15.8	39.2	66.2	5.4	86.4	27.3	13.6
paper production	56.6	24.6	28.4	0.0	89.0	70.9	52.7	92.7	65.5	23.7
steel and light metal products	17.8	0.0	64.4	17.8	64.1	61.7	33.0	74.5	44.7	12.8
machinery	18.9	15.2	92.9	26.0	54.3	74.2	40.9	83.3	56.6	41.1
automatic data processing industry	.	.	.	.	.	.	.	.	.	.
road vehicles	43.1	17.6	83.7	20.4	63.3	72.8	36.3	97.1	44.0	27.9
ship building	.	.	.	.	.	.	.	.	.	.
electrical goods	23.9	29.6	70.7	46.0	69.0	77.4	26.9	79.8	37.8	39.3
precision engineering, optics	22.0	8.4	70.7	38.3	62.7	75.1	22.4	81.1	28.2	25.2
steel construction	27.7	23.6	51.3	62.4	69.1	68.5	25.4	66.6	33.9	28.7
iron, sheet metal and metal goods	31.0	22.8	77.4	30.2	68.6	70.5	27.1	85.1	44.3	24.3
fine ceramics	14.6	17.1	19.5	74.8	82.2	28.2	10.4	36.0	78.5	13.3
glass production and processing	0.0	5.9	94.1	48.9	48.6	54.6	3.4	75.1	17.1	53.6
leather industry	66.0	12.2	34.5	23.3	76.3	9.9	38.5	82.9	6.4	39.4
textiles	51.1	19.0	52.9	34.2	68.4	62.1	29.4	65.3	44.1	30.6
clothing	76.9	15.5	43.2	35.8	77.9	72.7	56.1	75.7	35.6	48.6
paper processing	37.7	16.6	28.1	55.2	34.1	83.4	21.9	91.8	74.5	83.0
wood processing	36.5	19.1	75.5	34.2	64.2	58.9	34.3	87.4	41.8	27.5
plastics production	50.7	9.3	60.0	31.7	59.4	77.6	15.2	89.1	24.5	40.6
toys production	0.0	42.1	35.1	31.7	42.3	37.6	53.2	100.0	0.0	0.0
printing	41.9	6.6	54.5	42.1	67.9	46.3	23.5	86.8	38.6	28.9
food, drinks, tobaccos	21.2	6.3	71.9	41.9	69.0	76.5	12.5	69.7	14.0	1.5
manufac. ind. (branches covered by survey)	30.5	23.2	73.6	36.0	73.2	74.7	34.0	87.4	46.6	37.1

Note: "." = value registered and included in aggregates but not published for reasons of data protection
more than one answer possible

Source: Ifo Innovation Test 1986.

Table 5.4: The Technological Areas of Future Innovative Activity

in ... % of firms the following technological areas will be significant for	product innovation				process innovation (production)			process innovation (administration)		
branches	new materials	new inputs	new functional solutions	fundamentally new products	new production technology	further automation	new techn.-organ. structures	data processing	word processing	cmmunication technology
stone and clay	55.9	21.9	36.3	42.5	82.7	71.5	13.8	89.6	36.9	14.5
non-ferrous metals (incl. foundries)	53.0	53.0	23.3	23.7	64.0	82.3	17.7	76.7	37.7	93.9
crude oil processing	.	.	.	.	.	.	.	.	.	.
special chemical products	49.7	38.0	57.8	98.0	90.8	89.8	62.6	58.9	54.7	92.1
rubber processing	.	.	.	.	.	.	.	.	.	.
wood working	25.6	2.0	90.3	39.1	44.3	79.1	32.7	63.0	60.0	14.2
paper production	47.5	4.3	23.4	42.4	82.4	70.3	30.7	100.0	72.4	44.7
steel and light metal products	29.9	15.7	77.8	43.2	60.9	67.6	81.2	77.6	30.5	66.4
machinery	31.4	18.0	81.7	52.8	64.1	80.1	58.8	81.9	45.5	57.9
automatic data processing industry	.	.	.	.	.	.	.	.	.	.
road vehicles	69.6	24.5	83.9	41.7	74.7	80.0	66.8	87.0	43.2	66.1
ship building	.	.	.	.	.	.	.	.	.	.
electrical engineering	42.6	30.3	80.0	63.3	72.2	68.2	49.2	72.9	45.3	76.4
precision engineering. optics	36.5	10.7	52.4	73.7	67.9	71.2	57.6	62.3	59.5	54.5
steel construction	36.6	22.5	58.0	75.5	71.6	76.7	42.7	74.9	25.0	46.2
iron, sheet metal and metal goods	34.0	25.8	69.9	52.1	65.7	53.8	42.4	66.6	29.1	53.6
fine ceramics	78.5	8.2	20.3	75.1	87.0	84.9	15.2	45.0	17.3	79.3
glass production and processing	6.1	8.1	91.9	12.1	78.7	65.3	5.7	90.7	0.0	90.7
leather industry	91.7	10.1	44.5	31.7	76.8	49.1	27.0	98.0	35.0	4.0
textiles	59.8	18.6	32.5	59.6	60.4	73.8	44.1	87.6	53.0	71.0
clothing	76.3	17.3	35.5	46.1	76.5	76.2	48.7	86.8	33.2	51.8
paper processing	37.4	29.3	69.2	56.2	48.4	94.9	47.9	93.7	63.7	91.5
wood processing	37.4	19.2	66.4	49.4	63.7	67.8	35.6	78.5	42.0	44.3
plastics production	54.6	18.8	61.8	66.6	70.6	72.1	33.5	79.2	44.6	49.9
toys production	14.0	42.1	30.7	31.5	44.8	44.7	46.4	32.3	16.1	56.9
printing	32.7	3.3	66.4	29.3	72.1	77.1	53.4	79.5	36.6	64.1
food, drinks, tobacco	31.3	13.1	57.0	68.0	67.9	76.3	35.1	69.6	40.5	45.7
manufac. ind. (branches covered by survey)	44.8	25.5	71.4	57.8	72.0	75.0	51.6	77.0	45.7	66.5

Note: "." = value registered and included in aggregates but not published for reasons of data protection
more than one answer possible

Source: Ifo Innovation Test 1986.

Product innovations based on new materials or which lead to fundamentally new products will come to dominate innovative activity, at least according to the information provided by participants in the Innovation Test. Process innovations, on the other hand, seem likely to require to a greater extent than in the past new technical/organizational structures based on modern information and communication technologies. Against this background the empirical results of Module 1 (micro-analyses) of the Meta Study are to be borne in mind, as they provide some answers to the question whether a fundamental change is already occurring in the structural pattern of innovation, or whether the changes in innovative activity form part of a continuous, gradual trend which would tend to confirm rather than cast doubt on the relationships revealed by our analysis between innovation, growth and employment and imply that the direction of the effects of technical progress remains relatively constant.

The Effects of Research and Development on Employment, Prices and Foreign Trade

Georg Erber and Gustav A. Horn

1. Introduction

According to a common hypothesis concerning the structural changes which occur between sectors as a result of technological change jobs are lost in industries in which rationalization is proceeding rapidly while simultaneously new employment opportunities are created in less productive sectors, especially in the service sector (Baumol 1967 and Baumol et al. 1985). Technological change is thus seen as leading to a restructuring towards a service society. Two fundamental causal chains originating from the technological change induced by research and development can be distinguished. Changes in factor productivity lead to changes in factor intensity and may lead, for example, to a fall in the demand for labor (substitution effect). At the same time technological change also affects prices and wages. This in turn influences demand which in many cases induces effects which run counter to the substitution effects (income effects).

In the first part of the study presented here the substitution effect at sectoral level stands at the center of the analysis, with demand determined exogenously. At this stage the hypothesis that technological change leads to a considerable displacement of labor cannot be rejected. The subsequent empirical analysis of price effects reaches the conclusion that technological change has an anti-inflationary effect, primarily by way of reduced unit wage costs. In order to identify effects which may partially compensate the labor-saving effects the impact of R&D on the foreign trade of the Federal Republic was analyzed by sector. Our results show that while increased innovation often improves chances on export markets, they are by no means sufficient to compensate fully for the negative substitution effects.

Such empirical analyses are only possible within a framework of an econometric model disaggregated by sector. We used parts of the FIND Model developed by the DIW which models the behavior of 51 sectors in order to identify those effects which appeared relevant.

In our study the R&D capital stock was used as the indicator of technological change; our analysis is thus input-oriented. Qualitative changes in output such as completely new product ranges are not included in this presentation.

The following section discusses the hypotheses and the measurement of technological change. In subsequent sections the results of the models on employment, price formation and foreign trade are presented.

2. The Employment Effects of Technological Change

2.1 Econometric Estimates

Two fundamental values are taken for the measurement of technical change. On the one hand it is assumed that the cumulative expenditure for R&D makes a decisive contribution to changes in production technology. In certain important sectors it is necessary to distinguish between R&D expenditure which takes place within the sector itself and that which is "bought in" embodied in investment goods,. On the other the so-called "degree of modernity" of productive plant, i.e. the ratio of current investment to the capital stock of productive plant, is also a factor which is taken into account.

In most cases the econometric estimates of labor demand do not lead to a rejection of the hypothesis that R&D expenditure results in a reduction in the volume of labor-demand (DIW 1988, Part IV). Purchases of investment goods with a high R&D content, in particular induce labor-saving effects. While the extent of these effects may be somewhat exaggerated by the "trend" nature of the R&D capital stock, the direction of the effect is clearly negative in all cases where the estimated coefficients are statistically significant.

However, in the case of (retail and wholesale) trade and financial institutions no statistically significant effect was registered. In the transport and insurance sectors labor-saving technological change was observed, although in most cases it was of a marginal nature.

R&D expenditure from within the sector and that purchased from other sectors tend to pull in different directions. With the important exception of mechanical engineering purchased embodied R&D leads to labor displacement. The purchase of high-tech investment goods, in particular is thus clearly seen as rationalization-oriented. R&D expenditure within the sector, on the other hand, tends to induce an increased labor input. Product innovations or changes in production technology by means of own research thus have a positive employment effect. This finding is confirmed by similar conclusions obtained in the recent studies by Flaig/Stadler (1987) and Scherer (1985).

An additional explanation for this result is that to a great extent R&D serves the aim of facilitating the absorption of new technologies: one of the primary functions of R&D activity is to acquire the necessary skills and knowledge. In such cases rationalization is not the main aim of R&D efforts within a sector.

2.2 Rationalization Effects

On the basis of our results it can now be determined whether the rationalization effects induced by modern technology are compensated e.g. by means of an increase in demand. The model does not contain any causal relations between the exogenous variables. This leaves the question open as to whether the compensation was caused by technological change itself or was due to other influences such as foreign trade or state intervention. This represents a significant limitation to the analysis presented here which can only be overcome by estimating the sectoral demand functions. With the help of econometric estimates changes in employment can be attributed to different factors.

Our analysis shows that changes in the R&D capital stock between 1970 and 1983 - considering here only the rationalization effect - exerted a relatively considerable influence on changes in employment. If all other factors had remained constant since 1970 and given constant wages, prices and effective working time per employee, the output of the economy as a whole (excluding the public sector, other services and rented accommodation) of the year 1970 could have been produced in 1983 with a work force reduced by approximately 3 million. Actual employment during this period decreased by only 2.7 million so that clearly certain factors have pulled in the opposite direction. Of these the most important are the reduction of effective working time (an increase in employment of 2.3 million) and demand effects (approx. 2 million). In this way the effects of labor-saving new technologies have not always made themselves felt on the labor market. However, as has already been mentioned, at the current stage of the analysis it is not possible to establish causal relationships between the exogenous variables.

Between 1970 and 1983 the rationalization effect on the labor market was even over compensated (e.g. via working time reduction and rising demand) so that from these three factors the net employment effect was positive. The fact that employment nevertheless fell dramatically is due to changes in (relative) wage rates which led to a considerable reduction in employment, particularly in the first half of the 1970s. It is important to recall, however, that changes in wages and demand are exogenous to the factor demand model. Thus in contrast to the macroeconomic approach taken by Blazejczak (1988) the increased purchasing power resulting from higher wages is not taken into account. Wages are merely considered in their role as a cost factor and their increase results only in a decrease in profitability and an increase in the relative price of labor as a factor of production, leading to substitution. It can therefore not be concluded from these results that a policy of wage moderation would lead to higher employment as the reduction in labor costs may be over compensated by a fall in demand. The DIW model, on which the simulations conducted by Blazejczak and described in this volume were based, indicates a far more modest influence of relative prices and wages. This is primarily due to the inclusion of the influence of purchasing power in the DIW's model, whereas in the FIND approach used here demand, relative prices and wages are treated as exogenous variables.

Since 1980 the strength of those factors which compensate for the effects of rationalization have declined sharply. In particular the fall in demand between 1981 and 1883 had a negative effect on employment. It was no longer possible to compensate for the rationalization effects and actual employment fell (tables 2.1 and 2.2).

In order to determine the structural changes induced by technical progress it is necessary to analyze the rationalization effect in the different sectors. In absolute terms the largest effects were to be found in the electrical engineering industry, in which a considerable part of the observed fall in employment is due to labor-saving technological change. This development was not compensated by working time reduction or an increase in demand in any of the sub-periods studied. The same applies, although in a somewhat milder form to the automobile industry. However, between 1976 and 1980 rising demand provided a compensatory increase in employment.

In order to obtain a more precise picture of the significance of rationalization effects a "rationalization quotient" was determined for each sector in addition to the absolute changes in employment. This coefficient indicates the relation between the share of a sector in total job losses due to the rationalization effect of technological change and its share of total employment. If a sector's quotient is greater than one that sector has a higher proportion of job losses than its share of employment. In other words, in this sector the technologically induced fall in employment is above average. The higher the value of the quotient, the stronger the rationalization effect. For example, the rationalization quotient for steel construction between 1970 and 1975 is 1.01. R&D activity in this sector thus leads to a fall in employment which is approximately proportional to its share of total employment.

Table 2.1: Employment Effects - FRG except public sector
- thousand employees -

	change (total)			demand			wages			R&D			other		
Sectors	1970-75	76-80	81-83	1970-75	76-80	81-83	1970-75	76-80	81-83	1970-75	76-80	81-83	1970-75	76-80	81-83
7 chemical industry	16	-26	-25	47.2	26.1	-1.1	-185.7	-97.2	-33.3	82.1	32.6	16.7	72.4	36.8	8.7
15 non-ferr. metal	-15	-10	-7	11.2	0.0	-0.9	-9.0	2.4	1.0	-15.8	-15.6	-8.6	-1.4	21.5	2.3
18 steel constr.	-13	0	-26	5.2	18.7	-20.0	-5.9	7.6	-2.5	-12.6	-11.5	-6.4	0.3	-6.2	-8.3
19 machinery	-13	-43	-76	87.8	42.0	-30.4	-19.3	-51.1	-3.5	-182.1	-49.4	-13.0	100.6	49.2	2.3
20 office mach., ADP-7	-13	-1	-1.6	5.6	3.6	-11.6	-5.3	-4.2	1.0	1.3	1.0	5.2	4.6	1.4	
21 road vehicles	13	132	-36	130.4	229.7	39.5	-87.8	48.3	-26.4	-139.2	-151.0	-106.4	109.6	-5.1	50.6
24 electrical engin.	22	-33	-105	115.4	205.5	-7.1	-94.5	77.9	45.5	-261.5	-219.3	-160.3	262.6	-37.3	-9.8
33 textiles	-169	-63	-66	-17.5	7.4	-11.9	-10.1	10.3	-2.6	-69.6	-49.6	-26.3	-71.8	18.0	5.1
40 wholesale trade	-30	34	-70	75.7	59.4	-9.6	-128.9	-32.2	0.8	-8.8	-8.9	-5.7	32.0	2.0	-5.0
41 retail trade	60	111	-113	19.8	22.9	-1.3	-51.4	21.9	-5.8	-11.5	-12.6	-8.3	103.1	12.4	14.6
46 financial inst.	100	44	23	44.8	62.7	11.7	-44.1	-5.1	-13.5	-14.4	-17.2	-11.4	113.7	4.4	8.2
47 insurances	16	7	0	9.3	7.5	2.1	-0.2	3.3	-5.2	-13.9	-16.0	-10.9	20.8	2.2	1.8
other sectors	-1454	-328	-778	372.7	717.0	-349.3	-1003.0	-225.7	75.9	-664.5	-570.1	-304.3	-159.2	-103.3	78.3
total	-1474	-188	-1280	900.4	1404.5	-374.7	-1651.5	-244.9	26.2	-1310.8	-1087.3	-643.9	587.9	-0.8	150.2

"Other" refers to employment changes induced by changes in working time, age structure of plant, capacity utilization and adjustment processes.

Table 2.2: Rationalization Quotient by Sector

sector	1970-75	1976-80	1981-83
agriculture	-0.78	-0.91	-0.92
energy supply (electricity)	3.52	4.22	3.49
gas supply	23.40	32.10	29.31
water supply	3.33	3.51	2.49
mining	1.40	1.25	1.05
chemical industry	-2.24	-2.21	-0.84
crude oil processing	3.44	4.46	3.44
plastics processing	3.40	3.47	3.28
rubber processing	1.49	1.42	1.30
stone and clay	2.08	2.17	2.06
precision ceramics	1.86	1.80	1.68
glass	2.79	2.91	2.86
iron	0.49	0.60	0.52
non-ferrous metals	2.57	3.72	3.32
foundries	1.94	2.06	1.82
steel drawing	1.84	1.88	1.64
steel construction	1.01	1.14	1.01
machinery	2.45	0.61	0.35
office machines, automatic data proc.	-0.15	-0.31	-0.39
road vehicles	2.43	2.95	3.16
ship building	-0.50	-0.53	-0.52
aeronauticals	3.32	3.28	3.67
electrical engineering	3.32	3.39	4.23
precision engineering	0.27	0.29	0.27
iron, sheet metal and metal goods	2.72	2.81	2.59
musical instruments	2.30	2.25	1.79
mill working, wood products	0.41	0.44	0.40
wood processing	-0.22	-0.23	-0.20
pulp	0.75	0.88	0.89
paper processing	1.71	1.75	1.66
printing	0.76	0.85	0.82
textiles	2.18	2.57	2.46
leather	0.87	0.82	0.73
clothing	3.29	3.56	3.35
food	1.21	0.66	-0.90
tobacco	0.88	1.04	0.98
construction	0.20	0.23	0.24
interior design	0.45	0.42	0.44
wholesale trade	0.10	0.12	0.12
retail trade	0.09	0.11	0.11
railway	0.37	0.47	0.44
shipping	0.84	1.16	1.10
transport (road)	1.62	2.32	2.97
Federal Post Office	1.94	2.61	2.25
financial institutions	0.48	0.64	0.60
insurances	1.10	1.50	1.47

The rationalization quotient is defined as the ratio between the share of a given sector in total job losses due to technological change to that sector's share of total employment.

The values for the rationalization quotient which are set out in the following table clearly show that the claim that technological change leads to rationalization effects primarily in manufacturing industry cannot be rejected.

Within manufacturing industry the rationalization quotient was particularly high in electrical engineering and the automobile industry, confirming the view based on absolute values. The share of jobs lost due to rationalization is higher by as much as a factor of four than their share of total employment. The value usually is slightly over two.

The effects are also particularly high in the chemical sector (although not in the chemical industry itself) and the entire electrical sector (excluding precision mechanics).

There are however a wide range of effects in several of the most important branches of manufacturing industry. For firms in the chemical industry and the sector "office machines", for example, employment gains from technological change were registered. As this industry is one of the most important suppliers of investment goods technological change has clearly led to increased output (and labor input) rather than rationalization.

In the chemical industry it is the labor-using effects of R&D efforts within the industry which predominate, so that R&D activities do not lead to rationalization. The quotient for the mechanical engineering sector is rather more surprising. The rationalization effect in recent years is on this figure very low, although the estimate had indicated at least a clear correlation. As the firms in this branch are the most important producers of investment goods it is to be expected that this sector would also introduce labor-saving technology. A labor-saving impulse from technological change was observed amounting to approx. 240 000 jobs between 1970 and 1983. This effect was, however, partially compensated by demand effects.

The calculation of the rationalization quotients also shows that the effects of technological change in the firms in the engineering sector weakened considerably between 1976 and 1983. Whereas until 1975 labor-saving rationalization cost a large number of jobs, more recently there has been a fundamental change in the situation: the quotient is now approx 1/3, indicating that this industry is shedding labor at a below average rate. This is due to the fact that R&D efforts within the sector, which in this branch lead to a fall in employment have increased more slowly than R&D purchases from other sectors (which increase labor demand) in the last subperiod. Clearly the remaining potential for rationalization in the engineering industry itself was low between 1976 and 1983: modern machinery developed in this sector is now being purchased by other sectors and is leading there to technology-induced changes in employment.

Outside manufacturing industry the greatest job losses are to be found in the energy and transport and communications sectors. During the period studied trade and service sectors, on the other hand, were characterized by very low job losses as a result of technological change. Whether this is due to a lack of creativity and innovation in these sectors or is a structural phenomenon can not be decided on the basis of a quantitative analysis.

It is interesting to test whether those sectors in which in recent years employment has decreased particularly sharply are the same as those where rationalization effects are especially high. A look at the textile and clothing industry, construction and retail trade shows that this is by no means generally the case. In the construction industry it is primarily a lack of demand which led to large job losses, while in retail trade the reduction of surplus capacity played an important part, i.e. the negative effect was again due to problems on the demand side.

In the case of the textile and clothing industry, on the other hand, it was indeed attempts at rationalization, reinforced by falling demand which was the main factor. It seems that strong international competition which exerts a particularly strong influence on this sector led to hightened efforts to reduce product prices.

The structural change induced by technical progress has resulted in higher employment in the service and trade sectors compared with manufacturing industry. Within manufacturing industry it is electrical engineering, the automobile industry and the textile industry which show the most severe job losses whereas in the chemical industry and - more recently - (mechanical) engineering no or only slight losses were registered.

3. The Price Effects of Technological Change

If price formation contributes to market clearing (even if imperfectly) product innovations lead to price increases since at any price an increased demand faces an unchanged supply. Suppliers are thus able to increase prices and profits without having to worry about sales problems.

Here too the R&D capital stock serves as an indicator of the intensity of product innovation. While this poses no problems for R&D activity conducted in the sector itself it is more difficult to take account of purchased R&D capital. Because this value mainly reflects process innovations, and in particular because the investment-good exchange was used to determine it, it is not suited to our approach and object of study. The results of the estimation confirm this view: in only a small number of sectors in which purchased R&D capital was used were significant or theoretically plausible results obtained. In most sectors therefore estimations were also conducted without these variables. Only those sectors for which intra-sectoral R&D efforts could be determined are included in the following interpretation.

The price effects of process innovations are determined indirectly. The decisive variable is unit costs which include the costs of both labor and intermediates (inputs). By definition labor cost is the product of the hourly wage rate and working hours demanded, both of which are influenced, in different ways, by modern technology. Thus in order to be able to quantify price effects it is necessary to study the effect of technological change on intermediate goods, the demand for labor and wage rates.

Improved techniques of production can, on the one hand, lead to higher overall factor productivity and so reduce unit production costs (Stoneman 1983, Nelson 1987) if firms reduce their expenditure on material or labor inputs due to the introduction of new production processes. If all other factors remain constant producers increase profits. In an economy based on competition, however, excess profits are a temporary phenomenon, as during the diffusion process of the more productive technology firms are in a position to aim for higher market shares by reducing prices without cutting profits relative to the *status quo ante*. There is thus a trend towards reduced prices due to reduced expenditure on intermediates and falling labor demand. These prices reductions are the higher the more competitive the markets in which firms are operating and the larger the fall in unit costs of production. At the same time the use of more productive technology creates scope for higher wages, scope which is often used. Higher wages, however, increase unit costs and thus tend to raise prices. The change in the wage bill, the decisive variable for changes in prices thus depends on the extent of wage increases compared with that of the reduction in labor input.

Significant and theoretically plausible results for the price effects of product innovations were only obtained for the electrical and mechanical engineering sectors. A high level of R&D activity within these sectors led to price increases which can be interpreted as a willingness on the part of customers to pay a higher price for products of higher quality and as the intention of the producers to demand this higher price. As these two branches are the most important suppliers of investment goods improvements in quality have raised prices primarily in the investment good sector.

It was not possible to verify this theoretical hypothesis for the other sectors. Price increases as a result of improvements in product quality are, inasfar as this can be studied within the framework of our analysis, not a general phenomenon.

A far more important causal chain for changes in prices is the effect which new technologies exert by means of reduced costs. Nevertheless, the relevance of input-cost and wage-cost effects are very different. This fact is shown by the results gained from the factor demand and wage determination models taking into account changes in labor costs and intermediates respectively.

Changes in expenditure on intermediates are so slight that it is scarcely possible to obtain a measurable price effect. Its explanatory value for price changes is less than one percent. Savings made in the use of material inputs obviously play only a minor role as an impulse for price reductions.

Changes in total wages as a consequence of changes in the capital stock, on the other hand, explain a not inconsiderable part of price movements between 1970 and 1983. The direction of the effect differs from sector to sector. In certain important sectors (e.g the chemical and automobile industries, electrical and mechanical engineering) the increased R&D activity since 1970 has not led to a reduction in the wages bill. On the contrary, the (in comparison to other sectors) sharp increase in hourly wage rates induced by the increases in productivity was not compensated by a proportional fall in labor input (hours of labor demanded) so that the overall wage bill increased.

As a result technological change in these sectors was observed as having a price-increasing effect, which explains up to 12% (chemical industry) of the total price increase in this sector.

However, in this branch, a large proportion of whose products are intermediates, the increase in the wage bill is not due to higher wage rates but to the decrease in labor productivity as a result of increased R&D efforts. Thus total labor demand increased leading to higher labor costs in the period studied, part of which were passed on in the form of higher prices. Under such circumstances the economic advantages of technology which increases productivity does not benefit the purchasers of the products but rather the supplier firms (in the form of higher profits) or their employees (in the form of higher wages). This pattern of distribution is particularly pronounced in the investment-good sectors.

There are other sectors though in which R%D efforts have the effect of dampening prices. In the plastics, non-ferrous metal and textile sectors in particular prices would have increased much more steeply had it not been for technical progress. Although hourly wages have also increased in these sectors since 1970 there was such a labor saving that overall wage costs fell considerably. These cost advantages are reflected in real price effects. With the exception of insurance companies only relatively slight price and wage impulses as a result of technological change were observed in the service sector. The reason for this is the low rate of productivity growth in these sectors.

In the office machinery sector is the only sector in which prices have fallen since 1970. But it is only within the framework of a long-term analysis that the price effects of technological change explain a considerable proportion of this price movement. In the short-term it is not possible to attribute price effects to specific causes. However, following the period of considerable adjustment within this branch more than one third of the fall in prices is to be attributed to technological change.

4. The Effects of Technological Change on Foreign Trade

One of the central problems in analyzing the employment effects of technological change are the foreign trade effects arising from competition on world markets. This problematic is particularly relevant in the case of the Federal Republic which sells a large proportion of its production on foreign markets, both within and outside the European Community. This sets West Germany aside from the USA and to some extent from Japan which, in this respect, are in a more favorable position. To what extent and how quickly the completion of the European Single Market in 1992 will change this situation remains to be seen. In the FRG a loss of export markets due to a loss of technical competitiveness in industries in which key technologies are developed would have immediate consequences for employment (Rosenberg 1982).

In the discussion on the employment effects of technical progress it is often claimed that job losses caused by rationalization in domestic industry are compensated by increased demand from abroad due to the increase in international competitiveness which it brings. Sometimes the argument is put the other way: the loss of technological competitiveness in key high-tech areas would involve drastic job losses in the long run. For these reasons the influence of changes in R&D activity on the foreign trade relations in different sectors in which the goods exports of the FRG are concentrated was analyzed in a comparative approach with the USA and Japan.

Our work in this area is based on "neo technology theories" of foreign trade (Vernon 1966, Posner 1961) although it does not follow their product life cycle approach in all respects (Aquino 1981). A number of studies have shown the relationship between technological intensity and foreign trade to be empirically significant (Keesing 1967, Gruber/Mehta/Vernon 1967, Mansfield/Romeo/Wagner 1979, Horn 1977; for a summary cf. Hughes 1986). The study presented here provides further evidence that "neo-technology theories" offer a better explanation for the foreign trade of the FRG than traditional foreign trade theories using terms-of-trade effects (cf. also the results obtained by Erber (1986)).

With the help of an innovation scenario an attempt was made to determine the order of magnitude of employment effects induced via foreign trade, in which the indirect employment effects due to the increase in labor productivity which compensate the foreign demand effects are also incorporated.

Since technical progress in a country such as the Federal Republic is increasingly being dynamized by an international system of technology transfer both the import and the export of technology are of equal importance for the analysis. Foreign trade in goods and services accounts only indirectly and partially for this technology transfer; other important indicators are payments for patents and licenses and direct investment in foreign countries (cf. Börnsen, Glismann, Horn 1985). Neither the figures for absolute nor for relative R&D activity (relative, that is, to the share of West German GDP) provide sufficient information whether this activity is adequate to maintain the competitive position of the FRG in the long term. This applies all the more if West Germany continues as it has in recent years to grow more slowly than the USA or Japan.

4.1 R&D Activity in the Federal Republic, Japan and the USA

A look at the changes in total real expenditure on R&D in the three countries in the last five years shows that the rate of growth was highest in Japan, with particularly rapid growth at the beginning of the 1980s. In West Germany and the USA, on the other hand, the increase was at about the same average rate. During the period 1983 to 1987 the average rate of growth was 6.9% in Japan, 4.4% in West Germany and 4.2% in the USA. In all three countries a slowdown in the rate of growth is to be observed.

If R&D expenditure is measured relative to GNP no dramatic differences between the three countries are revealed. In all three the share of R&D expenditure in GNP has constantly increased (1977: USA 2.25%, Japan 1.98%, FRG 2.24%; 1987: USA 2.99%, Japan 3.03%, FRG 2.93%). These figures show unmistakably the faster rate of growth in Japan. In order to analyze the effects of R&D activity on foreign trade the most important causal relationships were simulated within the framework of a foreign trade model (DIW 1988).

4.2 Direct Effects of R&D Activity on the Foreign Trade Relations of Selected Sectors

In order to test the innovation scenarios developed with the aggregated model an aggregated foreign trade model was estimated which simulates the influence of changes in R&D capital stock in the FRG, Japan and the USA on the export and import of goods and services and their prices.

The results at the aggregate level for the FRG can be summarized as follows:

- an increase in R&D expenditure leading to an increase of the West German R&D capital stock increases the real net exports of the FRG.
- The increase in R&D expenditure also leads to higher export prices, especially where the accelerating growth of the German R&D capital stock occurs in comparison with the "technological leader", the USA.
- Export prices increase more rapidly than import prices as a result of a rise in R&D expenditures. In other words nominal net exports increase more rapidly than real net exports.
- Rising expenditures on R&D lead to a more rapid integration of the FRG in the process of the international division of labor. Its volume of trade with the rest of the world both in real and nominal terms increase as a result of more intensive R&D activity.

The foreign trade model was then linked to an employment model which simulates the influence of R&D capital stocks and their effects on total factor productivity and labor productivity.

4.3 The Employment Effects of Increased R&D Activity on Foreign Trade

4.3.1 The Assumptions of an Ex-post Scenario of Increased R&D Activity

The scenario investigates the employment effects of increased R&D activity in the FRG on the foreign trade relations of four selected sectors. The analysis is based on the foreign trade model. It is assumed - in accordance with the assumptions made in the previous section (DIW 1988) - that R&D expenditures in 1980 prices increase by 10% for five years (1973 - 1977) in all sectors. The cumulative additional expenditures for the four sectors for the entire period are DM 7.12 billion (in billion: chemical industry DM 2.396, mechanical engineering DM 1.262, automobiles DM 1.041, electrical engineering DM 2.421). At the end of this "R&D push" the total R&D capital stock has increased to DM 5.362 billion in 1980 prices. This represents an increase of 6.2% in the automobile industry and up to 6.8% in chemicals and electrical engineering.

This leads to relative increases compared with the R&D capital stocks in the USA and Japan. For the chemical sector this means, for example, an increase in the technology index of 6.97% compared with Japan and 6.43% compared with the USA. For mechanical engineering the difference is 6.61% (Japan) and 6.29% (USA). In the automobile industry the additional R&D activity leads to an increase of 6.74% and 5.49% and in electrical engineering by 7.07% and 5.87% compared with Japan and the USA respectively.

The R&D scenario was extended beyond 1977 to 1984; for the period 1978 to 1984 the same assumptions were made as in the reference scenario. This period thus provides a demonstration of the consequences of the previous R&D program. Due to the rate of depreciation of 15%, the R&D capital stock is more than one percent higher than in the reference scenario in all sectors even in the final year of the simulation. Because of their cumulative nature R&D activity exerts a long-term influence on the R&D capital stock of a given country.

4.3.2 Employment Effects Induced by Foreign Trade

Tables 4.1 - 4.4 provide a summary of the results of the scenarios of the employment effects in the four sectors. The presentation of the employment effects consist of a comparison of two variations: In the scenario with no productivity effects it is assumed that no productivity raising effects occur and that only the induced additional foreign demand has a positive effect on employment. In the chemical industry this "pure" demand effect would initially (in 1973) induce an increase in employment of 2 000 people, this figure rising to 6 000 by 1977.

In the scenarios with productivity effects the compensatory effect of the rationalization of production is taken into account as well to determine a net effect. The additional employment induced by the increase in productivity is now much lower in the chemical industry, with less than 1 000 additional jobs.

In contrast to the chemical industry, in (mechanical) engineering the additional R&D activity has a labor-displacing effect. In the scenario without the productivity effect labor displacement occurs only at the end of the R&D program in 1977 and is very low, with 1 000 employees. However, if the productivity effect is also taken into account there is a cumulative displacement effect which increases from 3 000 to 9 000 employees. After the R&D program has been completed the displacement effects continue for several years.

Table 4.1: Chemical Sector incl. Production and Processing of Fossile Material
- thousand employees -

	reference (1)	+10% R&D with productivity effects (2)	+10% R&D without productivity effects (3)	difference with productivity effects (4)	difference without productivity effects (5)
1973	632	632	634	0	2
1974	637	638	640	1	3
1975	629	630	634	1	5
1976	601	601	606	0	5
1977	605	606	611	1	6
1978	599	600	603	1	4
1979	595	596	599	1	4
1980	592	593	595	1	3
1981	570	570	572	0	2
1982	567	567	569	0	2
1983	532	532	534	0	2
1984	521	522	522	1	1

Legend: (4) = (2) - (1); (5) = (3) - (1); estimation method: SUR - Seemingly Unrelated; regression using the foreign trade model for the chemical sector

Table 4.2: Mechanical Engineering
- thousand employees -

	reference (1)	+10% R&D with productivity effects (2)	+10% R&D without productivity effects (3)	difference with productivity effects (4)	difference without productivity effects (5)
1973	1164	1161	1168	-3	1
1974	1115	1110	1124	-5	0
1975	1076	1069	1087	-7	0
1976	1090	1082	1102	-8	0
1977	1060	1051	1075	-9	-1
1978	1028	1020	1027	-8	-1
1979	1042	1035	1041	-7	-1
1980	1065	1059	1064	-6	-1
1981	1013	1009	1012	-6	-1
1982	969	966	968	-3	-1
1983	954	951	953	-3	-1
1984	933	931	932	-2	-1

Legend: (4) = (2) - (1); (5) = (3) - (1); estimation method: SUR - Seemingly Unrelated; regression using the foreign trade model for the mechanical engineering sector

Table 4.3: Automobile Industry
- thousand employees -

	reference (1)	+10% R&D with productivity effects (2)	+10% R&D without productivity effects (3)	difference with productivity effects (4)	difference without productivity effects (5)
1973	882	884	886	2	4
1974	879	883	888	4	9
1975	843	848	854	5	11
1976	813	818	825	5	12
1977	841	847	856	6	15
1978	875	880	887	5	12
1979	893	897	903	4	10
1980	926	930	935	4	9
1981	881	885	888	4	7
1982	853	856	858	3	5
1983	883	885	888	2	5
1984	899	901	903	2	4

Legend: (4) = (2) - (1); (5) = (3) - (1); estimation method: SUR - Seemingly Unrelated; regression using the foreign trade model for the automobile industry

Table 4.4: Electrical Engineering
- thousand employees -

	reference (1)	+10% R&D with productivity effects (2)	+10% R&D without productivity effects (3)	difference with productivity effects (4)	difference without productivity effects (5)
1973	1204	1200	1208	-4	4
1974	1221	1215	1227	-6	6
1975	1158	1151	1166	-7	8
1976	1102	1093	1112	-9	10
1977	1105	1096	1116	-9	11
1978	1084	1077	1093	-7	9
1979	1087	1081	1094	-6	7
1980	1093	1088	1098	-5	5
1981	1077	1073	1082	-4	5
1982	1022	1018	1025	-4	3
1983	1004	1001	1006	-3	2
1984	1004	1002	1006	-2	2

Legend: (4) = (2) - (1); (5) = (3) - (1); estimation method: SUR - Seemingly Unrelated; regression using the foreign trade model for the electrical engineering sector

In the automobile industry positive employment effects are to be observed due to improved price competitiveness. Without the productivity effect an additional 15 000 persons would find employment due to increased foreign demand and import substitution. The productivity effect reduces this to 6 000 in 1977. However, the improvement in international competitiveness would secure a positive employment effect for a number of years after the end of the simulated R&D program.

In the electrical engineering industry the "pure" demand effect is positive while the net effect including the rationalization effect of increased R&D activity is negative. Instead of the additional 11 000 workers, 9 000 workers less find employment as a result of the rationalization effect. The situation is similar to mechanical engineering where an inadequate expansion of demand leads to an employment deficit; the demand effects are insufficient to maintain the level of employment in electrical engineering in the FRG.

At the aggregate level the increase in R&D activity in the four sectors does not lead to additional employment. This is due to the greater magnitude of the employment losses in mechanical and electrical engineering compared with the gains in the chemical and automobile sectors. The "pure" demand effect of 31 000 additional jobs is more than fully compensated by the rationalization effect leading to the displacement of 42 000 workers. The net effect for 1977 is a fall in employment of 11 000.

5. Conclusion

Starting from the theoretical considerations mentioned at the start of this presentation regarding the effects of technological change it should now be possible to evaluate the results obtained here from the perspective of the different approaches.

The empirical analysis shows clearly that the price reactions usually assumed in neoclassical approaches in fact only occur to a very limited extent. Consequently, technological change does not necessarily lead to lower product prices. At best newer products, possibly of better quality, are supplied; a measurable, direct price reaction in the existing supply of goods could not be identified, at least not at sectoral level within the framework of the FIND model.

Wage reactions to technological change, on the other hand, are most definitely measurable. The increases in labor productivity as a result of technological change have frequently led to proportional wage increases, so that the reduction in the costs of production through the more productive use of resources was at least partially compensated by wage increases. However, the positive demand effects of these wage increases have not made themselves felt, particularly in recent years.

The primary reason for this was shown in our analysis to be the fact that firms shed labor as a reaction to wage increases. In the simulations conducted here the increases in nominal wages nearly always represent an increase in relative wages. As the rate of interest is determined exogenously and investment does not react in a predictable way to technological change, capital costs remain relatively constant, so that the relationship between wages and capital costs moves in the direction of the changes in wage rates. This implies that labor becomes relatively more expensive than capital as a factor of production. This leads to a fall in employment due to substitution which reduces total wage income and thus the demand for consumer goods.

Although in our approach the causality between changes in wages and demand cannot be satisfactorily modelled it would seem plausible to conclude that the demand-increasing effect of the wage increases will not compensate the demand-dampening effect of the fall in employment. Further insights in this regard can only be provided by a model of the entire macroeconomy including all feedback effects (cf. Blazejczak 1988). The results of simulations using a macroeconomic model indicate larger compensation effects.

Demand is not only influenced by changes in wage income: investment demand and the activities of the state are also major determining factors. The different results for the periods before and after 1980 are a good indication that the framework set by economic policy changed and that between 1980 and 1983 it was no longer oriented towards expanding effective demand. For this reason rationalization during these years led to a drastic deterioration in the employment situation, although sectoral differences were considerable. The service and trade sectors exhibit only negligible rationalization effects whereas in particular the electrical engineering, automobile and textile industries required considerably less labor as a result of technological change.

On the basis of this it is possible to sketch the "winners" and "losers" of the developments induced by technological change. Innovative firms benefit from an increase in profits. Those workers which remain in such firms also benefit from at times considerable wage increases. Customers of such firms profit only to the extent that new products of higher quality are offered which, in the case of investment goods, may enable them also to cut costs. In addition technological change leads to a higher share of total employment for firms in the service and trade sectors.

The costs of these developments are borne by those previously employed who lose their job as a result of rationalization and those looking for work, for whom the chance of finding a job has deteriorated due to the fall in labor demand.

The negative employment effects identified in this study as resulting from technological change should not lead to the premature conclusion that it would be better to do without technical progress altogether, or at least to desist from trying to accelerate it. Our study did not address the question of how employment would have been affected if technological change had not taken place. The danger is that the loss of competitiveness especially on international markets would have had a far more deleterious effect on employment.

At the same time the study has shown that innovation is likely to exacerbate rather than solve employment problems, at least in the short-run, if measures are not taken to stimulate employment. Economic policy must provide conditions in which the negative consequences of innovative activity are "cushioned" in order to prevent the establishment of a state of long-term underemployment.

The scenario presented here of the employment effects of R&D activity in foreign trade represents an attempt to evaluate the significance of international competitiveness of key export industries for the labor market of the FRG using an econometric model. Our results indicate that following the "neo-technology theory" concepts used in the study no increase in employment could be registered for the most important sectors for the export of goods and services (mechanical and electrical engineering). On the contrary, displacement effects occurred in these sectors. In the other two sectors (the chemical and automobile industries), however, positive employment effects were observed, although they were not sufficient to reverse the overall result.

However, it does not follow that R&D expenditure should be channelled into those sectors with a positive employment effect in order to obtain a positive overall result. In this context more study is required to determine whether or not there is a relationship between the elasticity of demand and the level of technology which has remained unidentified in our analyses. It would also be of interest to test whether the more rapid expansion of trade in services, the growing importance of intra-firm trade and direct investment would have to be incorporated more fully into the analysis in order to be in a position to consider technology policy more adequately as a constituent part of employment policy.

The Labor Market Effects of New Technologies - an Econometric Study for the Federal Republic of Germany

Peter Kugler, Urs Müller and George Sheldon

1. Introduction

Against the background of the persistently high level of unemployment in the Federal Republic of Germany the question as to the effects of technological change on the labor market has become a mayor issue in public debate. A wide range of opinions have been put forward. Following Blattner (1986), a spectrum can be drawn from, on the one extreme, the optimists, who view the introduction of new technologies as the beginning of a sustained economic upturn with full employment, to, on the other, the pessimists who believe that in the politically relevant medium term labor-displacing effects will dominate.

The divergence of opinion on the employment effects of new technologies rests on the answers given to the following fundamental questions:

- To what extent is technical progress labor-saving/labor-using? The introduction of new technology can alter the relative productivity of capital and labor in the production process and thus the optimal factor intensities. This so-called "bias" of technical progress, increases the demand for labor at the expense of that for capital (labor-using), or that for capital at the expense of labor (labor-saving).
- How elastic is the employment system in its reactions to changes in optimal factor-input relations, i.e. to what extent does it adjust to movements in relative factor prices? This is primarily a question of the substitutability of the factors of production and their price and cross-price elasticities of demand.
- How quickly does technological change progress? How rapidly does total factor productivity increase? How long does the adjustment of production processes and structures to the change in optimal factor input proportions take? In the short term, factor substitutability is very limited for technical, organizational and to some extent for legal reasons (in particular due to the provisions of labor law). Thus the nature and extent of substitutability is heavily dependent on the time period under consideration.

- What are the effects of technical progress on prices and product quality and how does demand react to these changes? The increase in total factor productivity resulting from the introduction of new technology provides scope for positive income effects. These may occur directly, in the form of wage or profit increases, or indirectly, in the form of lower prices. This will result in increased sales, whereby the extent of the increase will depend on the income and price elasticities of demand of different products. If the elasticity of demand is sufficiently high, the increase in sales can be so great that the initial fall in the demand for labor as a result of rationalization is fully compensated, or even overcompensated.

Our contribution to this volume deals with the first three causal relationships.

An empirical analysis of the effects of technological change on employment can be conducted in many ways. Following Weißhuhn (1986), three main approaches can be distinguished:

- descriptive studies, e.g. case studies, often based on enterprise surveys,
- limitational analyses, such as input-output analysis and the manpower-requirements approach, which exclude the possibility of substitution between the factors of production (at least implicitly), and
- econometric studies based on production theory, which, founded on a theoretical model of producers' factor demand, attempt to quantify the employment effects of the introduction of new technologies with the help of highly developed statistical techniques.

All three approaches have their strengths and weaknesses. Indeed all provide interesting insights into the process of change in employment structure. In the case of descriptive studies, this may even be possible at the level of the individual firm. Yet this is the very weakness of case studies, as it is far from certain to what degree the results extend to branches and time periods not covered by the study. Limitational approaches can provide some important supplementary information in this respect.

In the limitational approach possible changes in the structure of employment in the course of technical progress are simulated by assuming a variety of different values for the labor coefficients. The result is a series of comparative scenarios. As such an analysis also takes account of the multiplicity of interrelationships between the branches of a national economy, the criticism made of cases studies - their partial equilibrium character - does not apply to them. However, limitational scenario analyses still leave something to be desired: the relationship between technical progress and the selected labor coefficients remains largely unexplained. A further source of complaint is that limitational analyses - like case studies - do not differentiate between changes in factor input relations which are due to technical progress and those resulting from the normal process of substitution typical of a market economy. It may be, for example, that reduced employment, identified in such analyses as due to new production technologies, is in fact merely the result of a substitution process directed by a change in relative prices.

The strength of the econometric/production-theoretical approach lies precisely in this attempt to distinguish between the effects of technological change and those of the substitution process. This advantage over the two other approaches is won at the price of having to make a series of *a priori* assumptions, which some view as too restrictive or unrealistic. These critics often seem to forget that, due to the complexity of the real world, no form of empirical social research can dispense with simplifying hypotheses. It should also be taken into account that during the decade since the critical analysis conducted by Bombach and Blattner (1976) the econometric/production-theoretical approach has been greatly improved allowing for a broader spectrum of behavioral modes. It should also be noted that the econometric/production-theoretical approach is not intended to explain the behavior of an individual firm, but rather that of a large group of enterprises. Experience shows that many of the factors which may play a large part in influencing decisions at the micro level cancel each other out at the macro level.

The study described in this paper is based on the econometric/production-theoretical approach. Its aim is to analyze the effects of technological change, in particular those of process innovations on the skill structure of employment in the Federal Republic of Germany. To this end disaggregated factor demand equations were estimated for the seven major branches of manufacturing industry. The factor capital was divided into structures and equipment; the factor labor into seven skill categories. The analysis covers the period from 1960 to 1984.

Three causal relationships stand at the center of the analysis:

- The direct causal relationship (i.e. at constant factor price relations) in the form of the factor-specific bias of technical progress. The shift in relative factor productivities resulting from technological change and the consequent change in the optimal factor combination induce profit-maximizing firms to substitute the factor whose relative productivity has increased for that whose relative productivity has fallen.
- An indirect causal relationship in the form of the substitution effects. The displacement effect of technical progress changes the relative scarcity in the factor markets and by extension the factor price relations. The change in relative factor prices in turn leads to substitution processes which run counter to the direct displacement effects (negative feedback).
- A second indirect causal chain relates to the speed of technical progress as determined by relative prices. Factor-price changes which outpace (counteract) the cost-cutting effects of technical progress reduce (increase) the urgency of technological change and hence slow (accelerate) the introduction of new tech-

nology. Thus price decreases (increases) of the displaced factor or price increases (decreases) of the factor whose use is intensified slows (accelerates) the growth of total factor productivity. This represents a second negative feedback effect.

The demand effect of the increase in productivity, which was mentioned above as fourth central question in the debate on the effects of modern technology, plays no role within the framework of our study.

The paper is constructed as follows. The second section describes the econometric/production-theoretical approach together with the model variants used in our analysis. The data used to estimate the parameters of the model, together with the estimation procedure are described in section 3. The results of our analysis are presented and interpreted in section 4. The paper concludes with a summary of the results and a discussion of their implications for economic policy.

2. The Theoretical Approach

Econometric/production-theoretical analyses of the employment effects of new technologies are based on a theoretical model of producers' factor-demand behavior. The point of departure for modern factor-demand models (Berndt 1981) is a quasi-concave, twice differentiable production function having the general form

$$y = f(x_1, x_2, \dots, x_J, t)$$

The equation relates output (y) to the minimum quantities of the different factors required (x_1, x_2, ... x_j) to produce it and to a technology index (t). Diewert (1974) has shown that under such conditions cost-minimizing producer behavior implies a cost function dual to the production function. Assuming that the production function of an industry describes a locus of long-run competitive equilibria, and thus exhibits constant returns to scale, one can depict the dual cost function as a unit-cost function (c) or as a price function (Jorgensen 1986):

$$q = c(p_1, p_2, \dots, p_J, t)$$

The output price (q) is determined by the factor prices (p_1, p_2, ... p_j) and the state of technology (t). The duality relationship implies that the price function is twice differentiable, linear homogeneous in factor prices as well as concave and monotonically increasing in input prices.

In order to make the theoretical approach suitable for empirical analysis, the behavioral model must be more closely specified. To start, a form for the production function must be chosen. The Cobb-Douglas or CES specification are not suited to production technologies with more than two input goods, as they restrict substitution possibilities. They imply that the partial elasticities of substitution are equal and constant for all pairs of factors. In order to avoid limiting the possibilities for substitution from the outset and thus prejudicing the results of the analysis, we choose to use so-called flexible functional forms (generalized Leontief function, translog function etc.).

A further matter of concern is the degree of factor disaggregation. In order to obtain as complete a picture as possible of the opportunities for substitution, it is obviously desirable to disaggregate the input factors as far as the data permit. Moreover, substitution studies from the USA (Hamermesh and Grant 1979) show that capital and highly skilled labor are complements, whereas capital and unskilled labor tend to be substitutes. It can therefore be assumed that labor cannot be considered homogeneous with respect to technical progress. However, there are practical limits to the degree of disaggregation as the number of parameters which must be estimated increases more than proportionately. A practical solution is to specify multi-level production functions, in which upper-level inputs are viewed as functions of lower-level or more elemental factors. This so-called weak separability assumption enables substitution possibilities to be analyzed among a large number of factors of production.

There are also several possible ways of modelling technical progress to choose from (Stoneman 1983, p. 3ff.). Technical progress can be modeled as either embodied (vintage model) or disembodied. A further distinction can be made between models in which technical progress is exogenous and those in which it is endogenous. Due to data problems vintage models were considered impractical; however in our econometric model disembodied technical progress has been endogenized.

Finally, the assumption that firms fully adjust their factor input relations to changes in prices and production levels within one observation period (quarter or year) must be questioned. Within the framework of a dynamic model, adjustment costs can be specified either implicitly or explicitly. Here implicitly means that the adjustment costs serve as a motivation for a partial adjustment specification. In the second case the costs of the adjustment process are explicitly specified and included in producers' optimization calculations. In addition to a static model a modified partial adjustment model is employed.

Having completed our general introductory remarks we now turn to the parameterization of the static model. For empirical analysis a specific form of the function for unit costs and prices must be chosen. In the following it is assumed that the price function takes the translog form, developed by Christensen, Jorgensen and Lau (1973):

$$\ln q = \alpha_0 + \sum_i \alpha_i \cdot \ln p_i + \frac{1}{2} \sum_i \sum_j \beta_{ij} \cdot \ln p_i \cdot \ln p_j$$

$$+ \sum_i \beta_{it} \cdot \ln p_i \cdot t + \alpha_t \cdot t + \frac{1}{2} \beta_{tt} \cdot t^2 \qquad i,j = 1,\ldots,J$$

Differentiating the translog price function with respect to the logarithm of factor prices and employing Shephard's Lemma yields demand functions for the factors of production in the form of factor cost share equations:

$$\frac{\delta \ln q}{\delta \ln p_i} = \frac{x_i \cdot p_i}{y \cdot q} = v_i = \alpha_i + \sum_j \beta_{ij} \cdot \ln p_j + \beta_{it} \cdot t$$

$$i,j = 1,\ldots,J$$

If the translog price function is differentiated instead with respect to the time index (t), one obtains the following equation for the negative growth rate of total factor productivity:

$$-\frac{\delta \ln y}{\delta t} = \frac{\delta \ln q}{\delta t} = -v_t = \alpha_t + \sum_i \beta_{it} \cdot \ln p_i + \beta_{tt} \cdot t$$

In the above equation the speed of technical progress (v_t) appears as a function of the factor prices and of time. The rate of growth of total factor productivity is thus endogenized. Nevertheless, it remains disembodied as it is treated as independent of the age structure of the factors of production.

In summary, the econometric model consists of J factor demand functions (share equations) and the equation for the velocity of technical progress. Partial differentiation of these functions with respect to logarithms of factor prices and the technology index yields the parameters to be estimated. We turn first to the bias of technological change.

According to Hicks, for a given production function and a given factor, technical progress is factor-using (factor neutral or factor-saving) if, with constant factor intensities, the marginal productivity of this factor increases (remains constant or falls). As this leads to a shift in the optimal input combination, the actual structure of factor inputs changes accordingly. This means that in the case of factor-using technical progress and constant relative factor prices the cost share of the "favored" factor increases. In the case of factor-saving technical progress, on the other hand, the cost share of the same factor falls (and remains constant in the case of neutral technical progress). This so-called Hicks' bias of technical progress is yielded by twice differentiating the logarithmic price function with respect to the logarithms of factor prices and the technology index:

$$\beta_{it} = \frac{\delta^2 \ln q}{\delta \ln p_i \; \delta t} = \frac{\delta v_i}{\delta t} = - \frac{\delta v_t}{\delta \ln p_i} = \beta_{ti} \qquad i = 1,\ldots,J$$

With reference to this last equation, it is possible to offer an alternative yet equivalent interpretation of the bias: based on assumptions of our model the bias also measures the change in the rate of technical progress to movements in relative factor prices. This reveals that not only the input coefficients and the factor cost shares, but also the speed of technical progress depends on factor influence prices and the current state of technology. This represents the third of the channels of influence relating to technological change mentioned in the introduction.

Twice differentiating the logarithmic price function with respect to the logarithms of factor prices provides information on factor substitution in the form of share elasticities:

$$\beta_{ij} = \frac{\delta^2 \ln q}{\delta \ln p_i \; \delta \ln p_j} = \frac{\delta v_i}{\delta \ln p_j} \qquad i,j = 1,\ldots,J$$

The cost share model derived from the translog price function provides a very flexible framework for the analysis of factor substitution. This can be demonstrated as follows. If the J factors of production are not at all substitutable, an increase in the price of one factor (p_i) will, at a constant level of output, have no effect on the quantity of inputs. In this case the cost share of the factor x_i, whose price has risen, increases, while the cost shares of the other factors, whose prices remain constant, decrease. If, on the other hand, the other factors can be substituted for the factor x_i, there will be a quantity effect in the opposite direction. This results in (at least) a dampening of the price-induced change in cost shares. If the degree of substitutability is very high, the cost share of the factor whose price has increased may remain constant ($\beta_{ii} = 0$) or even fall ($\beta_{ii} < 0$). If, after the increase in the price of x_i, all the factor cost shares remain constant ($\beta_{ij} = 0$ for all i and j), the translog price function reduces to a Cobb-Douglas function. If the production factor x_j is a complement to x_i (whose price has increased) the cost share of x_j will fall.

The above shows that the degree of substitution between the factors of production can be evaluated using the parameter β_{ii}. Combined with the cost shares v_i and β_{ij} they can be used to calculate Allen partial elasticities of substitution (AES) (Berndt and Wood 1975, p. 261):

$$\sigma_{ii} = \frac{\beta_{ii} + v_i^2 - v_i}{v_i^2} \qquad i = 1,\ldots,J$$

$$\sigma_{ij} = \frac{\beta_{ij} + v_i v_j}{v_i v_j} \qquad i, j = 1,\ldots,J \quad i \neq j$$

An AES measures the relative change in the demand for factor x_i in reaction to a relative change in the price of the factor x_i or x_j after adjustment of all inputs to the now least-cost combination for a given output level. Put formally it is a share-weighted price elasticity. The weighting has the effect of setting σ_{ij} equal to σ_{ji}. If σ_{ij} ($i \neq j$) is negative there is a complementary relationship between the two factors. Positive values, on the other hand, imply substitution.

Finally, if the logarithmic price function is twice differentiated with respect to the technology index a measure can be derived for the change in the rate of technical progress in response to a change in the current state of technology.

$$\beta_{tt} = \frac{\delta^2 \ln q}{\delta t^2} = - \frac{\delta v_t}{\delta t}$$

It should be noted that the translog price function used in this model assumes constant values over time for the bias of technical progress (β_{it}), the share elasticities (β_{ij}) and the change in the rate of technical progress.

The model presented above for the J cost shares and the negative rate of technical progress is static. It is based on the assumption that in each period the cost shares adjust fully to the changes in prices and technological conditions. This is likely to be a very unrealistic assumption, as experience tells us that the adjustment of production processes and structures to changed conditions is usually spread over an extended period of time. For this reason the static model was transformed into the following modified partial adjustment model:

$$v_{it} = \alpha_i(1 - \lambda) + \sum_j (\beta_{ij0} \cdot \ln p_{jt} + \beta_{ij1} \cdot \ln p_{j,t-1}) + \beta_{it} \cdot t + \lambda \cdot v_{i,t-1}$$

$$-v_{tt} = \alpha_t(1 - \lambda_t) + \sum_i \beta_{it} \cdot \ln p_{it} + \beta_{tt} \cdot t - \lambda_t \cdot v_{t,t-1}$$

$$i,j = 1,\dots,J$$

The added delayed price term in the share equation takes account of the fact that relative price changes may provisionally shift the cost share in the "wrong" direction. This is reflected in a change in the sign of the coefficient of the delayed and non-delayed price term. If, for example, relative wages increase, the cost share of wages will probably increase in the short term, because in the short run the possibilities for substitution are very limited. In the medium to long term this increase will be reduced or even overcompensated due to substitution towards capital, which, given the appropriate values for the share elasticities, finally results in a lower cost share than at the outset.

Put technically, the parameter λ measures the stickiness of the cost shares, whereas $1 - \lambda$ represents the constant rate of adjustment to changes in price constellations. The coefficients β_{it} and β_{ij0} represent the short-term bias of technical progress and the short-run share elasticities, respectively. The equivalent long-term values can be calculated as follows:

$$\frac{\beta_{it}}{1-\lambda} \quad \text{respektive} \quad \frac{\beta_{ij0} + \beta_{ij1}}{1-\lambda} \quad i,j = 1,\dots,J$$

In a similar way long-term elasticities of substitution can be derived from long-term share elasticities. The adjustment parameter λ also permits estimation of the duration of the adjustment process to changes in factor prices.

Up until now, the econometric/production-theoretical approach to the study of the employment effects of new technology has rarely been implemented in the

Federal Republic of Germany. Recently some studies have been made in this area (e.g. Friede 1978, Griffin and Gregory 1976, Hansen 1983, Jäger 1980, Nakamura 1984, Pindyck 1979 and Schulte zur Surlage 1985), but in none of these was the emphasis placed on the questions concerning us here. In addition these studies generally ignore the skill differences between different groups of workers, assuming instead that labor is a homogeneous aggregate. A notable exception are the studies by Jäger and Schulte. However, they are both based on a number of restrictive assumptions which do not shed any light on the question of the effects of new technology on employment. Jäger's approach excludes the possibility of substitution between labor of different skill levels, while Schulte does not consider the possibility of biased technical progress. Furthermore, both approaches implicitly assume that firms adjust factor input to changes in factor prices within one observation period. Both models are thus static.

In contrast to the studies by Jäger and Schulte, our analysis controls for the possibility of delayed adjustment. In addition a model specification was chosen which allows for both biased technical progress and complex substitutive relations. Furthermore, technical progress is endogenized. Last but not least, differentiating among skill groups enables the effect of technical progress on the skill structure of employment in the Federal Republic to be econometrically analyzed for the first time.

3. The Empirical Procedure

In order to put the theoretical model to empirical use data are required on output, producer prices, factor quantities and factor prices. It is also necessary to divide the inputs into categories which are as homogeneous as possible so as to achieve reliable results for the actual degree of substitution between the factors of production. The demand for data exceeds the available supply in West Germany. Implementation of our model thus required data specially prepared for this purpose. Two data sets were used in the study.

The first data set was collected in conjunction with research conducted at the University of Frankfurt (Sonderforschungsbereich 3) under the leadership of Prof. Hujer. This set has been employed in several recent econometric studies (Bauer 1985, Schulte 1985) on factor demand in the Federal Republic. The data set contains annual figures from the period from 1960 to 1981. The data on factor inputs is restricted to capital and labor. As there are no figures on intermediate goods, weak separability between gross value-added, used to measure total output, and intermediates is assumed.

The data from the University of Frankfurt, with its 22 observation periods, provides too narrow a data base with which to estimate the dynamic specification of our model. For this reason a second data set was prepared, which relies on the same sources which served as the basis for the figures from the University of Frankfurt. The differences are that intermediates are also considered as a factor of production, 1980 is taken as the base year instead of 1976, and the greater actuality (1970-1984 instead of 1960-1980) and periodicity (quarterly rather than annual figures) of the data. As a considerable number of the original data consisted of annual figures, it was necessary in some cases to interpolate using indicator models designed especially for the purpose.

The data were subdivided by economic sector. The factors of production include labor (divided into blue-collar and white-collar workers), capital (divided into structures and equipment) and intermediates (only in the second data set). In a further step the blue-collar workers and the white-collar workers were subdivided into three and four skill groups (based on a classification by the Federal Statistical Office), respectively. The higher the number of the skill group, the lower the level of qualifications of the workers in it.

The available data on labor are differentiated by industry. The seven major industries of the industrial sector (in the wider sense of the term) were used: utilities, mining, basic and semi-finished goods industries, investment goods industries, consumption goods industries, the food industry and construction. As no suitable indicator models could be found for utilities and construction, it was not possible to include these industries in the second data set. For a detailed account of the way the data were constructed the reader is referred to our final report (Kugler, Müller, Sheldon 1987)

Finally, information is required on the rate of growth of total factor productivity (v_t) in order to operationalize the theoretical model for empirical analysis. The problem here is that technical progress itself cannot be directly observed. Following Jorgensen it can, however, be defined as the difference between the weighted relative change in input prices and the relative change in output prices in two consecutive time periods. In accordance with our theoretical model, if the output price increases less than it would have, had the price function been homogeneous and linear, this difference is ascribed to technical progress.

The complete estimating model consists of J cost share equations and an equation for the negative growth rate of total factor productivity. A stochastic disturbance variable is added to each equation in order to take account of the unsystematic influence of determinants not included in the model and/or unsystematic optimization failures on the part of firms. The resultant statistical specification of the model is as follows:

$$v_i = \alpha_i + \sum_j \beta_{ij} \cdot \ln p_j + \beta_{it} \cdot t + \varepsilon_i \qquad i,j = 1,\ldots,J$$

$$-v_t = \alpha_i + \sum_i \beta_{it} \cdot \ln p_i + \beta_{tt} \cdot t + \varepsilon_t$$

In the dynamic specification the share equations are supplemented by lagged price variables and cost shares, while the lagged rate of technical progress is added to the equation for technical progress

Notice that as the J disturbance variables of the share equations sum to zero in each period, the covariance matrix is singular. For this reason, one of the share equations is (arbitrarily) dropped. In addition certain parameter restrictions must be fulfilled so that the model conforms to our theoretical model. For one, all estimating equations must be homogeneous of degree zero. This implies that if all factor prices change proportionally, relative prices and the cost structure remain unchanged. Further, the cost shares must always sum to unity and, due to the interchangeability of the second order partial differentials, the coefficient matrix ß must be symmetrical.

Because of the endogeneity of factor prices at the level of an industry factor prices must be instrumented, a procedure which also permits a general test of the model's specifications (Hansen 1982). Once the parameters have been identified (i.e. sufficient instruments have been used) the model can be estimated using a three-stage least squares method (Jorgenson and Laffont 1974).

As has already been mentioned in section 2, the number of parameters increases disproportionately to the number of factors of production. To limit the parameter space factor demand is modeled in two stages. It is assumed that firms first choose the factor intensities for white-collar and blue-collar workers, capital equipment and buildings (in the second data set, also intermediates). Then they choose factor combinations for the four skill groups within the aggregate "white-collar workers" and for the three skill groups within the aggregate "blue-collar workers", respectively. In total then, three systems of cost share equations are estimated for each of the industries named above using two data sets and a static and a dynamic specification, respectively.

Estimation is considerably more complicated when the equation for the rate of technical progress is also considered. Estimation using two-period averages leads to a moving-average error, the structure of which is known, but the correction of which requires complicated data transformations (Jorgensen 1986). Furthermore, it requires that not only the current price variables, but also those lagged by one time-period be instrumented.

The calculation of the rate of technical progress (v_t) requires a data series for the output price. As no price series which is independent of factor prices exists for the contributions to production made by blue and white-collar workers, an index of the rate of growth of total factor productivity cannot be determined in these cases. Thus in the estimation models for blue and white-collar workers the function for technical progress must be deleted, with the result that no estimate can be made of the rate of change of the velocity of technical progress (β_{tt}). This of course lightens the estimation problem in these cases, as it is not necessary to work with two-period averages.

4. The Results of the Estimation

It would exceed the limits of this paper to present all of the estimation results. The following represents an attempt to provide an overview (in qualitative form) of the most important findings. It is preceded by a a few summary comments on the general fit of the models.

The data set with 22 annual values proved to be too narrow for the estimation of dynamic models. The estimates were often characterized by convergency problems and by very inexact parameter estimates. The results of these estimations are excluded from the following summary. The test of the overall specification of the model with the help of overidentifying restrictions (Hansen 1982) mentioned in the last section, showed that the dynamic specification performed better in all cases, with the exception of the sub-aggregate "white-collar workers". Frequently the static models had to be rejected, pointing to a misspecification of the model. Moreover, the autocorrelation of the residuals turned out to be a problem with the static, but not with the dynamic models. On the other hand, the large number of parameters in the dynamic models led to a standard error of individual coefficients which in some cases was very high. In view of these facts and with the desire to achieve robust results, it seems sensible to consider both the static and dynamic models separately.

As far as the explanatory value of the individual cost share equations are concerned the coefficient of determination is in most cases greater than 0.9, which may be considered very high. It is a different story, however, in the case of the equations for the rate of technical progress, where the coefficient of determination is in all cases close to zero. This means that the price elasticity of our selected measure of v_t

is very weak, so that the endogenization of technical progress is empirically unsatisfactory.

Our primary concern is the bias of technical progress, i.e. the changes in factor cost relations which arise solely from technological innovation. Given the translog specification of the unit costs equations, the bias of technical progress can be judged from the sign in front of the coefficient (β_{it}) of the trend variables in the cost share equations. A positive sign indicates that technical progress is factor-using (positive bias) whereas a negative sign points to factor-saving technical progress (negative bias).

Tables 4.1 to 4.3 provide an overview of the estimation results for the bias of technical progress. The columns numbered (1) contain the results based on annual values from the first data set (static specification). Columns (2) and (3) present the results of the dynamic model specifications, based on quarterly data. The distinction between them is the exclusion/inclusion (respectively) of intermediates at the highest level of aggregation.

We shall first consider table 4.1. A double sign (++ or --) in this and the following tables indicates that the assumption of neutral technical progress (no bias) can be rejected at the 5% significance level. In the static model with annual values, the results point to a significant labor-saving bias of technical progress. A similarly clear picture emerges for both sub-aggregates of capital: with one exception (utilities) technical progress has a positive bias towards capital in all branches. For white-collar workers, the picture emerging from the static model is ambiguous. In some branches technical progress is factor-using, in others factor-saving.

As can be seen from column (2) in table 4.1, with the change of model specification (static to dynamic) and the time period (from 1960-1981 to 1970-1984) the "bias pattern" becomes less clear. While the bias can still be considered to be labor-saving and machine-using, in the case of structures the results are highly ambiguous and there are signs of a factor-using bias for white-collar workers.

In contrast to columns (1) and (2) the bias in column (3) refers not to shares of value-added but to those of gross production. Even in this case technical progress is shown to be clearly labor-saving. No definite pattern can be discerned for equipment and white-collar workers while the negative signs for structures and the positive values for intermediates are all the more robust. The latter is an sign of increased externalization of production in manufacturing industry independent of factor prices.

To summarize the results of all three columns: technical progress is clearly labor-saving; there is evidence that technical progress leads to increased inputs of capital equipment and intermediates; no clear direction of bias could be identified for white-collar workers and structures.

Table 4.1: The Bias of Technical Progress

	white-collar workers			blue-collar workers			equipment			structures			intermediates
main group	(1)	(2)	(3)	(1)	(2)	(3)	(1)	(2)	(3)	(1)	(2)	(3)	(3)
utilities	--	.	.	--	.	.	--	.	.	++	.	.	.
mining	-	++	+	--	--	--	+	-	-	++	++	+	+
basic and semi-finished goods	++	-	--	--	+	--	+	+	-	++	-	--	++
investment goods	+	-	--	--	+	--	++	+	+	++	+	--	++
consumer goods	+	+	+	--	-	--	++	++	+	++	-	-	++
food industry	-	++	-	--	--	-	++	++	+	++	-	-	+
construction industry	++	.	.	--	.	.	++	.	.	++	.	.	.

Approach: (1) = static, annual values from 1960 to 1981, excl. intermediates
(2) = dynamic, quarterly values from 1970 to 1984, excl. intermediates
(3) = dynamic, quarterly values from 1970 to 1984, incl. intermediates

signs: + = factor-using technical progress
- = factor-saving technical progress
. = insufficient data

A double sign (++ or --) means that the assumption of neutral technical progress can be rejected at the 5% significance level.

As can be seen from tables 4.2 and 4.3, the ambiguity in the case of white-collar workers contrasts with the clarity of the negative bias of technical progress towards lower skills within the group of both white-collar (table 4.2) and blue-collar workers (table 4.3), and the positive bias towards better qualified workers.

Table 4.2: The Bias of Technical Progress: White-Collar Workers

	SG II		SG III		SG IV		SG V	
main group	(1)	(2)	(1)	(2)	(1)	(2)	(1)	(2)
utilities	++	.	--	.	-	.	--	.
mining	++	+	-	-	--	+	-	-
basic and semi-finished goods	++	++	--	--	--	--	--	--
investment goods	++	++	++	--	--	--	--	-
consumer goods	++	+	-	+	-	-	++	-
food industry	++	+	--	+	-	--	-	-
construction industry	+	.	-	.	+	.	-	.

SG: skill group (cf. section 3)

Approach: (1) = static, annual values from 1960 to 1981
(2) = dynamic, quarterly values from 1970 to 1984

signs: + = factor-using technical progress
- = factor-saving technical progress
. = insufficient data

A double sign (++ or --) means that the assumption of neutral technical progress can be rejected at the 5% significance level.

As has already been shown, the bias of technical progress also measures the sensitivity of the velocity of technical progress to factor prices. Looked at from this perspective, the overall results on the bias of technical progress indicate that an increase in the price of low-skill labor (i.e. the costs of the input of such labor) accelerates technical progress (ceteris paribus), whereas an increase in the price of highly qualified labor and/or of capital leads to a slowdown of technical progress. This constellation seems reasonable if it is borne in mind that the financial incentive to increase the rate of technological change is increased (decreased) when the factor whose input share falls (rises) becomes more (less) expensive. In light of this, the displacement effects of factor-saving technical progress can be slowed if the price of the displaced factor falls relative to other factors. According to this causal chain, technical progress thus contains the seeds of its own reversal. However, the low empirical explanatory value of the equation for the rate of technical progress v_t (see above) indicates that the speed of technological change is only marginally affected by changes in relative factor prices.

Table 4.3: The Bias of Technical Progress: Blue-Collar Workers

	SG I		SG II		SG III	
main group	(1)	(2)	(1)	(2)	(1)	(2)
utilities	++	.	--	.	+	.
mining	+	++	-	--	-	--
basic and semi-finished goods	++	++	-	--	--	--
investment goods	+	+	-	-	-	+
consumer goods	--	+	++	-	+	-
food industry	+	++	++	-	--	--
construction industry	++	.	-	.	-	.

SG: skill group (cf. section 3)

Approach: (1) = static, annual values from 1960 to 1981
(2) = dynamic, quarterly values from 1970 to 1984

signs: + = factor-using technical progress
- = factor-saving technical progress
. = insufficient data

A double sign (++ or --) means that the assumption of neutral technical progress can be rejected at the 5% significance level.

This section has dealt with the direct effects of technical progress on the structure of factor demand and with the negative feedback on technical change arising from its factor-price sensitivity. In addition to the latter channel of influence there also exists a further indirect causal chain: factor substitution as a result of changes in relative factor prices. Those factors which, due to the positive bias of technical progress in their favor, increase their share of total factor input, will sooner or later become more expensive relative to those factors displaced by technical progress has. This can lead to a subsequent increase or decrease in the demand for the displaced factor, depending whether the factors are substitutes or complements.

In this context the hypothesis that low-skill labor and capital are substitutes while high-skill labor and capital are complements, a hypothesis which has its origins in the USA (Hamermesh and Grant 1979), is of particular interest. If such a constellation is also applicable to the Federal Republic of Germany it would mean that the increased scarcity and hence cost of equipment affected by the capital bias of technical progress would contribute to a stabilization of employment for the less highly skilled blue and white-collar workers displaced by technical progress. In table 4.4 the validity of the US substitution pattern for the Federal Republic is tested.

Table 4.4: Substitution between White-Collar Workers (W), Blue-Collar Workers (B) and Equipment (E)

	W - B			W - E			B - E		
main group	(1)	(2)	(3)	(1)	(2)	(3)	(1)	(2)	(3)
utilities	+	.	.	+	.	.	+	.	.
mining	+	+	+	-	--	-	+	-	+
basic and semi-finished goods	+	-	--	+	-	--	+	+	++
investment goods	+	-	-	+	+	--	+	-	++
consumer goods	+	+	++	-	--	-	+	++	+
food industry	+	+	-	-	-	--	+	+	++
construction industry	+	.	.	+	.	.	+	.	.

SG: skill group (cf. section 3)

Approach: (1) = static, annual values from 1960 to 1981, excl. intermediates
(2) = dynamic, quarterly values from 1970 to 1984, excl. intermediates
(3) = dynamic, quarterly values from 1970 to 1984, incl. intermediates

signs: + = factor-using technical progress
\- = factor-saving technical progress
. = insufficient data

A double sign (++ or --) means that the assumption of neutral technical progress can be rejected at the 5% significance level.

The table gives the signs of the estimated Allen partial elasticities of substitution (AES, cf. section 3). A positive sign means that an increase in the price of one factor leads to an increase in the use of the other factor (positive quantity effect), indicating a substitutive relationship between the two inputs. A negative sign means the opposite: a negative quantity effect, a complementary relationship. A double sign indicates that the estimated value of the elasticity of substitution is at least 1.7 times greater than the relevant standard factor. If a normal distribution of elasticity estimates is assumed, this corresponds to a 10% level of significance.

The model specifications compared in table 4.4 are the same as in table 4.1. Standard errors were not calculated for the elasticity estimates based on the static model, hence the statistical significance of these values remains unknown.

The results drawn from estimation of the static model using annual values clearly indicate a substitutive relationship between blue-collar workers and equipment as well as between the former and white-collar workers. A complementary relationship between white-collar workers and equipment, a result obtained in studies in the USA, was confirmed for certain branches.

The results in columns (1) showing that blue-collar workers and equipment are substitutes, are by and large confirmed by the figures in columns (2) and (3), which result from estimation of the dynamic models. However, the substitutive relationship between blue and white-collar workers found in estimating the static model must, in the light of the dynamic models, be treated with caution. With regard to the complementary relationship between white-collar workers and equipment, the dynamic model specifications show even clearer results than the static models. To summarize: the results clearly indicate a substitutive relationship between blue-collar workers and equipment and, at least in the majority of cases, a complementary relationship between white-collar workers and equipment. The quality of the relationship between the demand for blue-collar and for white-collar workers, however, remains ambiguous.

For the dynamic model specifications it is interesting to note that the results of the model based on shares of value added and those based on shares of gross output are almost entirely uniform. This tends to confirm the assumption of a weak separability between intermediates and value added. The substitutive relationship between blue-collar workers and equipment and the complementary relationship between white-collar workers and equipment (cf. table 4.4) indicate that blue and white-collar workers cannot be used to construct an aggregate labor index. Studies which ignore this fact and instead attempt to do so run the risk of producing distorted results. On the capital side, a look at the estimation results shows that this warning also applies to attempts to combine equipment and structures. According to our results, equipment is a substitute (complement) to blue-collar (white-collar) workers, whereas the demand for structures is complementary (substitutive) to that for blue-collar (white-collar) workers. Within the subaggregates blue and white-collar workers, the relationship between neighboring skill groups are mostly complementary, whereas those between the groups at opposite ends of the skill spectrum tend to be substitutive. This pattern is particularly pronounced in the case of white-collar workers, where the highest and lowest skill groups represent substitutes in all the branches studied.

In the dynamic models it is not only the direction of the elasticities of substitution which are of interest, but also the speed of adjustment to changes in factor prices. Using the adjustment coefficient (λ) of the dynamic specification it can be determined how long it takes cost shares to adjust 90% to a factor price shock. This degree of adjustment was found to take slightly less than two years (somewhat less for white-collar workers and rather longer for blue-collar workers).

It has now been shown how the bias of technical progress and the elasticities of substitution influence cost shares. The question remains which of these two factors has had the greater effect in recent years. Simulations were conducted to determine the relative importance of price-led factor substitution and the bias of technical progress for changes in the structure of factor costs, which take into account the actual development of factor price relations and time trends. To this end *ex post* estimates were made of the factor-cost shares of blue and white-collar workers in the upper level of the dynamic model for the period from the first quarter of 1982 to the fourth quarter of 1984. This served as a control scenario. This was followed by two further scenarios in which only the trend or only factor prices were allowed to vary in accordance with the real world development while the other variable was kept constant. A comparison of these two scenarios with the reference scenario reveals which of the two causal factors was the stronger during this period. The results are summarized in table 4.5: a P means that price-led substitution was dominant, a TP appears where the bias of technical progress was the more important factor. The results appearing in the "total" column show that in all branches the (negative) bias of technical progress exerted a stronger influence than price changes, whereas in the case of white-collar workers the results are very heterogeneous.

Table 4.5: The Relative Importance of Changes in Factor Prices and the Bias of Technical Progress for Changes in the Structure of Labor Cost Shares

	White-collar workers					Blue-collar workers			
main group	total	SG II	SG III	SG VI	SG V	Total	SG 1	SG 2	SG 3
mining	P	P	P	P	P	TP	TP	TP	TP
basic and semi-finished goods	TP	TP	P	TP	TP	TP	TP	TP	TP
investment goods	TP	TP	TP	TP	P	TP	TP	P	P
consumer goods	P	P	P	TP	P	TP	P	P	TP
food industry	P	P	P	TP	P	TP	TP	P	TP

As technical change cannot be considered merely a "thing of the past", the question arises as to future developments: By how much would (ceteris paribus) the relative prices of the factors of production with a negative bias have to be reduced in order to compensate the factor-saving effects of technical progress? To put it another way: on the basis of the results presented here, for which factors is the greatest downward pressure on factor prices to be expected? This analysis is restricted to the blue and white-collar labor aggregates. In order to hold the share of an aggregate constant, the fall in wages will have to be greater the higher the speed of the negative bias and the more inelastic demand are. On the basis of these considerations, the greatest pressure for downward wage adjustment is to be expected in the basic and semi-finished good industries. Overall, the relative wage cuts to compensate for the negative bias of technical progress will have to be greater for white-collar than for blue-collar workers, due primarily to the difference in their price elasticities.

5. Summary and Conclusions

The aim of our study was the empirical analysis of the effects of technical progress on the skill structure of employment in the Federal Republic of Germany. The point of departure was a general model of producer factor-demand behavior. With the help of this approach three major causal relationships could be studied:

- A direct causal relationship in the form of the factor-specific bias of technical progress. Technological change may alter the relative productivity of the factors of production and thus the structure of inputs chosen by firms. This favors certain factors while displacing others. The former case is described as a positive bias or factor-using technical progress, the latter as negative bias or factor-saving technical progress.
- An indirect casual relationship via the substitution between the factors of productions following changes in their relative prices. The factors increasingly employed in line with a positive bias of technical progress become more expensive relative to those factors which are displaced. This can lead to a subsequent increase or decrease in the demand for the displaced factor, depending on whether it is a substitute or a complement to the increasingly more expensive factor.
- A second indirect casual chain via the factor-price sensitivity of the speed of technical progress. Factor price changes which anticipate (run counter to) the cost-cutting effects of technical progress create positive (negative) incentives for firms to proceed with the introduction of new technology.

These three causal relationships stood at the center of our study. The employment effects of an increase in the demand for goods - a further likely result of technical progress - was excluded from the analysis.

In order to operationalize the model its general structure had to be more precisely specified. The result was an econometric model of factor demand characterized by the following elements:

- non-neutral (biased) technical progress (Hicks)
- an endogenization of the speed of technical progress
- a functional form which offered a very flexible framework within which the substitutability of factors of production could be studied, and
- a dynamic specification which takes account of the delayed adjustment of firms to changes in factor prices.

The econometric model has a hierarchical structure. On the upper level the model consists of factor demand equations for blue-collar workers, white-collar workers, buildings, equipment and intermediates, respectively, and one equation for the rate of technical progress. On the lower level the demand equations were differentiated by dividing blue and white-collar workers into seven skill groups. In total 20 systems of equations were studied. They were constructed for the seven main industrial branches.

Two data sets were used to estimate the model's parameters. The first was based on data for the period from 1960 to 1981. It was put together by a research group at the University of Frankfurt and has been used in earlier econometric studies on factor demand in the Federal Republic of Germany. The second set was prepared by the authors. It consists of quarterly data for the period from 1974 to 1984 and is based on the same data sources as the set from the University of Frankfurt.

The main results of our econometric study can be summarized as follows:

- During the period under study technical progress was clearly labor-saving for blue-collar workers. In the case of white-collar workers the results were ambiguous, i.e. in some branches technical progress was factor-using, in others factor-saving.
- Within the subaggregates of labor (i.e. blue and white-collar workers) the bias of technical progress was directed away from low skill labor. The bias towards the better qualified groups in both subaggregates was clear.
- The speed of technical progress was only marginally affected by changes in relative factor prices: it could not be shown that the rate of technical progress is factor-price sensitive.
- The results clearly indicate a substitutive relationship between blue-collar workers and machines and at least a clear trend towards a complementary relati-

onship between white-collar workers and equipment. This implies that changes in the price of equipment produce a parallel (opposite) quantity effect in the demand for white-collar (blue-collar) workers. This result indicates that in the past factor substitution ran counter to the displacing effects of the bias of technical progress.

- Substitutive relationships between blue-collar and white-collar workers could not be unambiguously determined. However, within both subaggregates, the relationship between labor of a high and of a low skill level was clearly substitutive.
- The results of the dynamic model specifications indicate that on average the adjustment of factor inputs to changes in factor prices was spread over a period of approximately seven quarters. The one significant deviation from this result was for blue-collar workers where the average adjustment period was about half a year longer.
- At the upper level of our two-tiered model the quantitative extent of the reaction of factor inputs to changes in factor prices was modest. The price and cross-price elasticities of demand for blue and white-collar workers have an absolute value of less than one. The values within both subaggregates, however, are considerably higher: between blue-collar workers of different skill groups absolute values of between two and six were recorded, i.e. factor-price changes within the group of blue-collar workers bring about quantity effects which are twice to six times as great. It must be emphasized, however, that some of the estimates of elasticity are subject to a large margin of error.
- An analysis of the relative importance of factor substitution and the bias of technical progress in the allocative process showed that for the period 1982 to 1984 in the Federal Republic it was the bias of technical progress was the predominant factor for determining the factor shares of blue-collar workers and their different skill groups. For white-collar workers neither of the two market forces appeared to be dominant.
- As far as the future is concerned, our results indicate that the downward pressure on wages will be higher for white-collar workers, if job loss due to technical progress be solely counteracted via the wage mechanism.

This study is only one of the constituent parts of the Meta Study. For this reason it would certainly be premature to draw conclusions for employment policy from the limited number and partial equilibrium nature of the results of this study. The most important factor which was excluded from our analysis is the compensation effect on labor demand due to the increased demand for goods and services coming from technical progress. Nevertheless, some of the results do appear sufficiently robust to hold up to any evidence from other analyses within the Meta Study. This applies particularly to our results on the bias of technical progress and the substitutability of factors of production.

The findings on factor substitution, however, show that the effects of the bias of technical progress can be mitigated if the prices of the displaced factors (low skill labor) fall relative to the price of those factors whose cost share has increased (equipment and skilled labor). Given that, it can be assumed that any factor-price changes resulting from the displacement effects of technical progress will work to the benefit of the factor which had initially been displaced. This implies that a policy of subsidizing wages in order to ease the reintegration of redundant workers into the labor market would succeed. However, in view of the structure of substitution between the different factors of production, such a policy would only lead to a reduced demand for skilled labor and discourage investment. Hence, a policy of subsidizing low skill labor would end up being a policy for "bad jobs". On the basis of our results and in view of the need to maintain international competitiveness it would appear to be more sensible to alleviate social hardship for workers displaced by technical progress not by interfering with the price mechanism but rather by direct income transfers. On the other hand, the bias of technical progress towards highly qualified labor shows that by improving the skills and qualifications of redundant workers, their chances of finding employment can be increased considerably. These proposals, which are frequently put forward in the unemployment discussion in West Germany, are given further support through our findings.

Technological Change and Employment Structures

Jürgen Warnken and Gerd Ronning

1. Introduction

There is in academic circles a general consensus that the direct employment effects of modern technology are, on the whole, negative, i.e. the use of new technology tends to increase productivity and so (ceteris paribus) reduce employment in the areas in which it is employed. At the same time it is scarcely disputed that direct job losses are (at least partially) compensated viá a variety of mechanisms at both the sectoral and macroeconomic level. The source of disagreement lies in the relative importance of these compensatory effects and the time lags involved before they are felt. A summary of this debate can be found in Mettelsiefen and Barens 1987. Consequently the focus of the debate on the overall effects of the use of new technology has switched to the effects of new technology on the structure of labor input, i.e. the shifts between different categories of "labor" as a factor of production (e.g. age, gender, skill level, occupation). (This is also emphasized in the contribution of Ewers/Becker/Fritsch in this volume). The aim of our research team (the ISG together with Prof. Ronning) within the framework of the Meta Study was to determine the most important factors which, in the context of technological and organizational innovation, influence the substitution of these different categories within the enterprise. The role of technology in the system of production is studied at two levels:

- the determinants of the planning and operationalization of the use of technology and the diffusion process at the level of planning and decision-making processes within the enterprise;
- the relationship of the different categories of labor to one another in the context of product and process innovation at the level of the use of concrete forms of technology.

This implies that the use of modern technology is conceived of as a "quasi-endogenous" component of the economic system, i.e. at enterprise level at least, as one element in a firm's planning and decision-making procedures. If, unlike in many approaches to the analysis of productivity, technical innovations are no longer seen as a sort of "manna" which falls from heaven, the phenomenon of "technology" becomes more complex, whereby not only the (skill-specific) employment effects of innovation but also the economic and institutional factors which determine technological change must be taken into account.

The questions posed above were investigated by the project group consisting of the ISG and Prof. Ronning using econometric models and so-called "analytical-interpretative" methods. The latter were applied because a considerable number of the relevant influential factors could not be quantified in a satisfactory way. Our contribution to this volume concentrates primarily on two aspects of our work:

- the comparative analysis of different restructuring processes in horizontal (i.e. activity-oriented) skill structures, in the context of the use of modern technology in the 1980s (in section 2);
- the influence of industrial relations on the planning and the actual implementation of new technologies including their effects on employment in the 1980s and the attempt to quantify them (in section 3).

Only the fundamental trends of the results of our comprehensive econometric model calculations concerning the shifts between different skill levels will be presented here, as similar analyses with comparable results have been carried out by other project groups within the framework of the Meta Study (Kugler et al. 1987). The interested reader is referred to the full-length version of our project report. A further omission in this summary is an exact description of the taxonomy used to differentiate between sectors, occupations and activities, and the grouping of branches.

2. Activity Structures and the Diffusion of New Technology in the Enterprise

A common distinction in academic discourse on technology and employment is that between horizontal and vertical skill changes. This study has attempted to deal with both aspects. Results from the econometric model show that the (vertical) composition of the labor force (i.e. the grouping of workers according to skills in the wage and salary statistics), has been subject to some long-term variation, but that the innovative activity of the 1970s and 1980s has played a comparatively minor role. The overall trend for white-collar workers (and to a lesser extent for blue-collar workers) in manufacturing industry is from lower to higher skills. Innovative activity, in particular process innovations, has reinforced this process, but only to a fairly limited extent (cf. fig. 2.1). The effects of modern technology on the composition of the labor force in terms of skill groups is concealed to a considerable extent by changes in sector-specific, demand-side factors.

Table 2.1: A Branch Typology by Innovative Activity

Dynamic innovators	a	b	Hesitant innovators	a	b	Weak innovators	a	b
automatic data proc. ind.	1	I	paper production	16	V	state	35	--
electrical goods	2	II	health services	17	--	restaurants, hotel services	36	--
non-ferrous metal prod.	3	--	chemicals	18	III	clothing	37	IV
glass processing	4	III	printing	19	V	foundries	38	--
crude oil processing	5	III	insurances	20	--	stone and clay	39	V
energy	6	--	financial institutions	21	--	transports	40	--
communications	7	--	rubber processing	22	III	retail trade	41	--
shipping	8	--	foods, drinks, tobacco	23	IV	railway	42	--
aeronautical production	9	--	textiles	24	IV	wood processing	43	IV
musical instruments etc.	10	II	machinery	25	II	leather	44	V
road vehicles	11	II	drawing and rolling mills	26	IV	ship building	45	II
steel production	12	--	plastics processing	27	IV	prod. of steel and light metal	46	V
science/education	13	--	precision ceramics	28	I	construction	47	--
iron, sheet metal/metal goods	14	II	wholesale trade	29	--			
precision engin., optics	15	II	paper processing	30	IV			
			millwork/wood products	31	V			
			agriculture	32	--			
			mining	32	--			
			remaining services	32	--			

a: values based on expenditure on innovation
b: number of the branch cluster based on shares of innovators

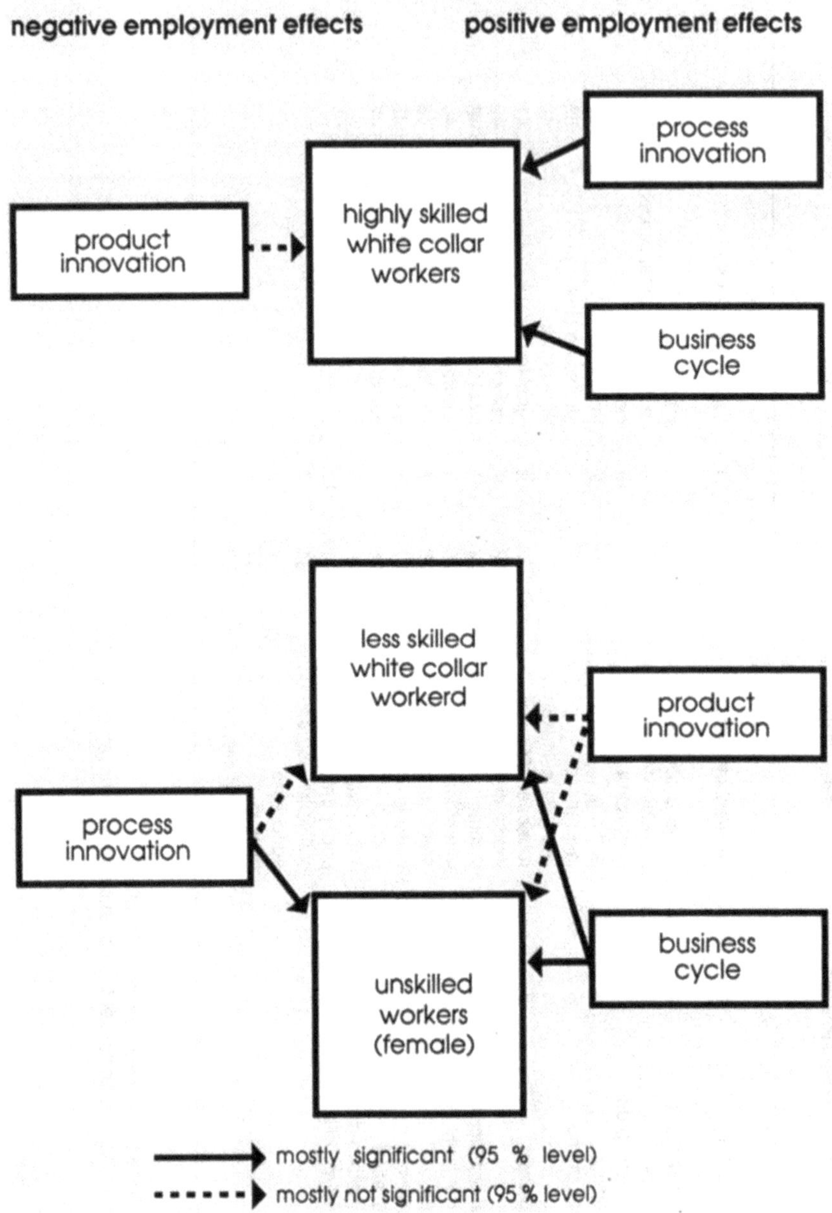

Figure 2.1: The Determinants of Changes in Employment in Manufacturing Industry
- selected skill levels -

Parallel to these "vertical" changes in skill requirements there has also been movement in the "horizontal" structure of the requirements made of the work force with respect to occupational, functional and so-called extra functional skills and qualifications. The analysis of these changes indicates which concrete educational and training measures will facilitate work with new technology, is an absolute prerequisite for its successful implementation. In order to construct indicators for such a "horizontal" shift in skills and qualifications, developing parallel to and dependent on both technological and organizational change and movements in demand, it is clearly not sufficient to judge a worker's qualifications merely by the occupation or profession which he or she has learned. Occupational skills, once accumulated, can be applied in a variety of different activities within the employment system.

This fact is taken into account in the relevant statistics available for the Federal Republic of Germany: normally the job or profession actually performed by a worker is recorded rather than that which he or she originally learned. Furthermore, in the course of the micro-censuses information on the actual characteristics of a worker's activity is collated, providing, it may be assumed, a more exact indication of skills and qualifications than simply a worker's "profession". Finally, by combining the educational and training characteristics of all employees subject to social insurance with their actual positions in the enterprise (i.e. information determined primarily by the supply of labor, with that derived from the demand for labor) it is possible to gain added insights into a specific aspect of the development of skill structures.

2.1 The Occupational Structure of Labor Input

A first approximation to the skills and qualifications of a worker is his or her profession or occupation. The interpretation of results achieved on this basis is, however, fraught with difficulties because technical and socio-economic change and its effects on the employment system can only be partially ascertained from changes in occupational structure measured in this way. The terms used to describe occupational groups have been determined as a result of an historical process, and are often based on a system of craft work which bears little resemblance to modern industry. It is therefore very difficult to draw conclusions about the actual activity performed from such a "description". Activity profiles are often adjusted in the face of changes in technological and organizational changes in working conditions, without this leading to a change in the term used to describe the worker performing these functions. This is partly because the use of programmable equipment is spreading to so-called "irregular" users. The number of jobs in which modern equipment is predominant and is used the whole day long is not growing nearly as rapidly as those where such equipment is used intermittently (Stooß and Troll 1988, p. 31).

Our study attempted to take account of this increasing obscurity in occupational classifications in the following way. In the sector-occupation matrices which form the basis of the analysis for 1980 and 1985 employees were divided into 51 occupational categories which, while they were derived from official classifications of occupations and are compatible with them, were classified into groups following a "functional" concept developed by the IAB (Institute for Labor Market and Occupational Research). This produces a more practical aggregation of individual occupations than that provided by the official statistics as they are divided into a series of hierarchical levels.

As is to be expected, the influence of modern production and communications technologies on occupational structures is relatively easily identified in technical occupations, whereas others (such as office workers) are less clearly defined. Workers in technical occupations can be said to have benefited from technological and organizational changes in the employment system because:

- against the background of falling overall employment between 1980 and 1985, they have usually managed to maintain their absolute position,
- and workers in technical occupations benefited to a more than proportional extent from the increase in employment in those branches in which the level of innovative activity was above average (see fig. 2.2).

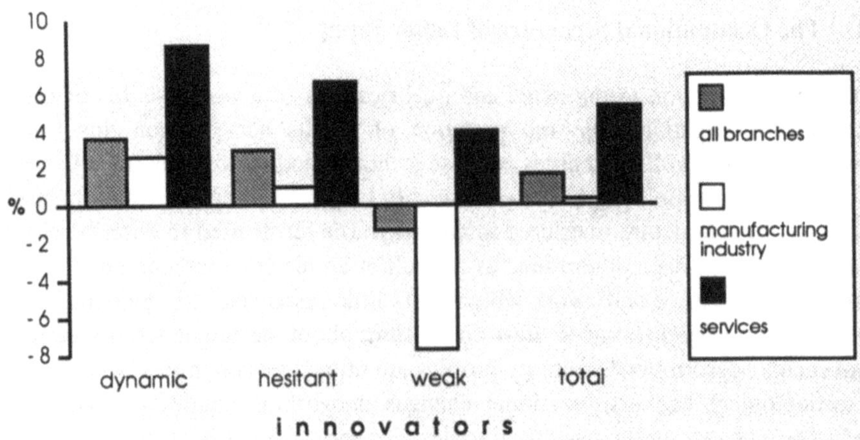

Figure 2.2: The Changes in Employment in Technical Occupations
- 1980 to 1985 -

The group of "dynamic innovators" is dominated by the capital-good sectors of manufacturing industry (excl. engineering), but also includes crude oil processing, communications and publishing. On the other hand, the following branches were characterized as "weak innovators" during the first half of the 1980s: consumer-good branches (in particular the leather and clothing industries and wood processing), but also steel and light metal construction, shipbuilding and the construction industry. The allocation of the different branches to the three types of innovators was based on a number of analytical grouping procedures including a cluster analysis. A comparison of the changes in employment in the three types of branches with regard to technical occupations reveals a wide disparity. Similar results are obtained for the analysis of manufacturing industry alone, in which in 1985 two thirds of all workers in technical occupations could be found. For this sector of the economy there was also confirmation of the tendency towards higher qualifications as a result of the use of technology which had been identified in econometric model calculations:

- By 1985 the number of workers in technical occupations had increased by 3.5% (+17 000) in branches with a high level of innovative activity compared to 1980. In branches with a comparatively low level of innovative activity, on the other hand, employment in these groups fell by 1.8%, or slightly less than 6 000 (fig. 2.2).
- The main beneficiaries of these developments were those with a higher level of academic training, at least in the area of "design" (table 2.2). Electrical and mechanical design engineers found an additional 18 600 jobs in highly innovative branches (out of a total increase of 24 000). This represents an increase of 19.3%.
- Finally, the extent to which the use of modern technology is changing working methods in the areas of product development and design becomes obvious. Technical draughtsmen were best able to maintain their position in branches with a low level of innovation, whereas in the dynamic branches there are already some indications of the effects of computer-aided design in the structure of the work force. In certain branches of manufacturing industry there was even an absolute reduction in the employment of technical draughtsmen, indicating not only that design is being carried out with the help of high-level computer programs, but also that this activity is to some extent being contracted out of manufacturing industry. This can be deduced from the rapid increase of draughtsmen in the service sector (+3 300 or 5.9%). However, this was accompanied by computer-aided technology, and its therefore highly probable, that some change in job requirements have occurred.

Table 2.2: Changes in the Employment of Technical Occupations in Selected Branch Groups - 1980 to 1985 -

	dynamic innovators		hesitant innovators		weak innovators	
	abs.	in %	abs.	in %	abs.	in %
technical occupations total	17,239	3.5	17,200	2.9	-5,938	-1.8
of which:						
engineers (design)	18,659	19.3	6,015	6.4	29	0.0
of which:						
- mechanical engineers	7,171	24.5	3,876	8.0	361	2.4
- electrical engineers	11,339	17.9	4,804	30.7	977	16.3
mechanical technicians	1,242	5.3	1,545	3.1	-514	-3.7
electrical technicians	3,439	6.2	4,662	18.2	1,143	8.5
other engineers	310	3.8	1,188	6.6	397	4.4
of which:						
- production engineers	434	10.0	566	4.7	482	17.1
other technicians	2,261	3.0	1,683	1.5	-3,676	-5.2
production controlling occupations	-10,218	-6.7	-6,609	-5.8	-4,830	-13.4
of which:						
- industrial/works foremen (Meister)	652	1.2	-3,904	-5.9	-2,782	-11.9
technical draughtsmen	440	1.5	2,211	3.4	1,386	3.9

The growing use of office and communications technology (data processing equipment, personal computers, word processors and Teletex, Telefax and BTX) means that the employment structures of office workers have also begun to be severely affected, after many years of comparative "resistance" to the effects of automatization and technical progress in general in administrative occupations. As a consequence the expansion in employment of both skilled and semi-skilled office workers has come almost to a standstill, at least among the dynamic innovators in manufacturing industry. Job losses in this area normally affect workers with lower skills, performing routine work and lacking decision-making powers. In branches with a high innovation dynamic overall employment of such groups of workers fell by almost 16% (fig. 2.3), in manufacturing industry by as much as 20%. As is to be expected, many of the routine tasks are no longer required due to the use of office and communication technology, or have been transferred to more highly skilled workers. Employment of the latter group increased slightly in the course of this process in the more dynamic branches, while in manufacturing industry it sank but not by more than the average amount.

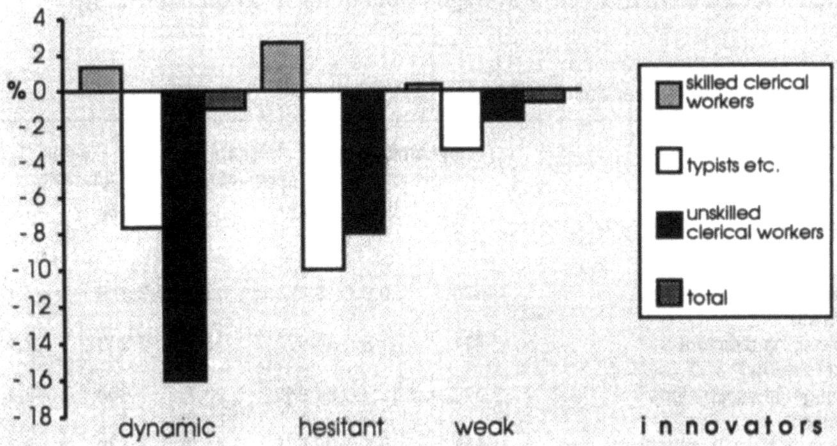

Figure 2.3: Clerical Workers and Innovation
- changes between 1980 and 1985 -

One remarkable trend was that (shorthand) typists, whose functions were considered to be in great danger from the spread of word processing and enterprise-level information systems, have, to judge by the overall employment figures, so far been only marginally affected by the use of these and other new technologies. While their absolute numbers did decrease (by some 30 000 or 6.5% between 1980 and 1985) their share of total employment decreased almost negligibly, and the difference between the dynamic and the weak innovators is very small. One is forced to the conclusion that the use of word processors with their storage capacity, efficiency and practicality in the office environment has tended to increase the quantity of written material. A further mechanism is presumably that clerical workers who were previously responsible merely for writing texts (esp. letters) now perform additional activities.

Employees performing managerial tasks in the wider sense of the term may also be considered under the rubric of office and administrative professions. They consist of two occupational categories, one concerned with planning and decision-making within the firm (enterprise management, advertising experts), the other ("external specialists") consisting of management, organizational and technical consultants.

Both occupational categories improved their position within the employment system during the first half of the 1980s, that is the overall fall in employment during this period affected them to a below average extent or not at all (cf. table 2.3).

Table 2.3: Employment Trends in Managerial Occupations 1980 to 1985

	dynamic innovators		hesitant innovators		weak innovators	
	abs.	in %	abs.	in %	abs.	in %
- all branches						
total "internal" managerial occupations	-526	-0.8	-3,371	-2.0	-7,479	-7.1
of which: - managing directors	-721	-1.3	-4,072	-2.6	-7,306	-7.1
total "external" managerial occupations	3,379	12.7	24,850	28.7	-606	-2.4
of which: - (organizational) consultants	1,810	25.1	22,528	41.2	130	3.2
- manufacturing industry						
total "internal" managerial occupations	-859	-1.7	-3,201	-4.5	-3,228	-8.5
of which: - managing directors	-1,096	-2.4	-2,842	-4.3	-3,223	-8.7
total "external" managerial occupations	2,901	12.8	943	6.8	-103	-2.9
of which: - (organizational) consultants	1,664	26.2	667	19.3	54	6.4
- service sector						
total "internal" managerial occupations	234	2.5	-1	0.0	-4,251	-6.2
of which: - managing directors	260	3.3	-1,064	-1.2	-4,083	-6.3
total "external" managerial occupations	163	7.7	23,833	33.2	-503	-2.3
of which: - (organizational) consultants	90	18.1	21,841	42.7	76	2.4

Technical and organizational change and the general uncertainty of the market environment have created additional tasks for management personnel, and existing tasks are now being pursued more intensively. As can be seen from the figures for "internal" management, these employment effects were not independent of innovative activity. The case of the "external" consultants for technical and organizational questions clearly shows the great importance of intensive outside support for corporate planning processes in a period of technical and organizational change: between 1980 and 1985 the overall employment of management personnel increased by 29 000. By far the largest increases were in the service sector. It seems plausible that this is primarily a reflection of the fact that management services for manufacturing industry are increasingly being "bought in" from firms in the service sector. Nevertheless, a glance at the figures for manufacturing industry itself shows that the direct hiring of management personnel by firms in these sectors is positively correlated to their level of innovative activity (table 2.3). It is rare for weak innovators to resort to the direct hiring of additional managerial staff.

Table 2.4: Changes in Overall Occupational Structure* 1980 to 1985

	dynamic innovators	hesitant innovators	weak innovators
all branches	0.509	1.298	1.112
of which:			
- manufacturing industry	0.492	0.976	1.234
- service sector	1.069	1.490	0.989

* as measured by the Shannon information measure on the basis of 142 occupational categories.

Against the background of the occupational restructuring sketched above one point emerges which is particularly worthy of note: using several different measures of structural change (e.g. the Shannon information measure) branches with a high level of innovative activity experience on balance a less far-reaching process of structural change than less innovative branches. On the basis of these figures there is evidence that dynamic innovators as a whole have changed the composition of their work force by occupation considerably less than hesitant or weak innovators (table 2.4). To a great extent this may be due to changes in the overall employment level in the different branches. In the course of their often drastic cuts in employment hesitant and innovators appear to have employed a staff selection process which had a considerable impact on occupational structure. Dynamic innovators, on the other hand, against the background of a relatively favorable employment trend, tended to retain workers within the enterprise; the use of new technologies led to a change of emphasis or an extension of their spectrum of activities rather than a redefinition of their occupational category.

Table 2.5: A List of the Activity Categories and their Aggregation into Occupational Groups

activity category	classification number	occupational group
1 cleaning, domestic functions	051, 052	
2 training etc.	053	
3 organization, surveillance administration of justice	055, 056	1 services (051 to 058)
4 physical/psychol. treatment/advice	054, 057	
5 publishing etc.	058	
6 those in training	059	
7 total services		
8 primary production	111, 112	2 goods production (111 to 116)
9 craft production	114 to 116	
10 machine-oriented prod.	113	
11 control and supervisory work in production	117, 630 to 636, 639	
12 production activities		
13 handling machines	2	3 maintenance (2, 3)
14 repair work	3	
15 storage and distribution	430 to 433, 435 to 438	
16 transport	434, 439	4 storage and transport (4)
17 general sales activities	530 to 532, 536	
18 product-oriented trade activities	533 to 535	
19 customer-oriented intermediaries	537 to 539	

Continuation of table 2.5

activity category	classification number	occupational group
20 trade activities		
21 research and development	.42	5 design and development (.42)
22 administrative activity in procurement	.41	
23 of which: office work	741	
24 production planning and guidance	.43	6 production planning and supervision (117, 630 to 636, 639, .43)
25 of which: office work	743	
26 administrative activity in sales/marketing	.44	7 trade (530 to 539, .41, .44)
27 of which: office work	744	
28 administrative activity in accounting	.45	8 administration (.45, .46, .48, .49)
29 of which: office work	745	
30 general administration	.48	9 in training (059)
31 of which: office work	748	
32 EDP activity	.46	
33 of which: office work	746	
34 organization, management	.49	
35 of which: office work	749	10 other groups
36 administrative activity	7 to 9	
37 of which: office work	7	
38 all activity categories		11 all occupational groups

* Classification number in the official statistics, cf. Statistisches Bundesamt (1981, p 114).

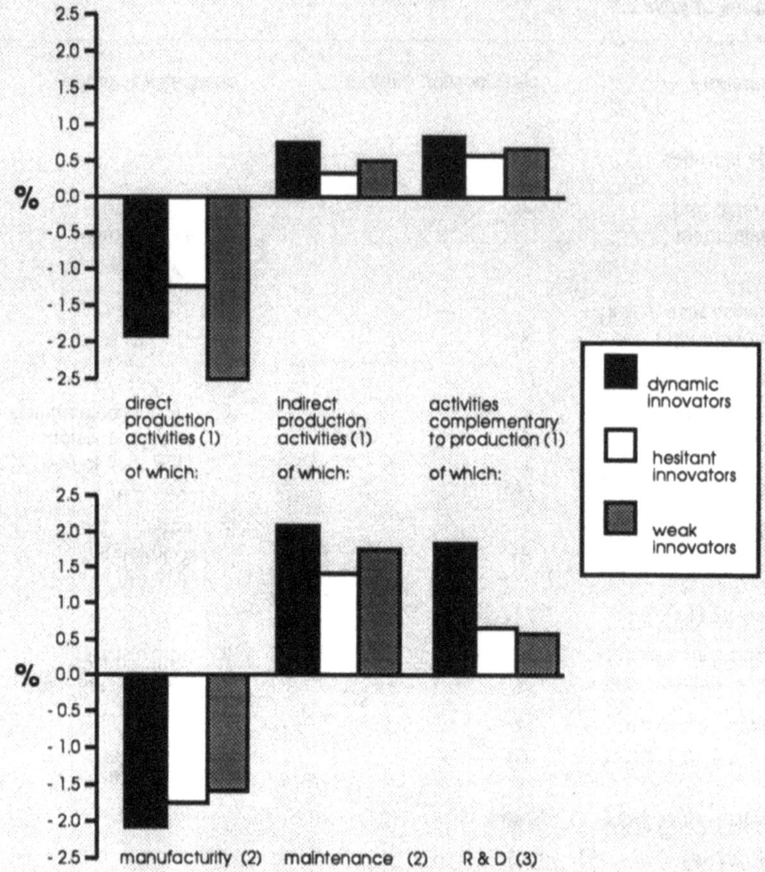

(1) with respect to all occupational functions incl. trainees and workers whose activities are not adequately defined
(2) with respect to production activities
(3) with respect to activities complementary to production

Key:

direct production activities:	**include all direct manufacturing activities (e.g. craft or machine-oriented adtivities) and service activities (e.g. cleaning, security, consultancy)**
indirect production activities:	**include all forms of maintenance activity (handling machines, repairs) plus warehouse and transport activities**
activities complementary to production:	**include design and R & D, planning and supervision of production, trade and administration**

Figure 2.4: Changes in Proportion of the Work Force Employed in Selected Occupational Functions of Manufacturing Industry, 1980 to 1985

2.2 Activity Structures in Different Branches

In order to obtain information on further aspects of the change in activity structures it is necessary to differentiate sectoral employment trends by various indicators of activity. An example of this technique is the concept developed by the IAB to improve the data on changes in enterprise-level task and activity structures (Stooß and Troll 1988). This necessitates the extrapolation of the activity structures of occupational groups - as determined in the 1970s with the help of the microcensus - until the year 1985 using logarithmic trends. In total 37 activity categories were chosen, some of which overlapped. These data were then linked to the appropriate sector-occupation-matrix for 1985. Table 2.5 presents the activity categories and shows how they were further collated in the course of the actual analysis.

This functional approach to the analysis indicates that the use of modern computer-aided technologies was accompanied by a slight "decoupling" of production processes in the enterprise from the deployment of labor. This is reflected in the fact that, for example, the use of modern technology, particularly in the innovative branches of the investment-good sectors and other branches of manufacturing industry, induced an accelerated reduction of staff with direct production activities (machine operators), while at the same time the use of labor increased in activities complementary to production and in the maintenance and repair of plant (indirect production activities) (fig. 2.5):

- Branches with a high innovation dynamic, at least those in manufacturing industry, reduced the amount of labor devoted to direct production activities considerably: the share of this activity category in total labor involved with direct and indirect production activities ("production-oriented labor"), which in dynamic innovators in manufacturing industry was by 1980 already relatively low (56%), fell by a further two percentage points. This amounted to a fall in employment of over 11%, compared with -6.1% for workers as a whole in dynamic innovative firms. By contrast, in branches dominated by hesitant innovators in 1980 a much larger proportion of the work force (68%) was employed in direct production activities. Their share of total employment in such branches subsequently fell by only 1.6 percentage points.
- In contrast to the development in direct production activities, workers involved with maintenance tasks fared better in terms of employment in branches with a high level of innovative activity than in those where the efforts at modernization were less intensive. Between 1980 and 1985 less than 10 000 (-1.4%) of workers performing maintenance activities lost their job (on balance) in innovative branches. In such branches more than 30% of all production workers were

assigned to maintenance activities in 1985. Job losses in branches dominated by weak innovators amounted to more than 50 000 (-9.7%).Thus, even in less innovative industries the share of maintenance activities increased, although it still amounted to only a little over 23% in 1985. In the course of the implementation of new technologies the importance of maintenance work thus increased, a result which also applies in large measure to the service sector.

The analysis of individual occupational categories indicated that the use of modern technology leads to a far-reaching restructuring of labor deployment, not least in those activities which are complementary to production. Between 1980 and 1985 branches with a high innovation dynamic increased their employment of workers in activities complementary to production compared with other branches. The major beneficiaries of these changes in activity structures were those involved with R&D. They can thus be considered an important factor supporting and promoting innovation. In 1980 workers in such fields already accounted for a greater proportion of activities complementary to production in the more innovative (14.4%) than in the weak innovator branches (12.8%), and their importance has since grown more rapidly (cf. fig. 2.4). In the innovative branches of manufacturing industry the number of such workers grew by about 7% between 1980 and 1985 and their share of employment increased by a further 1.5 percentage points.

Another significant development was registered in administration. In innovative branches the relative expansion of the number of workers engaged in administrative activities (accounting, general administration, organization, scheduling and the use of data processing equipment) applied overwhelmingly to skilled workers. In such branches workers in the administrative field are concerned to a greater extent with setting up projects, design, (cost) calculation, and coordination, organization, leadership and negotiation than in less innovative branches. The number of workers performing administrative tasks increased by roughly the same extent in all three types of branch (Warnken and Ronning 1988, ch. 2.6), i.e. their respective shares in total employment in those branches increased by the same amount (1.5 to 2 percentage points). However, in the course of this development technical and organizational restructuring has had a decisive effect on the proportion of routine, low-skill office work in total administrative activity (table 2.6):

- From the start of the period studied (1980) less workers were employed for such routine activities as bookkeeping, registration, typing/correspondence in the innovative sectors of manufacturing industry than in less innovative branches. In the former they accounted for less than 61%, whereas in less innovative branches 67% of all administrative workers were employed to deal with routine tasks.

Table 2.6: Changes in Office Activities 1980 to 1985

	dynamic innovators		hesitant innovators		weak innovators	
	employees in %	employee share (in % points)	employees in %	employee share (in % points)	employees in %	employee share (in % points)
- all branches						
total office activity*	-2.9	-1.7	-1.3	-1.5	-3.6	-0.8
of which:						
- trade activity**	-17.3	-3.0	-17.4	-2.2	-18.9	-1.7
- administrative activity ***	0.1	-2.2	1.4	-2.6	-0.3	-1.3
- manufacturing industry						
total office activity*	-8.0	-2.5	-7.0	-1.6	-8.8	-0.7
of which:						
- trade activity**	-21.9	-3.5	-21.3	-2.3	-24.5	-3.0
- administrative activity***	-4.9	-3.0	-3.7	-2.1	-5.4	-0.9
- service sector						
total office activity*	8.2	0.6	1.1	-1.7	-2.3	-0.7
of which:						
- trade activity**	-5.1	-2.2	-15.0	-2.1	-17.5	-1.5
- administrative activity***	10.8	-1.1	3.4	-2.9	1.0	-1.5

* with respect to all trade and administrative activities
** with respect to all trade activities
*** with respect to all administrative activities

- Furthermore, against the background of the rapid growth in administration and the correspondingly high demand for additional personnel the proportion of routine activities fell far more rapidly in the dynamic innovators of manufacturing industry (-3 percentage points) than in less innovative branches (-0.9 percentage points).

2.3 Human Capital and the Use of New Technology

In many cases the changes in occupational activity structures were linked to a change in skill level and the status of the affected workers within the enterprise. There is a particularly clear trend for dynamically innovative branches to employ a larger proportion of better qualified blue and particularly white-collar workers compared to the other branches. This has been shown in both the descriptive and econometric, although, as already mentioned, these effects are not due solely to the use of modern technology (innovation). A further trend is for occupational skill elements, usually gained through vocational training, to lose some of their value and to be replaced or supplemented by so-called extrafunctional skills and qualifications. This puts the spotlight on the nature and extent of human capital investment within the enterprise which is required for the efficient use of modern technology.

This aspect which has been of increasing importance (Staudt 1988) particularly under the conditions of economic development in the FRG in the 1970s and 1980s (stagnant markets, rising costs and changes in customer demand), is usually studied on the basis of formal "paper" qualifications. This is, however, very problematic as it excludes further training within the enterprise as a constituent element of the skills and qualifications of the work force. Nevertheless, even using this restrictive measure, it is apparent that dynamic innovators have a superior human capital base compared with branches dominated by hesitant or weak innovators, at least in manufacturing industry: here the share of higher education certificate holders in total full-time employment was 4.3% (1980), considerably above average. The figures for hesitant (2.8%) and weak (1.5%) innovators were substantially lower (cf. table 2.7).

Depending on the circumstances, practical and/or technical skills may be more important for the successful planning and implementation of modern means of production than the theoretical knowledge gained in universities and colleges by highly qualified staff (Kleine 1980). As was mentioned above, to concentrate attention solely on the share of "academics" as the indicator of human capital is to miss out important aspects of the level of human capital available to an enterprise and must therefore be supplemented by analyses which take account of occupational activities.

Table 2.7: The Status of Workers within the Enterprise in Manufacturing Industry in 1980 and the Changes between 1980 and 1985

	dynamic innovators		hesitant innovators		weak innovators	
	share in 1980 in %	change 1980-85 in % points	share in 1980 in %	change 1980-85 in % points	share in 1980 in %	change 1980-85 in % points
blue-collar workers to full-time employees	69.4	-2.4	67.9	-1.8	80.9	-1.7
skilled workers to blue-collar workers	39.6	4.2	44.5	3.1	64.8	4.6
skilled workers without vocational training to skilled workers	13.5	3.1	15.1	-2.4	10.8	-2.0
non-skilled workers with training to non-skilled workers	24.1	5.2	17.4	5.6	13.2	4.1
master craftsmen (Meister) to full-time employees	2.9	0.1	3.3	-0.1	3.6	0.2
white-collar workers to full-time employees	27.1	2.3	28.3	2.0	15.1	1.5
white-collar workers without training to white-collar workers	12.0	-3.3	8.9	-1.9	7.7	-1.2
higher education certificate holders to total employment	4.3	1.2	2.8	0.6	1.5	0.3
part-time workers to total employment	2.9	0.1	3.9	0.3	2.7	0.3

A look at the activity profile of the three groups of branches shows, for example, that R&D activities were about twice as important (in terms of employee shares) throughout the time period studied for dynamic innovators as for weak innovators (fig. 2.5). A comparatively large potential of skilled R&D personnel would thus seem to be one of the major prerequisites of the use of modern technology. However, it remains an open question whether this segment of employment actually deals exclusively with (direct) R&D activities, or whether a stock of such skilled labor is "hoarded" as potential human capital and in fact is employed to implement and adapt technical (external) knowledge purchased from other firms in the form of the exchange of investment goods.

The differences in occupational structure between the three groups of branches also indicate the necessity of highly qualified personnel with a broad range of technical and organizational knowledge in order to prepare and implement innovations.

This is true both of technical and organizational and commercial qualifications. To quote but one example in this context, of the 275 000 engineers responsible for design, i.e. primarily electrical and mechanical engineers, over 60% were employed in manufacturing industry, and over a half of these worked in dynamically innovative firms.

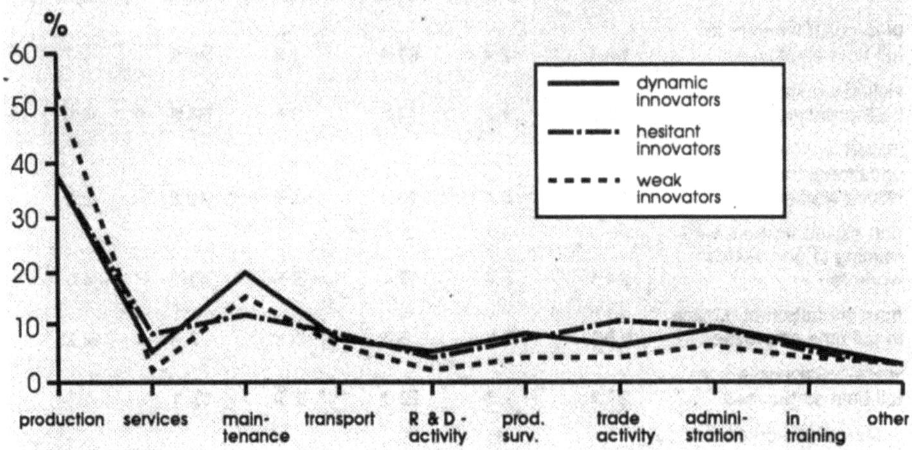

Figure 2.5: Activity Profiles in Selected Branch Groups
- manufacturing industry, 1980 -

So-called "external" management consultants, including organizational and administrative experts and engineers with advisory and/or controlling functions, on the other hand, are under-represented in manufacturing industry. They are employed primarily in the service sector and are concentrated there in branches at the medium innovation level, what we have called "hesitant" innovators. Thus the accumulation of technical and organizational knowledge is taking place largely in a group of branches within which engineering and consulting branches are to be found. This reiterates the importance of such qualifications in the course of the process of technological and organizational modernization but it also indicates that such know-how need not be "stored" in the same enterprises in which new technology is actually used. Small and medium-sized enterprises in manufacturing industry, in particular, may encounter organizational and financial barriers in obtaining the know-how required for innovation. Dynamic innovators in this sector, however, rely less on the help of external experts and know-how in order to carry out the technical and organizational restructuring of production: in most cases such firms have accumulated the required knowledge (human capital) internally.

An additional basic prerequisite for the successful implementation of new technology is the so-called operational knowledge. Employees whose work brings them in direct or almost direct contact with complex production plants need to be familiar with the technical processes involved, and be in a position, based on a relatively broad spectrum of knowledge, to meet all operating requirements. It is generally assumed that a practical vocational training or a combination of formal educational and practical training serves as an indicator of such capabilities. However, such a view of the importance of skilled labor for the use of technology is not confirmed by our analyses. The employment of trained and skilled labor, irrespective of whether it is considered to be "skilled" in terms of the hierarchy of workers within the enterprise, proved to be a relatively minor factor in determining the level of innovative activity in manufacturing industry.

Table 2.8: Changes in the Share of Skilled Workers by Branch Group
- dynamic, hesitant and weak innovators 1980 to 1985 -

	share of skilled employees*		share of skilled blue-collar workers**	
	1980 in %	change 1980-1985 in % points	1980 in %	change 1980-1985 in % points
dynamic innovators	54.4	+3.5	33.3	+1.4
of which:				
- manufacturing industry	52.3	+3.4	33.3	+1.6
- service sector	61.1	+4.4	28.5	+0.2
hesitant innovators	61.0	+4.1	24.8	+1.5
of which:				
- manufacturing industry	53.7	+4.7	31.5	+3.2
- service sector	71.4	+2.5	14.0	-0.2
weak innovators	61.9	+4.1	31.6	+1.3
of which:				
- manufacturing industry	58.8	+5.5	46.8	+4.1
- service sector	64.2	+2.9	20.6	+1.0
all branches	60.1	+3.9	28.6	+1.1
of which:				
- manufacturing industry	54.8	+4.3	36.8	+2.7
- service sector	66.8	+2.9	18.5	+0.4

* share of all blue and white-collar workers which have completed vocational of practical training in total full-time employees

** share of all blue-collar workers which have completed vocational or practical training in total full-time employees

In 1980 the share of workers who had completed their training as skilled workers in total full-time workers amounted to 47% for weak innovators, i.e. it was considerably higher than for dynamic innovators where only one in three full-time workers and less than one in two blue-collar workers had an occupational qualification as a skilled worker (table 2.8). The result is similar if white-collar workers are included in the comparison. In 1980 the share of skilled employees (blue-collar and white-collar workers) was slightly over 50% in the dynamically innovative branches of manufacturing industry, a share which increased to 55% by 1985. For weak innovators, on the other hand, the share in 1980 was already almost 59%, increasing in the first half of the 1980s to over 64%.

Once more, however, it must be emphasized that the formal level of education and/or training does not cover the whole spectrum of occupational qualification. Furthermore the slow rate of growth of the share of skilled workers may also be interpreted as being due to an acute shortage of skilled labor in the most innovative branches of manufacturing industry. Some confirmation of such a hypothesis is provided in the case of blue-collar workers by the following facts: in the more innovative branches of manufacturing industry

- a relatively smaller proportion of the total labor force (40%) are employed as skilled workers, whereas in the less innovative branches the figure is approx. two thirds, and
- a comparatively large proportion of the skilled workers have not completed vocational training, a proportion which, in contrast to the hesitant and weak innovators, has also increased during the 1980s.

We can summarize the above remarks by concluding that during the 1980s the most innovative branches, i.e. the majority of investment-good producers and certain other branches, have developed their modernization strategies on a comparatively broad base as far as human capital is concerned, i.e. on an above average proportion of highly skilled labor. The availability of qualified personnel with the necessary technical, commercial and organizational knowledge has proved to be an important factor in the planning and implementation phase of new technology. The role of formal qualifications for skilled labor for the successful, i.e. efficient use of new technology, however, was not confirmed by the data used in our analysis. This may be the result of a quantitative imbalance between the demand for and the supply of skilled labor, but also gives occasion to reflect on the content of training programs for skilled workers, in particular on the extent to which they are compatible with modern technology.

3. The Institutional Framework and the Use of New Technology

As was mentioned at the start of this paper the aim of our project was to analyze not merely the employment effects of innovation, but also the causes of innovative activity, as each process exerts feedback effects on the other. In order to sheed light on the causes of innovative activity econometric model calculations were conducted, the results of which can be summarized as follows (fig 3.1):

- Branches which are exposed to tough competitive pressure, particularly on foreign markets, are under greater pressure to innovate than those with less intensive foreign competitive relations - as measured by the export share.
- Tough import competition on the domestic market, on the other hand, proves to be a factor which tends to exert a negative influence on innovative activity. Domestic firms evidently react to increased pressure from imports not (or no longer) by increased efforts to improve efficiency, but are increasingly withdrawing from markets for such labor-intensive import goods.
- Firms competing with a small number of (generally large) firms exhibit a high inclination, or are under greater pressure to rationalize production processes.
- An innovative environment, resulting not least from favorable sales expectations and a high degree of capacity utilization, proves to be a factor promoting innovation, particularly production processes .

It was not possible in our quantitative approach to incorporate the institutional conditions under which economic processes occur and which, as mentioned earlier, are to be seen as a bundle of determining factors for the use of technology and its effects in the production and employment system. On the basis of available literature and a number of comparative statistical studies the role of industrial relations on the structure of the modernization process and the way in which it was conducted at enterprise level were, however, considered separately.

3.1 Industrial Relations and the Use of New Technology

The implementation of new technologies based on micro-electronics can be organized in a variety of ways. In contrast to the automatization processes of earlier periods they are not tied to certain forms of labor organization by limited technical potential or for cost and competitive reasons. Rather they open up a "corridor" of various forms of the division of labor and the design of individual jobs, all of which are compatible with a given technology (Lutz and Schultz-Wild 1986).

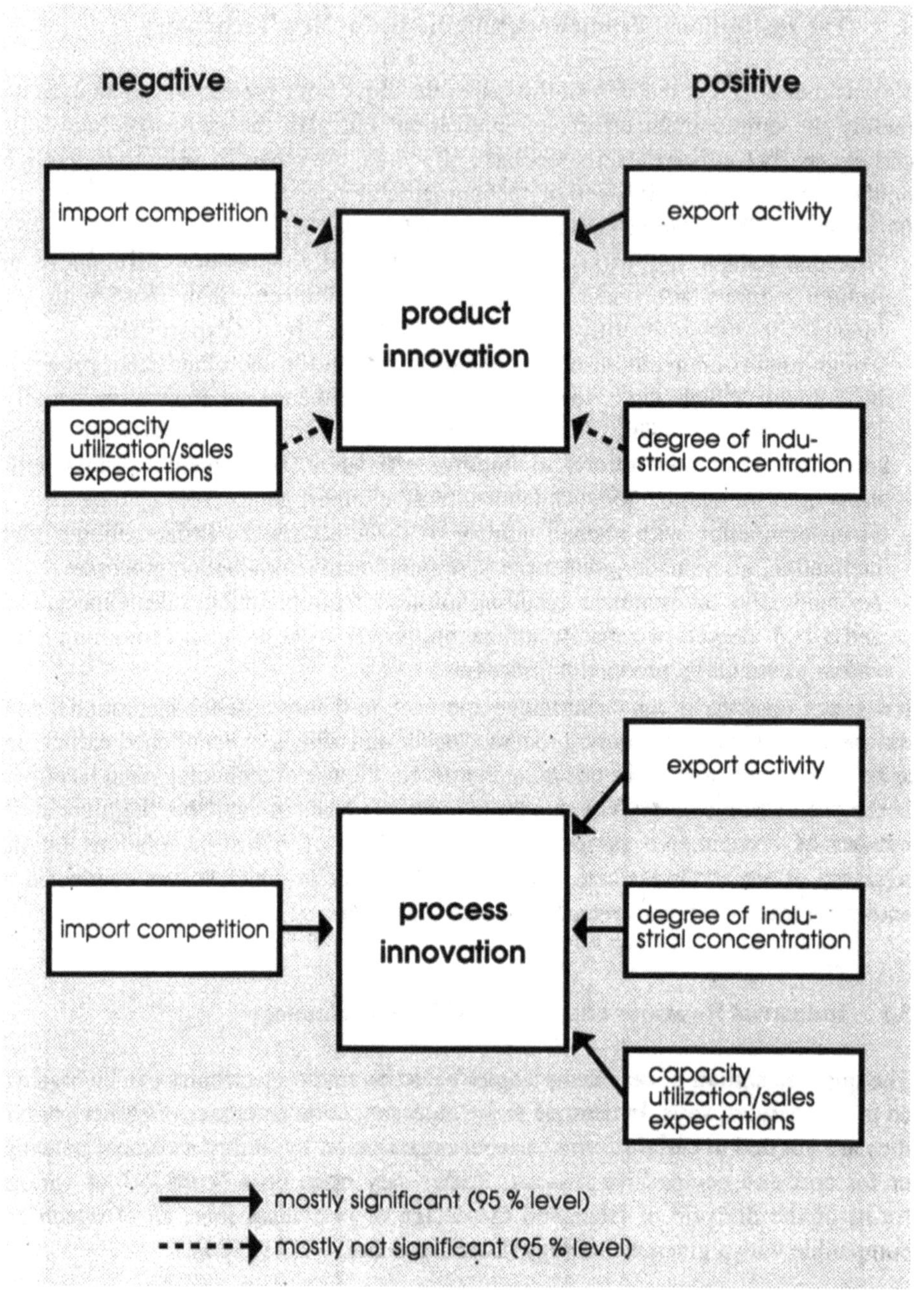

Figure 3.1: The Determinants of the Use of Technology with their Direction of Influence

A reorganization of the entire production process within the enterprise is nowadays often an integral part of the strategy of introducing new technology. The term "systemic rationalization" has been coined to describe this restructuring process (Baethge/Overbeck 1986), although it should be noted that innovations themselves (as has been constantly reiterated in the different projects making up the Meta Study) are not limited to the one aspect of rationalization, i.e. reducing costs and increasing efficiency. Against this background it is evident that industrial relations and training systems in which workers or their representatives (trade unions and works councils) have codetermination rights are important determinants of the selection, planning and implementation of new technologies and new products. The prerequisites for the ability of workers' representatives to exert an influence on the process of implementing new technology are considered to be relatively good in the Federal Republic compared with other countries (Sorge/Streeck 1987). This is due among other things to a broadly based system of vocational training, the largely cooperative relations between employers and work force, especially at enterprise level, and the fact that in recent years the employers' strategy has increasingly turned towards quality competition. Such factors make it possible and/or necessary to develop a form of organization which is decentral and therefore likely to meet the interests of workers (Helfert 1987).

Two basic levels are available for workers to participate in this process, in addition to the influence they can exert on the legal framework. At the collective bargaining level, collective agreements, which in Germany are mainly signed at branch and regional level, provide an additional overall framework for the introduction of new technology and its effects on the work force in a given branch/region, whereas at enterprise level it is the concrete use of new technologies by the firm and its consequences for the work force which is one of the most important tasks facing the works councils. However, at both levels the demands made of workers' representatives change in the course of the gradual shift away from the use of individual technologies towards "systemic" rationalization. This change implies that they have to focus their activities more and more towards questions of the overall organization of the enterprise (Bispinck and Helfert 1987). The different approaches currently available on both levels will now be briefly illustrated.

3.1.1 Trade Unions, Collective Bargaining and Innovation

The cooperative nature of industrial relations in the Federal Republic has favored the development of legal and collective regulations whose primary aim is to promote inter- and intrasectoral restructuring processes in a socially acceptable fashion and to cushion their effects on the labor market. These considerations apply not least to collective agreements on protection from the effects of rationalization, collectively negotiated measures which have existed from as early as 1968 in the metal industries.

Agreements on protection from the effects of rationalization are concluded at branch and/or regional level, and are thus insensitive to the specific conditions of individual enterprises. For this reason they tend to be of a general nature and often explicitly leave room for enterprise-level arrangements. It was not until the end of the 1970s (e.g. in the printing industry) that collective agreements also included special elements dealing with specific technologies and groups of workers. Nevertheless, the dominant factor continued to be the reactive, "cushioning" aim of maintaining employment levels, rather than the active aim of influencing the mode of implementation of new technology. Nevertheless, existing agreements do exert an influence, albeit an indirect one, on decision-making on questions of innovation and the work force. Insofar as such agreements - and the same applies to legal regulation of labor markets and social conditions - are seen by employers exclusively as implying additional costs, they will have the effect of reducing the propensity to innovate by lowering profit expectations. However, such a hypothesis, based on a fundamental antagonism between the "social partners", must be challenged by the alternative hypothesis of a fundamentally cooperative relationship. On the latter view, agreements of this type offer employers a degree of certainty in planning and decision-making as they facilitate a cost-benefit analysis of the use of technology. This means that even assuming an economically rational behavior on the part of employers agreements on protection from the effects of rationalization may exert a positive influence on the propensity to innovate.

By the mid 1980s about 50% of workers subject to collectively bargained agreements were covered by agreements on work-protection, i.e. are employed in branches or regions where such agreements were in force. So-called agreements on "protection from dismissal and income maintenance" (Kündigungsschutz- and Verdienstsicherungsabkommen) for older workers also apply potentially to about three fourths of workers covered by collective agreements. Due to the fact that in most of these agreements protection is linked to age and length of job tenure of individual workers it is clear that not all those potentially benefiting from these agreements are in actual fact protected. It is only recently in regional agreements in certain branches (engineering, printing) that protection is no longer limited to certain age groups or workers who have remained in a single firm for a long period. Even if it is assumed that the majority of older workers fulfil the condition of an adequate length of job tenure, only an estimated one fourth of all blue and white-collar workers in the Federal Republic are offered protection from the effects of rationalization under existing agreements, and only about 12% are covered by agreements on dismissal protection and income maintenance on the basis of their age and job tenure.

Table 3.1: Protection from the Effects of Rationalization in Different Branches

	employment	proportion benefitting from		
	in 1985 (in 000's)	protection from the effects of rationalization (in %)	dismissal protection (in %)	level of protection (in %)
agriculture, fishing	231.1	-	-	-
energy	245.1	23.5	13.3	20.1
mining	227.8	3.6	32.0	13.1
manufacturing industry	7,814.3	53.4	13.3	40.0
chemicals	574.7	46.6	23.5	38.9
crude oil processing	27.8	57.3	27.5	47.3
plastics production	238.6	27.7	19.0	31.1
rubber processing	98.8	32.9	9.8	25.0
stone and clay	190.4	22.3	8.5	17.7
fine ceramics	65.0	44.3	11.6	33.4
glass production and proc.	69.3	9.0	14.1	10.7
steel construction	206.2	-	9.0	3.0
non-ferrous metal prod.	60.0	-	-	-
foundries	105.6	75.5	10.8	53.9
drawing and rolling mills	287.6	74.9	8.7	52.8
steel/light metal production	184.4	76.4	8.5	53.7
machinery	936.8	76.8	10.6	54.7
automatic data proc. ind.	78.3	74.8	5.7	51.7
road vehicles	974.7	74.5	7.2	52.0
ship building	47.2	78.8	11.4	56.2
aeronautical production	53.0	77.2	7.6	53.9
electrical goods	997.0	74.9	8.9	52.8
precision engin.,optical goods	203.9	73.1	8.5	51.5
hardware and metal goods prod.	334.3	76.3	10.5	54.3
production of MSSS-goods[1]	50.7	75.8	11.4	54.3
millwork and wood products	60.3	16.8	12.1	15.2
wood processing	350.5	13.6	6.1	11.1
paper production	58.7	50.3	25.5	42.0
paper processing	97.3	42.6	-	28.4
printing	211.4	81.4	-	54.2
textiles	256.7	36.6	11.7	28.3
leather	85.5	39.2	9.2	29.2
clothing	215.8	1.9	7.8	3.9
foods	575.0	5.4	8.9	6.6
drinks	101.7	30.6	24.9	28.7
tobacco processing	17.2	57.2	25.5	46.6
construction, building	1,025.1	-	-	-
interior design	592.3	-	1.0	0.3
wholesale trade	1,113.8	-	10.6	3.5
railways	133.0	3.6	59.6	22.2
shipping	58.4	-	3.2	1.1
road transports	570.5	2.1	3.4	2.5
communications	233.3	42.0	42.0	42.0
financial institutions	581.8	19.5	7.9	15.6
insurances	222.7	39.5	16.1	31.7
real estate, housing	132.1	12.4	4.0	9.6
restaurants, hotel services	472.8	-	1.0	0.3
science, art, journalistics	257.5	3.2	0.3	2.2
health services	435.5	-	-	-
remaining services	1,106.6	-	0.2	0.1
public administration	2,088.0	42.0	42.0	42.0
social insurance	249.7	3.5	7.1	4.7
all branches	20,378.4	26.8	12.4	22.6

1 musical instruments, toys, sports goods, and jewellery

A relatively high proportion of workers is covered in the investment-good producing branches, in the printing and chemical industries, but also in the paper industry, the insurance sector and the communications industry (Federal Post Office) (cf. table 3.1). In these branches collective agreements usually apply to all regions and branches, and are often characterized by an above average proportion of older workers. This relatively high "level of protection" consists of a combination of agreements on protection from the effects of rationalization and on dismissal protection and income maintenance. On this basis the branches studied can be classified as follows:

- branches with a high level of protection (more than 30%)
- branches with a medium level of protection (between 10% and 30%)
- branches with a low level of protection (less than 10%)

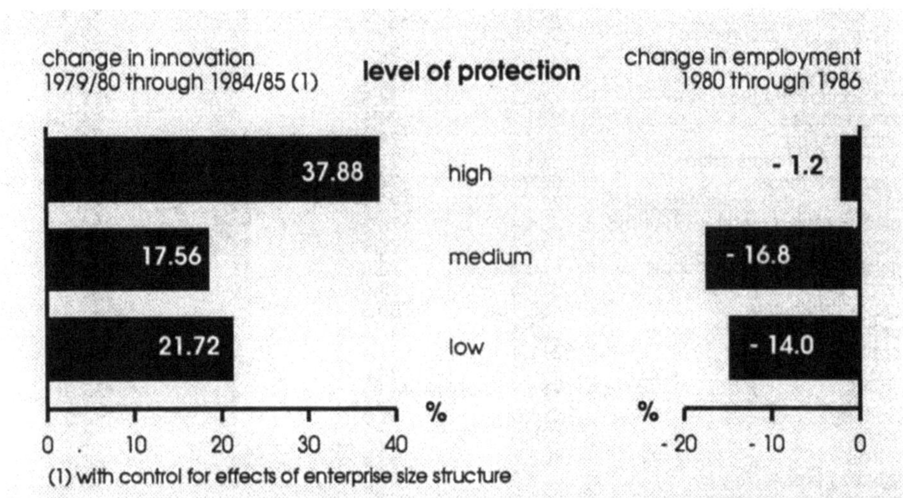

Figure 3.2: Groups of Branches with Different Levels of Protection - manufacturing industry -

Our results show no tendency for a more hesitant propensity to innovate in branches with a high level of protection. If anything the reverse is the case: in these branches investment in technology-intensive goods increased by 37% between 1979/80 and 1984/85. In branches with a low level of protection against the effects of rationalization (e.g. the steel and clothing industries, retail and wholesale trade and the transport sector) the increase in investment in such goods was only half as great. A similar, indeed even clearer picture emerges if the analysis focuses on innovative activity in manufacturing industry (where such agreements are concentrated).

For this sector of the economy it was also possible to control for the effect of firm size on innovative activity in different branches. This was considered necessary on the basis of the experiences gained in other studies regarding differences in innovative activity in various types of firms. Nevertheless, the differences between branches with a high and a low level of protection were not appreciably affected by controlling for enterprise size (fig. 3.2).

Overall, these results are not particularly surprising given that the trade unions presumably have concentrated their activities regarding protection from the effects of rationalization on those branches in which jobs were endangered by increased innovative activity. However, the actual impulses towards the signing of such agreements, particularly in manufacturing industry, occurred many years ago. Firms in the relevant branches have increasingly grown to view existing agreements as part of the overall framework under which they operate. Thus it can be tentatively concluded that the existence of provisions which protect workers from some of the negative effects of technological and organizational restructuring has not led to a measurable decline in the propensity to innovate, but indeed that the opposite would appear to be the case. The assumption would thus appear to be justified that such agreements have contributed to organizing the implementation of new technology and managing its consequences in a way which is both more transparent and more acceptable for employees.

However, due to the lack of comparable data for a period of innovation without protection clauses, it cannot be ruled out that innovative activity in dynamic branches might have been even higher in the absence of protection agreements. The fact that eight of the eleven branches of manufacturing industry which in the course of our research were identified as dynamic innovators (Warnken and Ronning 1988, ch. 2.6) exhibit a comparatively high level of protection from the effects of rationalization in the form of collective agreements would, nonetheless, appear to cast considerable doubt on such an assumption.

It is clear that there is a whole series of other factors exerting an influence on innovative behavior. Agreements on protection from the effects of rationalization are but a small part of the network of legal and collective regulation of labor market relations in general and the use of new technology in particular. The positive correlation between rationalization protection and innovative activity must therefore not be equated with a causal relationship, even though it can be seen as a reflection of cooperation between the two sides of industry and thus as exerting a positive effect on innovation. At the same time it would be equally inaccurate, based on the fact that many of the existing agreements were concluded several years ago and are now seen as a constituent part of the employers' planning framework, to interpret them simply as an automatic reaction on the part of the trade unions occurring wherever the use of new technology has been realized or is planned.

Similar considerations apply to the relationship between the degree of protection and employment. The following trends were observed for manufacturing industry:

- The branches of manufacturing industry in which workers enjoy a comparatively high level of protection from the effects of rationalization and which - as was shown above - are relatively innovative were able to maintain employment between 1980 and 1986 at approximately the same level (-70 000 or -1.2%).
- Branches with a medium level of protection, against the background of a level of innovative activity which, while lower, was still considerable, reduced employment in the same period by about 234 000 or 16.8%.
- Finally, branches with a low level of protection and a low level of innovative activity cut 420 000 jobs or 14% of their work force (fig. 3.2).

A degree of caution is of course to be observed when drawing general conclusions from such a descriptive comparison of different branches. A regression analysis of the correlation between the level of protection and relative change in employment shows a positive, albeit not particularly pronounced correlation. Changes in employment are, of course particularly susceptible to a wide range of other factors (e.g. changes in demand). Nevertheless, in view of the concurrence of a relatively high level of innovative activity, a comparatively favorable employment trend and a high level of worker security, it can be supposed at the very least that collective agreements protecting workers from some of the negative effects of rationalization are not a barrier to economic and technological development.

3.1.2 Industrial Relations at Plant Level and their Influence on the Use of Technology

Their very conception at regional or branch level precludes collective agreements in the FRG from containing more than a limited amount of detail. This is particularly true where they are concluded for a broad heterogeneous spectrum of branches, as in the engineering industry. But even in the printing industry it proved difficult to adapt the agreement on rationalization (RTS agreement) at plant level despite the fact that it was highly specific regarding both technology (the primary aim was to regulate the use of modern text processing technology) and for a well defined group of workers (type-setters) (Roback and Schlecht 1983). This suggests that trade unions will have to adapt their strategy to the changing requirements of the use of new technology (in particular the increased orientation towards the production process as a whole). From the point of view of the work force, it also emphasizes the importance of agreements at plant level, irrespective of whether these take a codified form or remain informal. Such agreements generally lead to additional costs for the firm in the first instance, as is the case with collective agreements, as employees will negotiate concessions with respect to wages, working time etc. However, these arrangements may also induce positive effects, seen now from the point of view of the employers, if they enable workers to be integrated into the innovation and rationalization process, and if they promote the acceptance and the interest of workers in determining the course of the implementation process. A positive attitude on the part of the work force to innovation is thus often reflected in an increase in labor productivity.

Overall the participation of works councils in the rationalization process is fairly high; at the beginning of the 1980s in the engineering industry they were involved in two thirds of cases. However, only in 35% of cases was there an agreement between management and the works council on personnel planning (IG Metall 1983). The participation of the works councils was thus often limited to the "social cushioning" of rationalization measures and exerted only a limited influence on quantitative and qualitative decisions on the use of new technology. To quote one (non-representative) example, an analysis of existing plant-level agreements on work with visual display units (Sedeno-Andres et al. 1981) has shown that only 45% of these agreements made provision for information and participation of the workers affected; moreover, this was often restricted to selected aspects.

The limitations of the ability of workers or their representatives to exert an influence on the way in which new technologies are implemented is largely determined by the definition of the rights of works councils as set out in the Works Constitution Act (Betriebsverfassungsgesetz). This legal framework clearly places the emphasis on (ex post) corrections to and "social cushioning" of the effects of new technology (Döbele-Berger et al. 1985). There are, however, a number of other reasons for the deficit of active participation in technological and organizational restructuring at plant level:

1) The spectrum of possibilities for participation is restricted if information on planned measures is not made available to those affected or their representatives until relatively late. It is only when comprehensive information is available at an early stage in development that workers will grapple with the problems posed by new technology. In many cases this will be reflected in concrete requests concerning working conditions, on the one hand, and in an increased willingness to participate in project groups or at least in the discussion within the enterprise on the effects of the new technology. In the course of active discussion on new technology workers have been observed to show increased sensitivity to changes in the content of their work (Hattry and Sydow 1982). A representative survey of

28 000 German workers showed that of those working with modern technology the acceptance level was, at 80%, very high, being positively related to the extent to which those affected were able to participate in decision making and received further training (Clauss 1988). This however occurred much more frequently in the case of staff with managerial responsibilities (52%) than with blue-collar workers (12%).

2) Another important factor is the traditional role played by the trade unions and works councils, which, under certain circumstances, may act as a barrier to the creative participation of the work force in the planning of changes in the production process. At both enterprise level and in the organization of the unions at higher levels, workers' representatives often see their primary, if not their sole task in protecting the workers from the adverse consequences of the employer's actions. There is no lack of voices calling for a change of direction in this regard (Bleicher 1987) and for a switch away from "social security" strategy (which finds its clearest expression in agreements on protection from the effects of rationalization) and towards an offensive strategy on innovation (Hinz 1986). However, at present it would appear that such a change of strategy will proceed only slowly and still affects intensive discussions within the trade union movement (Benz-Overhage 1985, Briefs 1986). At enterprise level, too, the majority of agreements deal with the results rather than the process of innovation (Report by the German Trade Union Federation cf. DGB-Abteilung Angestellte and WSI-Projektgruppe HDA 1984).

3) The fact that the necessary process of reorientation within the German labor unions and works councils has advanced only marginally in recent years is due in large measure to the unfavorable labor-market situation during the 1980s. In times of job losses and mass unemployment workers' representatives at enterprise level and beyond are forced to set other priorities than in periods in which there is an adequate demand for labor. Short-term efforts to save jobs and maintain income levels predominate over long-term strategies to improve and protect working conditions.

4) The fourth factor is that in conditions of general underemployment the institutionalized participation of the work force can be used by management as an instrument for realizing their aims because of the lack of interest of the affected workers in the participation rights which it in fact contains. In such cases management interprets these institutionalized forms de facto as a means of gleaning information from the knowledge and experience of the work force, rather than as a source of real participation (Hattry and Sydow 1982).

Particularly problematic for the ability of workers or their representatives to exert an influence on the process by which technology is introduced is the lack of information available to the works councils. In addition to the fact that information is often provided too late ("time" aspect) there is also a "qualitative" aspect. Clearly, early and punctual information is important, but it only makes sense if it is complete and comprehensive. The results of a study conducted for the machine tool industry would seem to be symptomatic for the majority of discussion processes at enterprise level. The study came to the conclusion that although works councils were always informed of planned modernization, they were rarely encouraged or in a position to participate in the actual planning of the measures (Hildebrandt 1987). Works councils often lack the required technical knowledge or the means to acquire information; new production and communication technologies represent new and difficult tasks for works councils as they often integrate different areas of the enterprise and usually "seep into" the firm in a process of successive decision-making and planning (cf. the distinction between traditional and "systemic" rationalization in Baethge and Overbeck 1986). This creates a number of barriers to effective strategic action on the part of workers' representatives within the firm:

- Due to their close ties to traditional labor union strategies, the activities of works councils remain dominated by certain groups of workers, in particular skilled (blue-collar) workers, while other groups (technicians, middle management, general white-collar workers) tend to be less involved in such work for career reasons.
- The wide range of innovative measures occurring simultaneously within the enterprise (an indication of which is provided by the creation of project-management departments within firms) require specialization within the works council. The creation of specialized "experts" which this implies can, however, lead to institutional problems and to the personal isolation of those members of the works council affected in this way.
- The information channels between all groups of the work force and the works council which are necessary to develop an overall strategy tend to be stochastic rather than of a permanent nature. In many cases this may reflect partial conflicts between individual worker interests and the group-oriented interests of the works council in dealing with new technology (Kern and Schumann 1984).
- The opportunities for gaining access to external sources of information and advice are often limited as the supply of such services is inadequate or is often of too general a nature to be applied to solving the concrete problems of meeting the interest of the work force in the firm's planning processes.

One aspect which cannot be ignored in this context is enterprise size. On the one hand it can be assumed that, as a rule, small and medium-sized enterprises face a greater degree of uncertainty in the use of new technology than large firms.

This is then reflected in the overall job situation within the enterprise, and makes it even more necessary for planning to be conducted jointly between management and the works council. On the other hand, the point made above that the knowledge gap of the works councils concerning the use of new technologies is particularly wide in the case of worker representation in small enterprises.

Works councils in small enterprises in which decisions on innovation are made under conditions of great uncertainty and usually with extensive help from external consultants (Ott 1985) are often not in a position - both in terms of the work load involved and the technical know-how required - to come fully to terms with the wide range of legal and technical problems surrounding the use of new technology, and to incorporate them into their work (Brötz 1983). This inability is compounded by the composition of the works council in terms of status groups (see above) and the more rapid "turnover" of works council members typical of small firms.

3.1.3 Some Concluding Remarks on Industrial Relations and the Use of New Technology

From our evaluation of the scientific literature and the empirical data on the influence of industrial relations on the use of new technology it can be concluded that the opportunities for worker participation in decisions on the use of modern technology in the FRG are decidedly limited. The partners in the process of industrial relations, on the one hand the employers and their organizations, on the other the workers represented by labor unions and works councils, are only slowly coming around to the view that a harmonization of their respective behavior may well bring considerable advantages in terms of more efficient use of resources and improved working conditions:

- The influence exerted by unions is at supra-enterprise level and as such is only able to set a framework for an entire branch or region. In this process they are not in a position to deal with the individual problems which arise in the planning and implementation of new product and process innovations at firm level. Collective regulation (agreements concerning protection from the effects of rationalization in the widest sense) can thus only provide a guarantee against the threat of the worst effects of rationalization on employment and working conditions. At the very least it can be claimed that the existence of such agreements has not hindered the innovation process in the past, presumably because it offers a degree of certainty for firms during planning.
- At enterprise level, on the other hand, workers' representatives (works councils) are much more familiar with the specific conditions under which new technology

are introduced, and could thus exert a much more differentiated influence on the actual planning and implementation of technical and organizational restructuring measures. In practice, however, there are a whole series of formal and informal barriers which stand in the way of the active participation of the workers affected by new technologies in the long-term process of planning their implementation.

3.2 Age-specific Displacement of Labor

In principle, agreements on protection from the effects of rationalization at both enterprise and regional/branch level aim to protect all workers from severe negative effects of the use of modern technology, however, older workers are implicitly considered to be most endangered in this respect. It is generally assumed that older workers are less adaptable to changes in job requirements and working conditions than younger workers, for this reason are likely to be perceived by management as a (potential) barrier to innovation, and thus require special protection.

This characterization of older workers, at least as far as their productivity is concerned, is doubtless largely a matter of prejudice, which only seldom stands the test of concrete analysis. The most difficult charge to answer is that older workers have inadequate or inappropriate skills and qualifications with respect to new and unfamiliar demands. A study for the Federal Ministry of Labor and Social Services (Naegele 1983), for example, has shown that in individual cases there may be a "skill risk" associated with older workers in the context of new technology. But this study also indicated that the process of introducing new technology is usually so evolutionary that older workers tend to remain (in important positions) in enterprises even if they have skill deficits and although younger workers are preferred in jobs which require workers to come to terms with new technology. The adjustment of an enterprise's human capital stock via an exchange of older for younger workers which may be necessary in some cases is usually achieved by natural wastage and only seldom through redundancies.

These observations have been confirmed by our analyses. Demographic influences on age structure were controlled for by using a "cohort analysis". However the statistical material available - evaluation of statistics on all employees subject to social insurance contributions - only allowed us to compare the number of workers in one age group (e.g. the 40 to 45 year olds) in the base year (1980) with the level of the next highest group (45 to 50 year olds) at the end of the period (1985). Entries into and exits out of these cohort groups which are not due merely to economic factors, cannot be accounted for. It was therefore assumed that these more or less cancelled each other out so that the comparison of employment levels gives at least an approximate picture of the extent to which age groups (cohorts) have been affected by a general increase or decrease in employment. Our starting point was the idea that almost all workers who were in the employment system in 1980 (excluding those over 65, for which no upper age limit could be determined, but which in any case are statistically negligible) must have been older than 20 in 1985. If the number of employees aged between 15 and 65 in 1980 is compared with the number over 20 in 1985, we can see an overall reduction of more than 10%. Of course this is partially compensated by the entry of new workers in the under-20 age group, so that the overall reduction in the number of those employed (and paying social insurance contributions) was, at 2.8%, much less. This change has affected the various cohorts to different degrees:

- Workers between 40 and 50 in 1985 suffered to a less than proportionate extent from the overall reduction in employment. Between 1980 and 1985 the number in employment in these age groups sank only by 6% to 7%.
- For workers in their fifties (1985) the employment risk and/or advantage is increasingly taken of the opportunities for early retirement (the two factors cannot be distinguished here). More than one tenth of workers who reached the age of fifty lost their jobs between 1980 and 1985; this corresponds roughly to the average figure. At the same time workers who were fifty or over in 1980 (55 to 60 year olds in 1985) the reduction of employment was by almost a quarter.

These figures represent the changes at the macro level and are to be taken as a yardstick for the evaluation of the structural changes in different groups of branches. Clearly the first question concerns the extent to which age-specific labor displacement patterns can be observed in connection with the introduction of new technology. The second relates to the effect of industrial relations: what has been the effect of the existence of collective and enterprise-level agreement on protection from the effects of rationalization on this displacement pattern. Even where this is not explicitly stated; in the following when a specific age group is mentioned the reference is to the group of workers which was in that age band *in 1985*.

Against the background of a slight increase in the average age of the work force in all branches and a less intense shift in age structure in the more innovative branches (the so-called dynamic innovators - cf. Warnken and Ronning 1988), it is apparent that the more innovative branches exhibit a clearer trend towards a rejuvenation of their age structure than those with an average or below average level of innovative activity:

- During the period studied (1980 to 1985) dynamic innovators took on workers in the age group 20 to 25 to a much greater extent than other branches. In 1985 58% more were employed in this age cohort than were in the 15 to 20 year-old group in 1980 (the increase in manufacturing industry - 51% - was almost as large, cf.

fig. 3.3). However, this shift is of course not to be interpreted as a "cohort development"; the main cause is the flow out of the education and into the employment system.

- In the most innovative branches the relative loss of jobs for workers in the 25 to 35 year olds cohorts was considerably less than the average for these branches. In weak innovative branches, on the other hand, these same cohorts experienced a reduction in employment which roughly corresponded to the average of all age groups in these branches, at least in manufacturing industry (23%).
- Older workers (50 years and over) which, as has already been mentioned, are usually considered to represent a particularly vulnerable segment of the labor market, both in general and specifically in the context of the introduction of new technology, saw their share of employment fall in the most innovative branches due to the large-scale entry of younger workers. However, in percentage terms they were not displaced to a greater extent that workers of the same age in less innovative branches. In the more innovative branches of manufacturing industry on balance 15% of workers whose 50th birthday occurred between 1980 and 1985 lost their jobs, whereas in the less innovative branches the figure was 21%. There were comparable differences for the 55 to 60 year olds: -32% in the more, -36% in the less innovative branches.

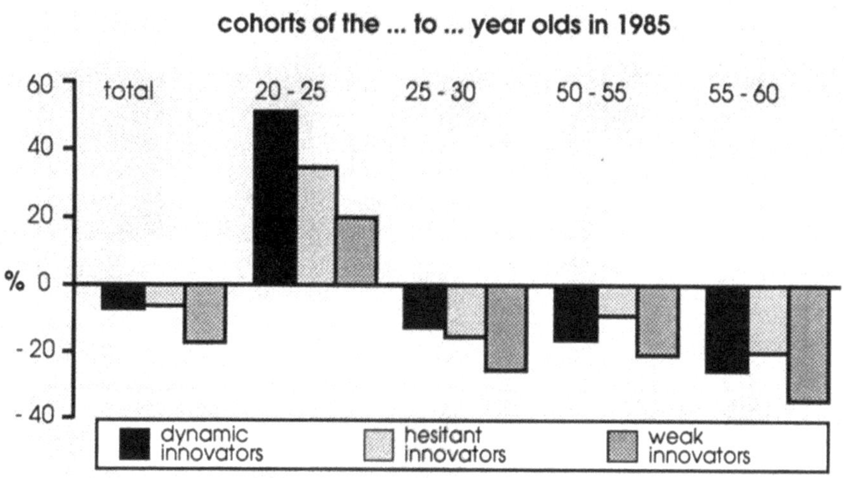

Figure 3.3: Employment Trends by Age Group and Innovative Activity in Manufacturing Industry, 1980 to 1985 in %

Overall the displacement effects for older workers in the most innovative branches of manufacturing industry and the service sector appear to be very slight. Our findings thus offer no support for the thesis of a conscious staff selection policy to the detriment of older workers due to their supposed lack of adaptability. Because of the above average entry of younger workers in more innovative branches the share of older workers fell by more than in less innovative branches but their numbers were not reduced to a greater extent than in the other branches.

In many of the more innovative branches, particularly in manufacturing industry, the industrial relations system has brought about protection from the effects of rationalization in both codified and informal form. As has already been mentioned this protection has not had much effect on the process of innovation and employment stabilization. On the other hand, these collectively bargained agreements have only fulfilled the task they were originally designed to meet (i.e. to offer older workers special protection) only to a limited extent. If the branches of manufacturing industry are grouped according to the level of protection we get a similar picture to the classification by level of innovative activity (fig. 3.4): The two methods of classification are highly congruent.

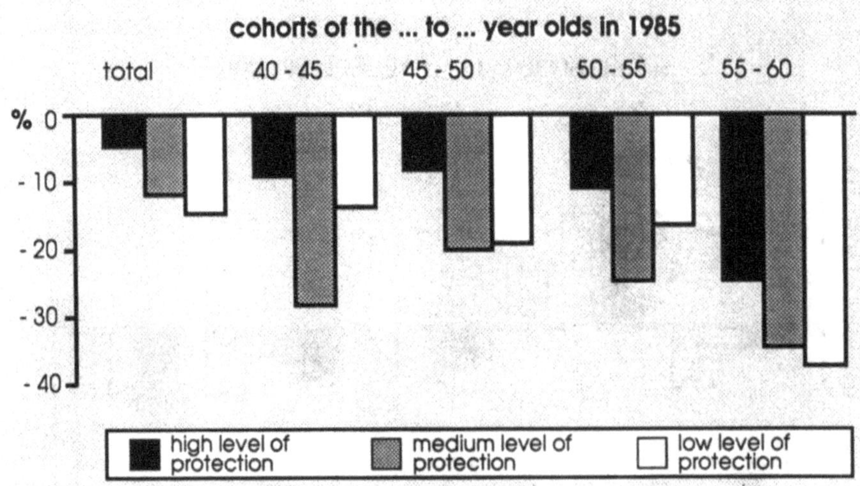

Figure 3.4: Employment Trends by Age Group and Level of Protection from Rationalization in Manufacturing Industry, 1980 to 1985 in %

- Branches with a high level of protection also exhibit a clear tendency towards the hiring of younger workers, particularly if they are also dynamic innovators. Branches such as crude oil refining, the production of analog-digital machinery and the aerospace industry have hired workers from the age cohort 20 to 25 at a rate which is clearly above the average, and have furthermore increased their employment of the next age cohort (25 to 30 year olds), in contrast to the majority of other industries.
- In view of the rapid increase in the employment of younger workers it comes as no surprise to discover that older workers have lost ground in those branches with a high level of rationalization protection. While their employment was not reduced to a greater extent than in branches with a lower level of protection they did not benefit from the more favorable overall employment trend in these branches which, in some cases, even led to an increase in total employment.

Thus collective agreements on protection from the effects of rationalization were at least partially successful in meeting their ("defensive") aims. In many cases they have contributed to a stabilization of employment and have kept structural shifts between different age groups within limits. On the other hand, they can hardly be labelled as successful with regard to more "offensive" aims such as the integration of older workers in new or changing structures of production and a more intensive use of their experience within the enterprise.

Innovation and Labor Market Dynamics

Ronald Schettkat and Bettina Bangel

1. Introduction

The use of new technologies in the production process is leading to constant changes in the skills and qualifications required of the work force. For many years the discussion on this subject was dominated by the 'de-skilling' hypothesis (Braverman 1977), according to which the skills required of the majority of workers will become more and more basic. In recent years the opposite trend, the 'reprofessionalization' of the work force is considered to be more plausible (Kern/Schumann 1985). There is a general concensus that changes in skill requirements are not so much technically determined but that they are rather the result of social processes (Doeringer/Piore 1971, Lutz 1986). Furthermore, new technologies can be applied much more flexibly than older technologies so that the freedom to organize work in different ways will increase rather than decrease (Sorge et al. 1982).

Irrespective of whether it leads to deskilling or reprofessionalization, technological and structural change entails a re-evaluation and often a devaluation of existing skills. In economics the slow adjustment of the labor supply to changes in labor market demand structures comes under the heading of 'structural unemployment', and has been repeatedly put forward as a cause of slow economic growth. Looking behind this concept of structural unemployment, the literature on the subject offers a wide variety of interpretations (Freiburghaus 1978) as well as various other concepts such as low-growth, technological, classical, and Keynesian unemployment (Schmid 1980).

Taken at its most general 'structural unemployment' implies that it is social and economic structure, in particular labor market structure which are responsible for unemployment. Such a view provides the background to the 'Eurosclerosis' discussion (Giersch 1985). At this level of abstraction, the concept of structural unemployment is certain to hit the mark simply by allowing for all possibilities, while at the same time and for this very reason it lacks analytical value.

In the current context of technological and structural change unemployment is often attributed to such changes *per se*, or to the lack of adjustment to them, the argument being that structural change - be it regional, sectoral or technological - requires processes of adjustment on the labor market to which the skills of the work force respond too slowly. If change occurs more rapidly on the demand side of the labor market than the supply side disequilibrium, in the form of a 'skill-profile discrepancy' will clearly ensue. The skill structure of the work force can thus become a constraint on economic development. Such a situation is similar to a lack of the physical means of production: an isolated increase in overall effective demand would lead to higher inflation but would create scarcely any new employment (cf. Meidner/Hedborg 1984). This will only occur, however, if the labor shed in declining sectors cannot find jobs in expanding sectors because of their specific skills or qualifications and if those entering the labor market for the first time are not oriented towards the new qualification requirements. The skill-structure explanation of unemployment does not hold if, for other reasons, there are simply no sectors expanding sufficiently rapidly (Abraham 1983). The obvious point should also be made that various forms of education and (re)training can facilitate adjustment to the new requirements (Schmid 1989).

For purists from both the neoclassical and Keynesian schools this change in the value of existing skills and qualifications does not pose a serious problem. Both sides ask merely that the mechanisms which each recommends, and only they, be applied - and the adjustment problem is solved. Skill imbalance is not a long-term problem for neoclassicals if only the provided market signals are clear enough: this is to be achieved by allowing greater freedom for 'market forces' (Soltwedel 1984). Price signals will then ensure an optimal allocation of resources. Clearly, even if the price signals are perfectly clear and strong, adjustment processes will take time, and even where it is possible to react appropriately to existing problems, the price mechanism fails when it comes to adjustment to future developments (Meidner/Hedborg 1984).

The supporters of demand management, on the other hand, do not regard the price mechanism as the most important means of equalizing skill imbalances on the labor market; they consider it sufficient to expand effective demand in order to overcome unemployment. If enough profitable opportunities are available, the profit interest of firms will see to it that they carry out the necessary training of the labor force. There is therefore little need for public training schemes which serve mainly to maintain a reserve army of the unemployed (Spahn/Vobruba 1986).

Is the labor market in the Federal Republic of Germany flexible enough to react to the adjustments required by technological and structural change? Is greater labor market mobility required or is the current degree of mobility sufficient to achieve a high rate of innovation? This paper addresses the question of whether a high level of innovation, which is seen as a precondition for sustained economic growth, necessarily implies greater labor market mobility. Firstly, the principle means of economic adjustment will be developed and statistical concepts will be discussed which can help to illustrate labor market processes. Because *changes* in the *stock values* of labor market aggregates only reflect the results and not the dynamics of labor market movements, analytical approaches are required which are based on *flow data* in order to take full account of the complexity of labor market processes .

The hypothesis that there is a relationship between the level of innovative activity in the various branches of manufacturing industry and their exchanges with the external labor market will be empirically tested. In both the academic and political discussions of the innovation-induced adjustment processes of firms and thus of the labor market as a whole it is the value for the flow out of employment and into unemployment which is of particular relevance. The flow from employment into unemployment is also one of the few labor market flows in the Federal Republic of Germany for which relatively highly disaggregated statistics are available. If firms primarily carry out skill adjustment via the external labor market, we would expect the flow into unemployment to be higher in more innovative industries as they will have a greater need for skill-adjustment.

This paper also analyzes the relationship between innovative activity and the branch-specific risk of becoming unemployed. Other factors such as the skill structure, the structure of enterprise size, the strength of worker representation and seasonal dependence of the products will also be considered for the different branches. If the diagnosis that the West German labor market is too rigid and hinders innovation is correct, then one would expect more innovative branches to make greater use of (at least) the existing possibilities for dismissal than less innovative ones. This article presents an empirical analysis of this hypothesis.

Finally, the elasticity of the supply of labor in the different occupational groups will be studied in order to examine the question whether the supply of labor reacts too slowly to changes in the skill requirements and whether this has resulted in a 'qualification gap' in the Federal Republic.

2. Labor Market Processes: Mobility and Flexibility, Flow and Stock Values

Technological and structural change requires flexibility and mobility either in the form of worker adjustment to new conditions or job-design adapting to existing skills and qualifications. Adjustment processes by workers result in a variety of labor market dynamics, depending on the level at which adjustment takes place.

These can be considered from the point of view of the supply of or the demand for labor.

If we initially assume a static situation, i.e. one in which there is a fixed allocation of workers to specific jobs, every change in the technology used in the production process (be it process innovation, product innovation or a combination of the two) will call for skill flexibility on the part of the work force. The flexibility which can be achieved by (internal) mobility increases with the level of aggregation at which adjustment is to take place. For example, an enterprise, even one with a constant workforce (both qualitatively and quantitatively), can adjust to new skill requirements through mobility within the enterprise, so that at the individual level only minor retraining is necessary.

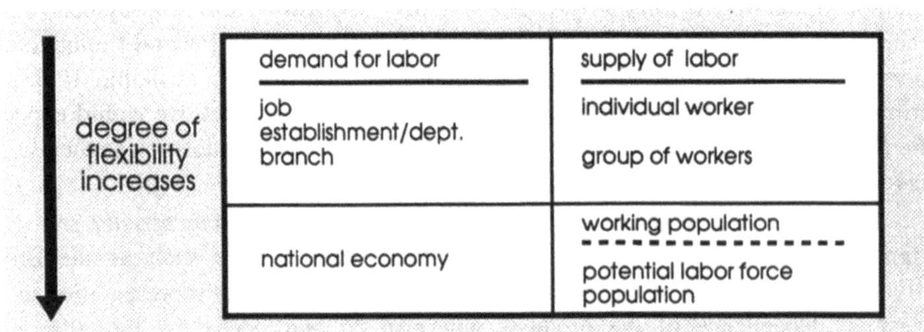

Figure 2.1: Flexibility and Mobility as Dependent on the Level of Aggregation

The flexibility which can be achieved by mobility is even greater in an entire industry or the economy as a whole. The larger the system under consideration, the greater is the skill flexibility which can be achieved through internal mobility with only marginal adjustment for each individual. At the macro level the adjustment processes to technological and structural changes will involve flows (mobility) within the economy as a whole. These can be both within firms, i.e. between different work places, between firms of the same branch or industry or between branches (the flows 1 and 2 in fig. 2.2). Adjustment can also lead to mobility out of the employment system with workers entering registered unemployment (flow 4) or leaving the active labor force (flow 3) i.e. retiring, returning to a role as housewife, reentering the education system or non-registered unemployment etc.

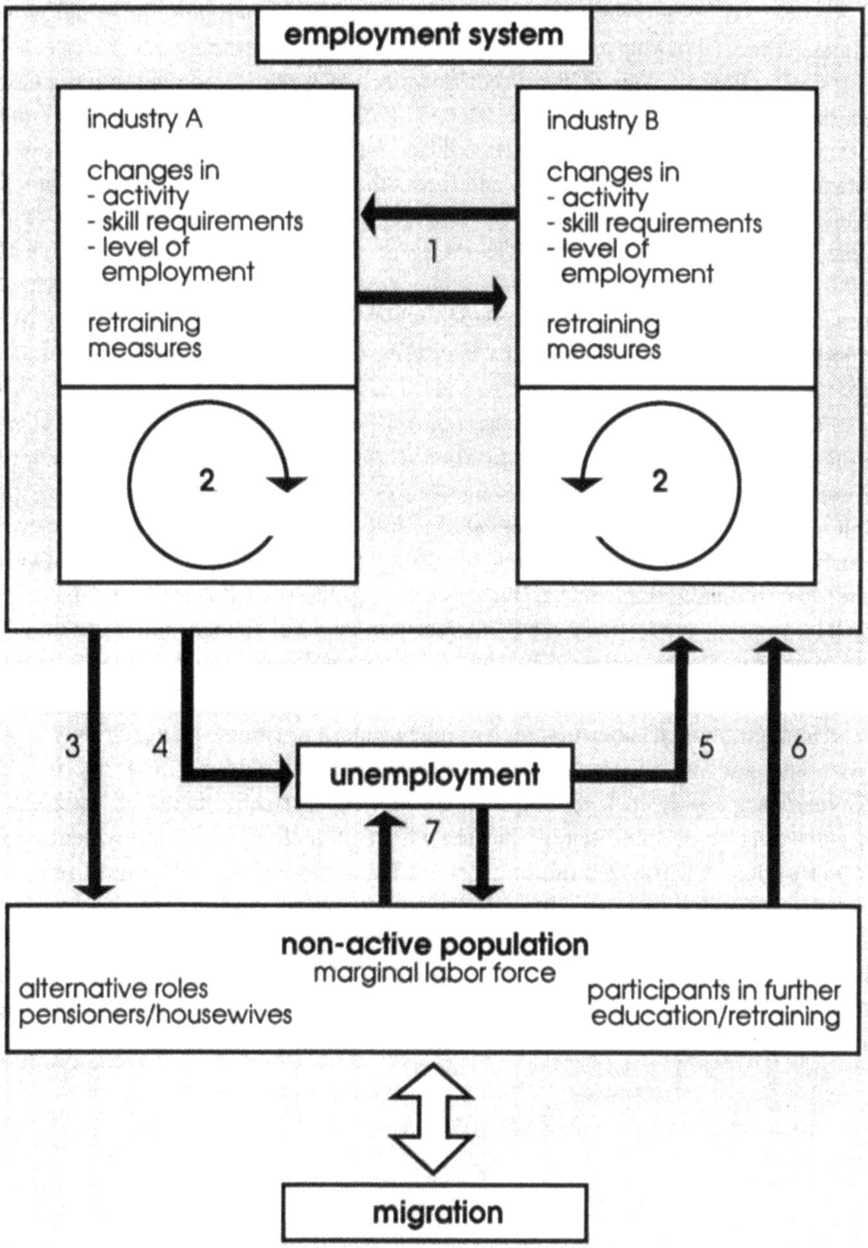

Figure 2.2: Technological and Structural Change and Labor Market Dynamics

Parallel to these flows out of employment are flows into the employment system, either from registered unemployment (flow 5) or non-participation in the labor force (flow 6). Due to the direct flows between registered unemployment and non-participation in the labor force (flow 7) a change in the unemployment figures does not automatically imply an equal and opposite change in the employment statistics . In addition, migration to and from other countries can provide not only for adjustment in the overall size of the working population but also for skill adjustment. To a large extent the skill flexibility of the employment system is due to a continuous exchange process between the employment system and the non-economically active section of the population which is partly natural (the 'passing of the generations') and partly regulated by institutional arrangements (provisions for early retirement etc.).

Flexibility can be described as *internal* when workers adapt to new skill requirements through training or education while remaining in the same branch, firm or even at the same work place (Sengenberger 1984 and 1987). At the same time skill flexibility can also be achieved by the direct exchange of labor between branches. This is known as *external* flexibility. Changes in the stocks of individual labor market categories conceal such dynamic processes (illustrated in fig. 2.2) on the labor market, since stock values represent merely *net* changes, ignoring the often very much larger flow values. The study of labor market processes therefore requires a methodology based on flow data.

The high level of labor market dynamics can be illustrated with reference to the flows into and out of employment. According to calculations made by the IAB (Reyher/Bach 1986), in 1985 employment in West Germany increased by 200,000; this however represents merely the net effect of a flow into employment of 5.2 million and an outflow of 5 million persons. These figures also allow for the possibility that an individual may have been in and out of employment several times during the year. This example indicates how changes in stock values merely reflect the sum of contrary labor market flows and thus conceal the actual dynamic movements taking place on the labor market, movements which are of great relevance for adjustment processes to new skill requirements.

3. The Risks of Unemployment

3.1 The Components of Unemployment

Constant figures for the average annual level and rate of unemployment can conceal a wide range of different unemployment dynamics (cf. Freiburghaus 1978, Clark/ Summers 1979, Egle 1979, Akerlof/Main 1981). An annual average level of unemployment can be composed of a given number of persons who remain unemployed throughout the whole year but also of a far greater number of persons each experiencing a relatively short period of unemployment. Situations with an equally high level of unemployment can therefore differ greatly both in the number of people experiencing unemployment and in its average duration. Whereas in the first case there are no exchange processes between registered unemployment and employment/non-participation, in the second case these flows will be considerable.

These two widely different types of labor market dynamics cannot be ascertained by the change in the stock data on unemployment between two points in time, as this merely represents the net result of flows into and out of unemployment. It provides no information on the dynamics, the gross flows lying behind it. The mechanisms behind a change in stock value can be seen from the following equation:

$$(1) \quad LU_t - LU_{t-1} = \underset{t-1}{\overset{t}{\mathrm{Sum}}}(FIU) - \underset{t-1}{\overset{t}{\mathrm{Sum}}}(FOU)$$

where:
LU = level of unemployment
FLU = flow into unemployment
FOU = flow out of unemployment
t = time index

Hence the unemployment risk can be divided into two components: a component of becoming unemployed, i.e. of entering registered unemployment; and a component of remaining unemployed, i.e. of not finding other employment or an alternative to unemployment. The risk of becoming unemployed is given by the number of persons entering unemployment relative to the number of persons who could potentially be affected by unemployment (wage and salary earners). The risk of remaining unemployed is given by the duration of unemployment.

The duration of unemployment is measured by the Federal Agency of Labor by analyzing the average length of unemployment experienced on a given date (usually

30th September). The duration of unemployment up to this date is subject to upward distortion because the probability of being unemployed on the date in question (stock value) is obviously greater for the long-term unemployed than for those experiencing a short period of unemployment. This is clear if we consider that for a person unemployed throughout the year the chance of being registered in the survey is equal to one, while for those unemployed for one month the chance is only one in twelve. Because of this distortion, the average duration of 'completed' unemployment is always considerably below the average duration of unemployment 'to date' (Cramer/Egle 1976, Freiburghaus 1978, Egle 1979).

In the Federal Republic both components of the unemployment risk - the risk of becoming unemployed and the average duration of unemployment can be established for all branches with the help of the unemployment statistics provided by the Federal Agency of Labor. The formal expression for the branch-specific net changes in the level of unemployment can thus be set out as follows:

$$(2) \qquad \sum_{j=1}^{n} \left(LU_t^{j} - LU_{t-1}^{j}\right) = \sum_{j=1}^{n} \left(\sum_{t-1}^{t} \left(FIU^{j}\right) - \sum_{t-1}^{t} \left(FOU^{j}\right)\right)$$

where:
j = index for branch (or non-participation)

By drawing this distinction in the labor market dynamic as shown in equation (2) an important component can be isolated in each branch, the risk of becoming unemployed, which is expressed by the ratio of the number of persons entering registered unemployment to the number potentially affected (wage and salary earners). This risk can be determined for each branch on the basis of available data: the number of those entering unemployment from a given branch is set in relation to the number of employees in that branch. Such data will indicate whether there are variations in the risk of becoming unemployed between the different branches and, more specifically for our purposes, whether the risk in innovative branches is above or below average.

The values for the risk of becoming unemployed are ratios (based on the number of workers in a branch), and as such are comparable between branches, forming a cardinal scale in which zero is the lowest value - the risk of becoming unemployed cannot be negative. By contrast there is, at least in theory, no upper limit as the flow into unemployment during one year can be higher than the work force in an industry due to the possibility of multiple unemployment.

While the flow from employment into registered unemployment accounts for a large proportion of the total flow out of employment and is statistically well documented, there is a dearth of statistical information on the reverse flow, from unemployment into employment. Although the flow out of unemployment and into employment accounts for 50% of the total flow into employment (Reyher/Bach 1986) the statistics provided by the Federal Agency of Labor only record those leaving registered unemployment if this occurs through the intermediation of the official labor offices. This, however, accounts for only about 50% of those moving from unemployment into employment. Only in an analysis based on a sample carried out annually (May/June) are more comprehensive studies made of labor market movements which also differentiate between the different flows out of unemployment. It is thus not possible to quantify the flow out of unemployment into employment or non-participation in a way which would allow us to differentiate between branches.

The second component of the unemployment risk is the risk of *remaining* unemployed. A long period spent in unemployment is a reflection of reduced chances for reintegration into the employment system, although it should be borne in mind that the flow out of registered unemployment can also result in non-participation in the labor force as well as employment (fig. 2.2). As was indicated above, the figures provided by structural analysis for the average duration of unemployment 'to date' are distorted as the long-term unemployed are overrepresented in the figures. Attempts were made in the 1970s to calculate the average duration of 'completed' unemployment in the Federal Republic by Cramer/Egle (1976) and Freiburghaus (1978) (for an overview cf. Buttler 1987) a calculation which proved highly complicated.

3.2 The Risk of Becoming Unemployed by Branch

The risk of becoming unemployed is a pivotal coefficient for determining the level of employment stability. It is determined by the ratio of the flow into unemployment to the number potentially effected, i.e. the wage and salary earners in a branch. While it is true that the flow into unemployment can also come from employment which is not subject to social insurance (e.g. the self-employed, family workers), so that the most comprehensive reference group for determining the branch-specific risk of becoming unemployed would be the labor force as a whole, wage and salary earners subject to social insurance make up the vast majority of those employed and provide a sufficiently accurate reference group for calculating the branch-specific risk of becoming unemployed.

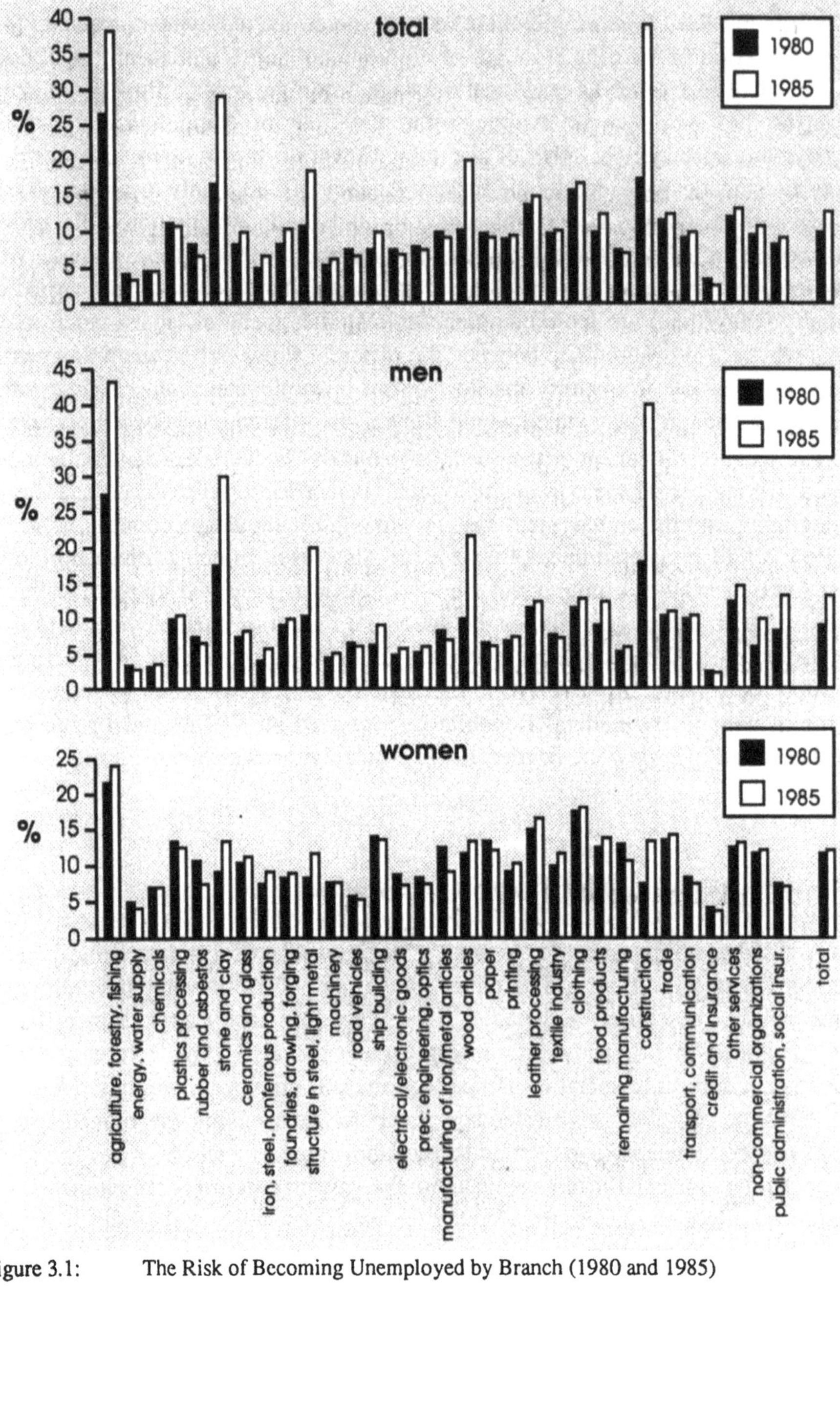

Figure 3.1: The Risk of Becoming Unemployed by Branch (1980 and 1985)

Table 3.1: The Risk of Becoming Unemployed by Branch (Flow into Unemployment in the Relation to Employees subject to Social Insurance)

Branch		1977	1978	1979	1980	1981	1982	1983	1984	1985	1986	1987
00-03	agriculture, forestry, fishing	30.0	31.4	29.6	25.9	30.9	34.4	34.1	32.9	38.2	35.7	36.8
04-08	energy, water supply, mining	4.2	4.1	3.9	3.7	4.3	4.3	4.0	4.2	3.3	3.2	3.5
09-10	chemicals	5.8	5.6	4.7	4.5	5.1	5.2	5.5	5.1	4.7	4.6	4.4
12	plastics	14.3	14.1	11.6	11.4	13.9	14.1	12.8	11.4	11.3	11.1	10.9
13	rubber and asbestos	10.6	9.3	8.8	8.4	10.7	9.0	7.9	6.9	6.8	6.4	6.8
14	stone and clay	20.9	20.3	21.6	17.0	23.2	28.0	24.1	21.6	28.4	25.1	23.2
15-16	ceramics and glass	9.8	9.4	8.8	8.6	10.7	11.1	11.0	9.0	9.5	9.3	9.3
17-18	iron, steel, non-ferrous production	5.7	6.4	5.1	5.0	6.0	7.7	9.3	10.2	6.5	6.7	9.8
19-22	foundries, drawing, forging	11.5	11.4	9.4	9.1	11.7	12.3	13.0	11.6	10.2	9.7	11.0
23-25	structure in steel, light metal	17.8	15.6	12.7	10.6	14.5	18.7	19.9	18.7	19.0	16.9	15.8
26-27	machinery	7.6	7.0	6.0	5.5	6.8	7.7	8.6	8.4	6.1	5.5	6.3
28-30	road vehicles	7.3	6.8	5.8	6.7	8.1	7.9	8.2	8.1	6.2	5.7	5.5
31	ship building	10.7	11.1	8.8	6.7	7.8	8.3	13.2	20.7	9.6	12.7	15.2
33-34	electrical, electronic goods	8.8	8.0	7.0	6.7	8.6	9.1	9.5	7.5	6.5	6.4	7.1
35-36	precision engineering, optics	8.7	8.2	7.3	7.1	8.3	9.4	11.4	9.0	7.1	6.9	8.2
37	manufacturing of iron, metal articles	12.1	11.7	10.3	10.1	12.7	12.6	11.3	9.0	8.0	7.8	7.9
40-42	wood articles	12.8	12.2	11.2	10.3	14.5	20.0	18.1	17.0	20.4	19.0	16.6
43	paper processing	11.0	10.8	9.5	9.2	10.4	11.5	10.3	9.2	8.0	8.3	7.7
44	printing	10.5	9.3	7.9	7.8	7.9	11.0	10.8	9.3	8.6	8.8	8.3
45-46	leather processing	16.0	14.9	13.4	13.9	16.3	15.7	15.8	13.9	14.9	15.1	16.7
47-51	textile industriy	12.3	11.2	10.1	9.3	12.5	14.4	13.6	10.4	9.9	10.3	10.4
52	clothing	20.3	18.3	15.8	16.8	20.6	21.2	20.9	17.1	17.4	17.1	18.1
54-58	food products	12.9	12.2	10.9	10.8	12.2	13.4	13.6	13.0	13.5	13.0	12.9
11, 32, 38 39, 53	remaining manufacturing	9.9	9.4	7.9	8.1	8.8	10.1	9.7	8.6	7.7	8.2	8.4
59-61	construction	28.0	22.9	18.3	17.4	23.1	33.0	32.8	33.2	38.4	34.2	33.2
62	trade	15.6	14.1	12.6	12.3	14.0	14.5	14.0	13.0	13.2	12.9	12.3
63-68	transport, communication	11.3	10.9	10.1	9.8	11.8	12.4	11.9	10.81	10.4	10.3	10.5
69	credit, insurance	5.1	4.6	3.9	3.7	4.1	4.1	4.0	3.8	3.5	3.3	3.3
70-86	other services	15.5	14.5	13.3	12.7	14.2	14.9	15.2	14.3	14.2	13.7	13.3
87-90	non-commercial organizations	12.5	11.6	10.8	10.3	10.8	11.1	11.5	11.4	11.7	11.7	11.8
91-94	public administration, social insurance	8.0	8.8	8.9	8.2	8.6	9.0	8.8	9.9	10.7	11.4	10.9
99	miscellaneous	-	-	-	-	-	-	-	-	-	-	-
total		13.2	12.2	10.8	10.4	12.4	13.9	13.8	13.1	13.1	12.4	12.3

There are wide variations in the risk of becoming unemployed in the various branches (fig. 3.1 and table 3.1). When analyzed over time it is clear that the risk of becoming unemployed closely follows the movement of the business cycle. At the last peak in the employment figures, in 1980, the overall risk of becoming unemployed was at its lowest, both for men and women. By contrast, in the trough in the employment figures in 1984 the overall risk of becoming unemployed was relatively high. This link between the risk of becoming unemployed and the general employment trend can be seen in almost all branches; it was only in a very small number of cases that the risk was greater in 1980 than in 1985. But the 1980's have also been marked by widely divergent developments in the various branches. The figures also illustrate the phases of crisis and consolidation in some branches; in ship-building, for example, where for 1983 and 1984 the risk of becoming unemployed was very high and where obviously in 1985 a phase of consolidation, at a low employment level set in. The risk of becoming unemployed was particularly low in two branches: "credit, insurance" and "energy, water supply, mining". Within the manufacturing sector the lowest risks by a long way are to be found in the chemical industry. The overall picture is one of great heterogeneity regarding the risk of becoming unemployed in the different branches (cf. table 3.1).

Remarkably high values for the risk of becoming unemployed can be seen in agriculture and the construction industry This is to be explained by the strong seasonal variation in the employment level. The seasonal influence on the risk of becoming unemployed or, conversely, on employment stability, can be most clearly seen from the wide divergence in the values for male and female workers in the construction industry: the risk of becoming unemployed in this industry is much lower for women than for men due to the different activities they perform. Women are employed almost exclusively in white-collar positions where employment is relatively stable, whereas men are occupied in the construction work itself which is heavily influenced by seasonal factors. In the trade sector the relationship is reversed, although the dimensions are not so dramatic. Here the risk of becoming unemployed is lower for men than for women as the latter are employed in a more unstable segment of the labor market.

4. The Risk of Becoming Unemployed and Innovation

If it is external rather than internal flexibility which is dominant in the adjustment of the work force to new skill requirements, i.e. workers with new qualifications are hired from the external labor market while some sections of the existing work force are made redundant, then it is to be expected that the risk of becoming unemployed will be higher in more innovative than in less innovative branches. If, on the other hand, *internal* flexibility is more important, if the existing work force is retrained in order to adjust to the new demands, then no connection would be expected between innovative activity and the risk of becoming unemployed.

If it is true that high rates of innovative activity primarily require external labor market adjustment (external or numerical flexibility), then we can expect a positive relationship between the branch-specific risks of becoming unemployed and innovative activity. By means of this analytical link between innovative activity and the risk of becoming unemployed in the different branches it is possible to address the question whether a high degree of numerical flexibility and/or extensive mobility on the labor market are a precondition for a high level of innovation.

4.1 Indicators of Innovation

For a long time economics regarded technical progress as an unexplained residual variable. That part of the changing relationship between factor inputs which could not be explained within the context of the model, i.e. by variations in factor prices, was declared to be technical progress. In the social sciences any explanation of a phenomenon using a residual category must remain unsatisfactory, and indeed attempts have repeatedly been made to measure innovative activity directly. Since 1979 the Ifo Institute (Munic) has collected data on innovative activity by firms in West German manufacturing industry. The data are based on a questionnaire similar to that used by the Ifo Institute for its investment and business cycle studies and to which firms are asked to supply the appropriate information. A broad conception of innovation is used, on the basis of which firms classify themselves as innovators or non-innovators. The Ifo Institute asks for information on new products or significant improvements to existing ones (product innovation) and new or improved assembly or production techniques (process innovation). These questions which form a part of Ifo's business-cycle test enable us to identify the branch-specific level of innovative activity representative for manufacturing industry as a whole (for a more detailed description cf. Schmalholz/Scholz 1985).

In addition to the values for the proportion of innovative firms in each branch, values which do not permit us to draw conclusions as to the extent of the innovations, indicators were also developed to take account of the dimensions of innovation. These indicators are based on extensive work carried out by the Ifo Institute within the framework of the Meta Study (for a detailed account cf. Scholz et al. in this volume). This work, which is based on data collected by Ifo as part of their innovation test but which also takes account of macro-economic input/output relationships, has made it possible to produce statistical data on the value of differentiated innovative activity in the various branches of West German industry.

In order to incorporate all the effects of innovative activity into the analysis the flow of innovation from earlier stages of production must also be taken into account (cf. Rosenberg 1979, Scherer 1982). An empirical analysis of innovative input/output relations between branches was successfully carried out by the Ifo Institute and forms part of the framework of the Meta Study (cf. Scholz et al. in this volume). Its results were then available for use in the labor market analyses.

The values established for innovative activity by the Ifo Institute encompass not only the innovation carried out in the branch itself (direct expenditure on innovation), but also innovation which flows into a branch from other industries at an earlier stage in the production chain. Agriculture, for example, develops scarcely any new products itself, i.e. its expenditure on direct innovation is very low; however, it does use machines and other inputs (fertilizer etc.) purchased from other branches, some of which have very high levels of innovative activity, with the result that agriculture is to be classified primarily as a *user* of innovation.

Our research was concerned with the connection between innovation and adjustment processes on the labor market and not with changes in the level of employment. It can be assumed that the necessity for skill adjustment increases with the level of expenditure on innovation. At the same time the absolute level of expenditure on innovation carried out in a sector depends to a considerable extent on the size of that sector. One way around this problem is to divide sectoral expenditure by the number of workers in that given sector (it is the effects on the employees which we are trying to establish). The basis for developing the indicators for the model (described in detail below) was therefore taken to be the number of wage and salary earners employed in the branch in question as revealed by national accounting statistics. For its analyses the WZB/AMB made use of a range of different indicators in order to establish how sensitive the results are to the way in which individual indicators are specified, and to exclude 'freak' results. Based on the data provided by Ifo, the following indicators were developed for the value of innovative activity:

- total direct expenditure on innovation with respect to the number of employees in the branch
- total direct expenditure on product innovation with respect to the number of employees in the branch (PDV)
- total direct expenditure on process innovation with respect to the number of employees in the branch (PCV)
- total indirect expenditure on innovation with respect to the number of employees in the branch

- total expenditure (direct and indirect) on innovation with respect to the number of employees in the branch
- direct expenditure on product innovation per employee multiplied by direct expenditure on process innovation per employee in the branch (INNV).

In addition to these 'value indicators' the 'share indicators' mentioned above (i.e. the proportion of innovative firms in a branch) were also used, again weighted by the number of employees in a branch. The correlation coefficients between the different indicators are presented in table 4.1 for each of the five years analyzed.

Table 4.1: Correlation Coefficients between Innovation Indicators

year	INN with PROD	INN with PROC	PROD with PROC	INN with PCI	PROC with PCI	PROD with PDV	INN with INNV	PCI with PDV
1981	0.90	0.90	0.64	0.50	0.70	0.60	0.50	0.55
1982	0.90	0.92	0.70	0.22	0.58	0.63	0.42	0.30
1983	0.91	0.94	0.74	0.48	0.76	0.68	0.61	0.51
1984	0.94	0.89	0.71	0.33	0.66	0.63	0.40	0.75
1985	0.94	0.94	0.80	0.27	0.56	0.55	0.31	0.70

INN = innovation variable, share indicator (PROC*PROD)
INNV = innovation variable, value indicator (PCV*PDV)
PROC = process innovation, share indicator
PROD = product innovation, share indicator
PCV = process innovation, value indicator
PDV = product innovation, value indicator

The correlation coefficients between the different innovation indicators show that there is a strong correlation both between product and process innovation and value and share indicators. While it is possible on the basis of available data to differentiate between product and process innovation, in practice the two forms correlate closely, so that here only the combination of product and process innovation will be presented. It is precisely in such cases that a high level of skill adjustment is to be expected and, given the predominance of external adjustment, this ought to lead to an increased risk of becoming unemployed.

4.2 The Analytical Model

The pooling of data from the 21 branches and 6 years which constituted the analysis, produced a combined cross-sectional and longitudinal data set of 126 values In order to carry out a simple regression analysis with this data set it must be assumed that both the regression coefficients and the absolute terms of the estimated equation remain constant for all branches and over time. In order to find out whether such an assumption can be justified, various covariance models can be constructed: firstly the assumption of constant absolute values over time can be relaxed, followed by that of constant absolute values in all branches; finally the absolute terms are allowed to vary both over time and in the branches (cf. Johnston 1984, Formby et al. 1984). The resulting models for use in the empirical analysis can be set out formally as follows:

Possible variations in the constant term in the combined longitudinal/cross-sectional analysis:

I. constant slope, constant absolute terms over time, constant absolute terms for branches
method: OLS-regression

$$RUE_{jt} = \beta_1 + \sum_{i=2}^{n} \beta_i X_{ijt} + \varepsilon_{jt}$$

II. constant slope, constant absolute terms for branches, varying absolute terms over time,
method: regression with dummy variables for the time

$$RUE_{jt} = \beta_{1t} + \sum_{i=2}^{n} \beta_i X_{ijt} + \varepsilon_{jt}$$

III. constant slope, constant absolute terms for the time, varying absolute terms for the branches
method: regression with dummy variables for the branches

$$RUE_{jt} = \beta_{1t} + \sum_{i=2}^{n} \beta_i X_{ijt} + \varepsilon_{jt}$$

IV. constant slope, varying absolute terms for the time, varying absolute terms for the branches
method: regression with dummy variables for the branches and the time

$$RUE_{jt} = \beta_{1jt} + \sum_{i=2}^{n} \beta_i X_{ijt} + \varepsilon_{jt}$$

with:
RUE = risk of becoming unemployed
X = independent variables
ε = error term
i = index for independent variables, i = 1 ...n
j = index for branches
t = index for the time

The use of dummy variables in the estimated equations has the advantage that different absolute values for the different years and for each branch can be taken into account. However the coefficients can only indicate the existence of these differences, not their extent. The explanation for the latter has to be left to qualitative analyses. The introduction of dummy variables into the regression equation also means the loss of degrees of freedom for the estimation. However, even in the model in which variation of the absolute value is possible both over time and for the branches 99 degrees of freedom remain. If, in addition to the dummies, a further variable is taken into account many more degrees of freedom are available than would have been the case with cross-sectional analyses of specific years.

4.3 Operationalization

In order to establish the effect of innovation on the risk of becoming unemployed and the risk of remaining unemployed (the duration of unemployment, not reported here, see Schettkat 1989) the WZB/AMB conducted regression analyses of 21 branches of West German manufacturing industry at different points in time. With the aim of making the fullest possible use of the available information, branches were grouped according to the available data on the flow into unemployment. The data on innovative activity in different branches provided by the Ifo Institute were then aggregated to the level of these branch groups. The share and value indicators were weighted by the number of employees in the branch as described above.

It would seem plausible to assume that the effects of innovation are not felt until after a time lag. In order to illustrate the lags built into the WZB/AMB model, the time periods covered by the different indicators will now be set out in detail. The change in the level of employment for each year is measured against the number of wage and salary earners subject to social insurance on June 30th of that year. June was chosen because at this time the influence of seasonal factors on both the structure and level of employment is at a minimum. The risk of becoming unemployed was measured by the flow into registered unemployment from October to September of each year compared with employment in the previous year. For example, for the flow into registered unemployment for the year 1981 the time period used was 1st October 1980 to 30th September 1981. The values for innovative activity refer to innovations carried out during the calendar year. It is apparent that the variables included in the analysis are based on different time delineations. This is illustrated in the following diagram:

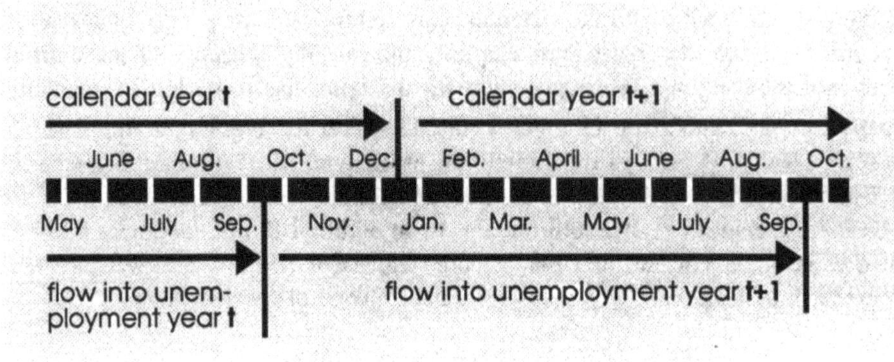

Figure 4.1: Time Delineation for Innovation and the Risk of Becoming Unemployed

If the same time scale were used for both the risk of becoming unemployed and the innovation indicators in a calendar year, some innovative activity would be included which could not possibly have exerted an influence on the risk of becoming unemployed. This is clearly illogical as the assumption is that the effects of employment *follow* changes in innovative activity. For this reason we specified the model so that the risk of becoming unemployed in the year t (i.e. from the October of t-1 to the September of t) is dependent on the innovative activity in the calendar year t-1 (January to December t-1). Thus the effect of innovation on the flow into unemployment is delayed by an average of nine months. During our analyses we also tested longer delay periods, but this did not affect the results obtained.

5. Empirical Analysis of the Relationship between the Risk of Becoming Unemployed and Innovation

5.1 Other Factors Influencing the Risk of Becoming Unemployed

In addition to the type of skill adjustment used as a response to innovative activity, the risk of becoming unemployed is also heavily influenced by the following factors:
- the employment situation (DB)
- seasonal changes in production and employment (SAIS)
- the skill structure (SS)
- the structure of enterprise size (RBGR)
- industrial relations (IR)

Both in the economy as a whole and in individual branches the risk of becoming unemployed is heavily dependent on the business cycle. It is therefore to be expected that in branches with a favorable overall employment situation the risk of becoming unemployed is less than in those where the trend level of employment is less favorable. An allowance is therefore made for the influences of the branch-specific, trend level of employment by including the figures for the change in employment level from one year to the next in the analysis.

To understand the importance of seasonal influences on the branch-specific risk of becoming unemployed it is important to remember the definition of the risk of becoming unemployed: it was calculated by relating the (continuous) flows out of employment into unemployment to the number of persons potentially subject to redundancy (wage and salary earners). Branches exposed to heavy seasonal influences on production and employment will experience a large flow out of employment into unemployment without the level of unemployment changing on the annual average, as workers are taken on and made redundant according to the seasonal cycle.

On the basis of quarterly employment statistics the seasonal influence on employment was established for each branch in terms of its deviation from the moving average (for detail see Schettkat 1989). The standard deviation of the relative differences between the actual employment figures and the moving average was taken as the intersectorally comparable measure for the importance of seasonal influences on each branch (SAIS).

It is well known that unskilled workers are at greater risk of being made redundant than skilled workers, a fact which can be explained in either terms of human capital theory, transaction costs and segmentation theory. One would therefore expect that branches with a greater than proportional share of unskilled workers would exhibit a higher average risk of becoming unemployed than those with a higher than average skill level. Blue-collar workers are also subject to a greater risk of being made redundant than white-collar workers due to the different provisions of labor law regarding dismissal applying in each case. The ratio of unskilled workers to skilled workers (SS) was used to construct an indicator to take account of the different skill structures in each branch. In addition, the ratio of unskilled workers to white-collar workers was also included in the analysis. This was done less to take account of the different skill structure than of the different protection from dismissal afforded by labor law.

The different structure of enterprise size in each branch can exert an influence on the flow into unemployment. Worker representation (e.g. the existence of full-time paid works councilors) and its influence is partly dependent on the size of the enterprise and, at the same time, larger enterprises usually have greater financial resources enabling them to overcome minor fluctuations in production more easily and with greater flexibility. Larger enterprises can also be said to have a more extensive internal labor market, i.e. the potential for internal flexibility is a function of, among other factors, enterprise size. In the analysis allowance was made for the structure of enterprise size by using the branch-specific, average enterprise size as an indicator.

The position of worker representation can exert an influence on the stability of employment and the dismissal policy of firms in the various branches. However, the influence of worker representation, like most socio-economic variables is very difficult to model. With the help of Wolfgang Streeck, also from the WZB, who has long and wide-ranging experience in the field of industrial relations research, an indicator for trade-union strength in each branch was developed by means of "expert rating". Obviously subjective factors play a part in this indicator so that from the beginning only a crude form of classification was attempted, in which the degree of trade-union influence can take values between 0 and 2.

It is not until account has been taken of these factors in the analysis that it becomes possible to analyze the specific influence of innovative activity, or, to be more precise, the specific form of skill adjustment resulting from it. This analysis is described below. It consists of a combined cross-sectional/longitudinal analysis of the 21 branches of manufacturing industry in West Germany between 1980 and 1985. This involves the use of various co-variance models with the main aim of examining whether cross-sectional relations are also valid longitudinally, or, to put it another way, whether a functional relation can be concluded from a cross-sectional one.

Numerous examples in economics and other social sciences show that all too often a functional relation is prematurely concluded from cross-sectional results. An example of this is provided by the explanation for female labor force participation: in cross-sectional analysis female labor force participation exhibits a negative income elasticity whereas over time it shows a positive income elasticity - with increasing national income the participation rates of women have increased. In this case the functional relation has to be explained using other variables (cf. Schettkat 1987).

5.2 The Results of the Analysis

The relationships as estimated using the combined cross-sectional/longitudinal approach are shown in table 5.1. Even the simple OLS estimation equation which contains no dummy variables explains approximately 70% of the deviation of the risk of becoming unemployed for men and about 60% for women. When the dummies are included, the deviation explained increases significantly, reaching about 90%. The vast majority of the regression coefficients of the dummies are significantly different from zero.

The innovation variable (INN) which refers to the proportion of innovative firms in a branch is statistically highly significant in all equations which are strongly cross-sectional in nature (models I and II in table 5.1) but loses importance in equations which include a longitudinal perspective (models III and IV) and the regression coefficient is statistically no longer significant. It is clear that varying levels of innovative activity can explain some of the differences in the risk of becoming unemployed between branches but that they are incapable of explaining these differences over time.

At first sight it seems that there is a contradiction between the effect of the innovation indicators on the risk of becoming unemployed in cross-section (between the branches) and longitudinally (within the branches). Two arguments can shed some light on this apparent contradiction: the first is a statistical argument based on the variations in innovative activity over time, while the second is a substantive one, based on those factors which lie behind and produce the relationship between innovative activity and the risk of becoming unemployed.

The analysis presented here is of labor market processes which are set in train by innovation and the skill adjustments resulting from it; the question dealt with is *not* whether innovation has a positive or negative effect on the employment level. A change in the level of employment clearly has an effect on the risk of becoming unemployed, as is confirmed by the analysis presented here, but this has little to do with skill adjustment processes which can influence the risk of becoming unemployed independently of changes in the employment level.

In cross-sectional analysis the risk of becoming unemployed was less in the more innovative branches, i.e. skill adjustment processes do not result in workers being made redundant; on the contrary, the negative sign before the regression coefficients for the innovation variables indicates that the flow of workers into unemployment from more innovative branches is less than from less innovative branches. It seems certain that highly complex, socio-economic, causal relationships lie behind this result and there is no reason to expect that the factors determining these causal relationships are subject to change in the short run. Rather it is to be expected that they are relatively stable over time and subject to change only in the long term.

Table 5.1: Regression Analysis of the Risk of Becoming Unemployed in a Combined Cross-Sectional/Time Series Approach Using Various Covariance Models

	Constant	INN	IR	PS	DE	SS	SAIS	TDUM	BDUM	R^2	R^2_{adj}
					Total						
I	13.0 (11.2)	-0.9E-3 (-4.7)	-0.7 (-1.9)	-0.2E-1 (-5.7)	-0.6 (-7.4)	0.3E-2 (0.9)	2.2 (8.0)	-	-	70.6	61.1
II	11.5 (8.9)	-0.9E-3 (-4.9)	-0.6 (-1.7)	-0.2E-1 (-5.9)	-0.6 (-6.3)	0.4E-2 (1.1)	2.2 (8.2)	most signif.	-	72.9	70.3
III	7.8 (4.9)	0.8E-4 (0.4)	- (0.0)	- (0.0)	-0.4 (-8.3)	-0.4E-1 (-0.3)	+ (0.0)	-	all signif.	91.2	89.2
IV	3.4 (2.8)	0.1E-3 (0.5)	- (0.0)	- (0.0)	-0.5 (-7.9)	+ (0.0)	+ (0.0)	most signif.	most signif.	92.3	90.2
					Men						
I	11.2 (11.5)	-0.9E-3 (-4.5)	-0.1 (-0.3)	-0.2E-1 (-5.6)	-0.5 (-6.4)	0.3E-2 (-0.8)	2.5 (10.2)	-	-	71.2	70.0
II	9.1 (8.5)	-0.9E-3 (-4.8)	-0.8E-1 (-0.2)	-0.2E-1 (-5.9)	-0.5 (-5.8)	0.3E-2 (1.2)	2.5 (10.6)	all signif.	-	74.7	72.3
III	9.9 (5.7)	0.7E-5 (0.3)	- (0.0)	- (0.0)	-0.4 (-6.1)	-0.8E-1 (-5.1)	- (0.0)	-	all signif.	91.4	89.4
IV	0.4 (0.3)	0.9E-4 (0.4)	- (0.0)	- (0.0)	-0.4 (-7.5)	- (-1.5)	3.8 (16.1)	all signif.	most signif.	92.4	90.3

Continuation of table 5.1

	Constant	INN	IR	PS	DE	SS	SAIS	TDUM	BDUM	R^2	R^2_{adj}
					Women						
I	11.4	-0.7E-3	-0.4	-0.4E-2	-0.6	0.8E-2	0.7	-	-	61.2	59.3
	(8.7)	(-4.5)	(-1.0)	(-1.5)	(-9.3)	(3.1)	(3.5)				
II	12.1	-0.7E-3	-0.3	-0.5E-2	-0.7	0.7E-2	2.5	not	-	63.5	59.9
	(8.3)	(-4.4)	(-0.8)	(-1.9)	(-8.1)	(2.6)	(3.4)	signif.			
III	6.6	0.4E-4	-	-	-0.5	0.1E-1	+	-	most	92.0	90.2
	(6.3)	(0.3E-1)	(0.0)	(0.0)	(-14.8)	(1.7)	(0.0)		signif.		
IV	8.2	-0.2E-4	-	-	-0.5	+	+	not	most	92.1	89.9
	(8.6)	(-0.1)	(0.0)	(0.0)	(-10.8)	(1.3)	(0.0)	signif.	signif.		

INN = indicator for innovations
IR = indicator for industrial relations
PS = plant size
DE = relative change of employment
SS = indicator for the skill structure
SAIS = indicator for the seasonal variations of employment
TDUM = time-specific dummies
BDUM = branch-specific dummies
in parentheses = t-statistics

It is equally unlikely that minor changes in innovative activity will have a significant effect on the risk of becoming unemployed within a branch over a period of time. The use of various value indicators, which vary more widely over time than do share indicators, for the innovative activity in different branches indicates a similar overall relationship between the risk of becoming unemployed and innovation to that given by share indicators. The innovation indicator is significantly negative for both cross-sectional models (models I and II) so that even using the value indicators the risk of becoming unemployed is less in the more innovative branches than in the less innovative ones.

Positive regression coefficients were registered for some of the covariance models in which the longitudinal aspect plays a role (models III and IV) but they were mostly not significant. Attention must, however, be drawn to one important exception: a significantly positive regression coefficient was found in models III and IV (i.e. over time) for process innovation and male workers. This means that increased expenditure on process innovation within a branch increases the risk of becoming unemployed. Even though the longitudinal analysis can have only limited validity due to the limited availability of data (since 1980) the results of the combined cross-sectional/longitudinal analysis show that one must be extremely careful about drawing conclusions from cross-sectional results and then applying them to the long run: a generalization over time is often invalid and in some cases the relationship can even be the exact opposite.

The indicator for the strength of worker representation in different branches (IR) indicates a similar relationship as the innovation indicators. Where there is no explicit control for the branch (equations I and II in table 5.1) the regression coefficient of this indicator is significant. The strength of worker representation can therefore partly explain the variation in the risk of becoming unemployed in different sectors. Within a sector (control for the branch, equations III and IV in table 5.1) trade union strength loses some of its importance as an explanation of variations in the risk of becoming unemployed. This is also highly plausible since trade union density and the influence which can be exerted by the work force change only slowly over time and given the indicators used in this analysis it was not possible to take account of such variations. Variations in the risk of becoming unemployed in a branch over time must therefore be explained by other factors. It should be noted that the differences between branches as regards the IR and the innovation indicators were "absorbed" by the branch-specific dummies (WDUM). The same is true for all variables which are branch-specific and which are almost constant over time. This does not mean that they lose their importance for the explanation of the differences between the sectors but rather that they "disappear" in the values for the dummy variables which were almost without exception significant.

Similar considerations apply to the seasonal dependence of employment in some branches. As described above, an indicator was constructed with the aim of accounting for those changes in the level of employment which are due to seasonal factors (for further details cf. Schettkat 1989). In cross-section, the seasonal dependence of the various branches offers a good explanation of the risk of becoming unemployed but, for the reasons mentioned above, the importance of this factor is much less over time. Again this is highly plausible as the seasonal dependence of a branch is unlikely to change from one year to the next. Seasonal dependence too is branch-specific and is also "absorbed" by the branch dummies.

Skill structure (SS), on the other hand, has a relatively low explanatory value: most of the coefficients are not significant. It is only in the cross-sectional equations for women that the regression coefficients take on significantly positive values, i.e. a higher ratio of unskilled workers to white-collar workers (who enjoy greater stability of employment due to the differences in labor law) in a branch increases the overall risk of becoming unemployed.

Changes in the level of employment exert a significantly negative effect on the risk of becoming unemployed in both cross-sectional and longitudinal perspectives, i.e. an increase in employment is linked to a reduced risk of becoming unemployed while a decrease in the level of employment increases the risk.

Overall it can be concluded that a high level of innovative activity in no way requires a high level of exchange with the external labor market. On the contrary, the reverse seems to be true: in more innovative branches the risk of becoming unemployed is less. The explanation for this clearly lies in highly complex socio-economic causal relations and it is unlikely that the factors which determine these relations are subject to change in the short term. For this reason it was not to be expected that the risk of becoming unemployed would change over time in response to variations in innovative activity.

6. The Flow out of the Active Labor Force

Skill adjustment processes also take place within the work force through inter-generational mobility (Stooß 1984, Berglind 1981). Younger workers tend to have more up-to-date skills and greater theoretical knowledge, whereas older workers are largely to be found in occupations which are currently less in demand. In view of this it is to be expected that more innovative branches would tend to have a work force of younger average age. The flow into 'non-participation in the labor force' is of particular importance in this process. If older workers tend to exhibit a skill profile which is no longer in great demand, one would expect greater use to be made of the various possibilities offered by the laws governing pensions and retirement. For firms, early retirement offers a very advantageous instrument for adjusting the skills of the work force because, on the basis of early retirement laws and redundancy agreements, it is possible to offer older workers a financially secure and socially acceptable 'alternative role'. A correlation between the level of innovative activity in a branch and the high rate of departure of older workers into non-participation would thus seem a reasonable expectation.

Other factors beside innovative activity exert an influence on the age structure of the work force. The structure of enterprise size in a given branch will probably play a part, although the direction of its effect is not something which can be determined *a priori*: on the one hand the seniority principle tends to be stronger in larger firms. On the other hand, personnel planning is more common in large enterprises and their better financial position enables redundancy agreements to be reached which encourage more workers to opt for early retirement. A further factor is likely to be the trend level of employment and the skill structure of the work force, the latter because the actual retirement age for blue-collar workers is considerably below that for white-collar workers (Hellberger 1977).

6.1 Age Structure

In our analyses the dependent variables were the average age of the work force and the proportions accounted for by specific age groups. The data base consisted of wage and salary earners (excluding trainees and apprentices) subject to social insurance in 1982 and 1986. The innovation indicator was constructed from an average of the years 1980 to 1982 and 1982 to 1986 respectively as the values for a single year may have led to distortions.

The regression analysis of the relationship between age structure, innovation and certain other variables in the 21 branches of West German manufacturing industry show for both years that the average age of wage and salary earners as defined above and classified in age groups spanning five years is not significantly influenced by innovation nor by the skill structure. The structure of enterprise size, however, has a positive influence on (i.e. increases) average age, a relationship which is significant at the 5% level for 1982. The relationship which had been assumed between innovative activity and age structure cannot be substantiated, at least as far as average age is concerned.

Again no significant regression coefficients were established between the shares made up by the different age groups and the innovation and skill-structure indicators. The regression coefficients of the structure of enterprise size are negatively significant at the 5% level for the share of the over 60 year-olds. This implies that in branches dominated by large enterprises the proportion of workers in their 60s is significantly lower, a fact which is possibly due to firm's pension schemes, compensation payments etc. leading to earlier retirement. A positive employment trend (DE) would appear to benefit younger rather than older workers: this is reflected in the positive regression coefficients for the share of under 55-year-olds (table 6.1) combined with a fall in the share of workers over 60.

6.2 Non-participation in the Labor Force

Besides the flow from employment into unemployment, external labor market adjustment can also occur via the flow into 'non-participation'. The non-economically active population consists of all those who are not in employment and are not registered as unemployed, i.e. those who have withdrawn from the labor market or have not yet entered it. (pensioners, housewifes and those in education, cf. fig. 2.2). The Mikrozensus, a 1% sample of the West German population, conducted in 1983 enables us to analyze the flows between employment and non-participation. Those interviewed in the Mikrozensus who were not economically active at the time of the interview were asked whether they had been in the active labor force during the previous three years and if so in which branch. In addition the reason for leaving the labor force was given.

On behalf of the WZB/AMB the Federal Statistical Office (Statistisches Bundesamt) constructed a special data set from the 1985 Mikrozensus differentiating the flow into non-participation by branch. At this point a note of caution must be sounded: Although the Mikrozensus is based on a relatively large sample (1% of the population of the Federal Republic) when the data are differentiated in some categories the individual cells have insufficient numbers to allow reliable extrapolation. For example, it proved impossible following differentiation by branch and age to then also differentiate by gender.

For the period between 1982 and 1985 the flow from employment into non-participation was analyzed by branch for the age groups 'under 55', '55 to 60' and '60 to 65'. These flows were set in relation to the initial level of employment in each branch in order to construct relative figures which were comparable between branches. These values were used as the dependent variables in the analysis and regressed to the branch-specific level of innovative activity (INN), change in the level of employment (DE), the structure of enterprise size (PS) and skill structure (SS). The results of the regression analysis for the flow into non-participation are set out in the upper part of table 6.2.

Table 6.1: Regression Analysis of the Link between the Age Structure and Innovative Activity, 1982 and 1986 (t-values in parentheses)

dependent variable (time period)	INN (1980-1982)	DE (1978-1982)	PS (1980)	SS (1979-1982)	R^2	R^2_{adj}
				1982		
average age	0.17 (0.79)	-0.32 (-1.36)	0.40 (1.78)	-	40.2	29.6
	0.21 (0.96)	-0.24 (-0.97)	0.43 (1.86)	0.19 (0.89)	43.0	28.8
proportion of age groups in total employment						
- under 55	-0.17 (-0.79)	0.58 (2.34)	-0.06 (-0.25)	-	34.1	22.5
	-0.19 (-0.79)	0.56 (2.06)	-0.07 (-0.27)	-0.05 (-0.23)	34.4	18.0
- 55 to 60	0.19 (1.00)	-0.57 (-2.67)	0.25 (1.20)	-	50.4	41.6
	0.23 (1.13)	-0.51 (-2.21)	0.27 (1.27)	0.16 (0.79)	52.2	40.3
- 60 and older	0.03 (0.13)	-0.30 (-1.08)	-0.49 (-1.84)	-	18.1	3.6
	-0.03 (-0.09)	-0.40 (-1.36)	-0.52 (-1.95)	-0.26 (-1.02)	23.1	3.9
				1986		
average age	0.13 (0.58)	-0.36 (-1.56)	0.38 (1.73)	-	37.6	26.6
	0.17 (0.73)	-0.35 (-1.46)	0.38 (1.66)	0.13 (0.61)	39.0	23.8
proportion of age groups in total employment						
- under 55	-0.15 (0.62)	0.55 (2.17)	-0.007 (-0.03)	-	26.5	13.5
	-0.21 (-0.83)	0.53 (2.06)	0.004 (0.02)	-0.19 (-0.85)	29.6	12.1
- 55 to 60	0.20 (0.93)	-0.59 (-2.59)	0.15 (0.70)	-	39.9	29.3
	0.28 (1.25)	-0.56 (-2.50)	0.14 (0.64)	0.24 (1.20)	44.8	31.1
- 60 and older	-0.08 (-0.34)	-0.19 (-0.74)	-0.46 (-1.87)	-	22.7	9.1
	-0.10 (-0.36)	-0.19 (-0.73)	-0.46 (-1.81)	-0.04 (-0.16)	22.8	3.6

INN = innovation variable, value indicator
DE = branch-specific change in the level of total employment
PS = structure of plant size
SS = skill structure
in parentheses: t-values

Table 6.2: Regression Analysis of the Link between the Relative Flow from Employment into Non-Participation and Retirement together with Innovation between 1982 and 1985

dependent variable (time period)	INN (1982-1985)	DE (1978-1985)	PS (1980)	SS (1979-1985)	R^2	R^2_{adj}
			flow into non-participation			
total	0.28	-0.31	-0.08	-	7.6	-8.7
	(0.11)	(-1.11)	(-0.30)			
	0.09	-0.28	-0.09	0.21	11.3	-10.9
	(0.34)	(-0.98)	(-0.32)	(0.81)		
- under 55	0.05	-0.31	-0.25	-	8.7	-7.4
	(0.18)	(-1.09)	(-0.96)			
	0.15	0.25	-0.26	0.33	17.9	-2.6
	(0.57)	(0.86)	(-1.01)	(1.34)		
- 55 to 60	-0.08	0.26	0.11	-	4.3	-12.6
	(-0.28)	(0.87)	(0.42)			
	-0.05	-0.17	0.11	-0.09	5.0	-18.8
	(-0.16)	(-0.64)	(0.40)	(-0.34)		
- 60 and older	-0.08	-0.17	0.36	-	21.4	7.5
	(-0.34)	(-0.64)	(1.48)			
	-0.09	-0.17	0.37	-0.01	21.4	1.8
	(-0.32)	(-0.62)	(1.43)	(-0.04)		
			flow into retirement			
total	-0.28	0.20	0.37	-	12.9	-2.4
	(-1.08)	(0.71)	(1.44)			
	-0.17	0.24	0.36	0.34	22.8	3.4
	(-0.64)	(0.88)	(1.44)	(1.43)		
- 55 to 60	-0.19	0.23	0.33	-	9.6	-6.4
	(-0.72)	(1.01)	(1.24)			
	-0.13	0.31	0.32	0.19	12.5	-9.3
	(-0.46)	(1.07)	(1.21)	(0.74)		
- 60 and older	-0.32	0.27	0.51	-	21.5	7.7
	(-1.29)	(1.03)	(2.05)			
	-0.23	0.30	0.50	0.26	27.3	9.1
	(-0.92)	(1.15)	(2.03)	(1.12)		

INN = innovation variable, value indicator
DE = branch-specific change in the level of total employment
PS = structure of plant size
SS = skill structure
in parentheses: t-values

No statistically significant relationship could be established between innovative activity, changes in the level of employment, the structure of enterprise size or skill structure and the (relative) flow into non-participation. There was, however, a positive regression coefficient between the structure of enterprise size and the outflow of older workers, but it was only significant at the 10% level. This means that the trend to early retirement is more pronounced in branches with a large average enterprise size. This finding corresponds to the results of the analysis of age structure in the previous section which showed a significantly lower share for the 'over 60s' in branches with a large-enterprise structure.

In a further step the status of 'non-participation' was specified and the analysis of non-participants was restricted to those who had stated 'retirement' as the reason for having left their last place of employment. In contrast to the unspecified flow into non-participation, by restricting the flow to that into retirement we obtained a significantly positive regression coefficient (at the 5% level) for the structure of enterprise size and workers over 60 (the lower part of table 6.2).

This result corresponds to the results of the analysis of age structure, and can be explained by firms' retirement schemes and personnel planning in large enterprises, but cannot be developed and verified further in this presentation. The effect of the innovation indicators on the other hand remained insignificant: hence, a significant relationship between the flow out of employment into retirement and the branch-specific level of employment could not be established. In conclusion it should be noted that the models did not provide a good explanation of the variances. This is probably partly due to the data base.

7. The Elasticity of the Supply of Labor by Occupational Group

The different occupational groups exhibit widely differing employment trends between 1980 and 1985. In some groups employment increased by up to 20% while at the same time other groups suffered losses of the same magnitude. Employment gains were registered mainly in service occupations (health, social services and educational occupations) and highly qualified technical occupations (engineers, industrial chemists, specialized technical personnel). At the same time those occupations directly involved with production suffered considerable job losses (cf. table 7.1), a result which has also been obtained in other studies (Baethge 1988, Warnken/Ronning in this volume).

Table 7.1: The Change in Employment, Unemployment and the Supply of Labor in 38 Occupational Groups

occupational groups	change 1980 to 1985 (%) employ-ment	unemploy-ment	labor supply
health service, diagnosing/treating occ.	+ 21.0	+ 166.9	+ 25.7
social occ., educationalists, researchers	+ 10.6	+ 170.6	+ 21.3
agriculture, animal XX, fishing	+ 10.1	+ 177.9	+ 20.0
journalists, interpreters, librarians	+ 9.0	+ 105.8	+ 13.0
engineers, chemicists, physicists, mathematicians	+ 8.8	+ 174.2	+ 12.7
goods and services, sales	+ 8.3	+ 131.4	+ 10.5
personal hygiene	+ 5.4	+ 184.8	+ 13.4
technical specialists	+ 4.8	+ 179.5	+ 9.9
hotel staff	+ 4.4	+ 176.0	+ 16.7
law and order, security	+ 3.4	+ 110.5	+ 10.2
electricians	+ 2.4	+ 233.1	+ 7.2
food, drinks and tobaccos related occ.	+ 2.0	+ 188.0	+ 10.8
trades goods occupations	+ 0.7	+ 151.8	+ 7.7
technicians	+ 0.6	+ 179.6	+ 3.2
organizers, administrative and clerical staff	+ 0.3	+ 122.1	+ 4.2
painters, varnishers and allied occ.	+ 0	+ 313.4	+ 12.8
fitters, mechanics and allied occ.	- 2.3	+ 281.4	+ 3.6
joiners, pattern making occ.	- 3.0	+ 521.5	+ 8.2
artists and allied occ.	- 3.0	+ 94.7	+ 5.1
chemical/plastic processing workers	- 4.4	+ 120.9	- 0.6
domestic services	- 4.7	+ 129.7	+ 5.6
miners - coal and mineral	- 5.7	+ 137.9	- 1.3
printers	- 6.0	+ 172.6	- 1.7
cleaning staff	- 6.6	+ 143.2	+ 0
building trade, fitting & furnishing, upholstery	- 7.2	+ 369.2	+ 5.0
transport occupations	- 8.5	+ 147.1	- 1.7
mechanic and metal fitting occupations	- 10.4	+ 67.7	- 3.6
makers of paper and paper products	- 11.3	+ 111.2	- 7.1
unskilled laborers without specified task	- 11.8	+ 33.3	- 4.3
checkers and packers	- 12.1	+ 96.9	- 4.0
producers of metal and metal goods	- 12.2	+ 198.2	- 7.3
potters and glass manufacturers	- 13.8	+ 146.9	- 8.6
machinists and allied occupations	- 14.9	+ 222.6	- 8.5
stone workers, masons, suppliers of build. mats.	- 16.7	+ 164.2	- 6.9
construction workers	- 18.1	+ 363.5	- 3.3
wood preparation, wood working	- 20.8	+ 182.9	- 11.0
leather and fur producers	- 21.4	+ 112.4	- 14.9
textiles and clothing	- 21.9	+ 73.8	- 16.1

Calculations based on employment statistics, wage and salary earners and registered unemployed subject to social security payments; the supply of labor consists of wage and salary earners subject to social security payments plus the registered unemployed.

However, a positive employment trend in an occupational group need not necessarily result in a reduction of the level of unemployment in that group. The relationship between the change in the level of employment and of unemployment is far from being as close as is usually assumed, and it becomes even more ambiguous if individual sectors of the labor market (e.g. occupational groups) are considered. If a fall in employment resulted in an increase in unemployment of equal magnitude and vice versa the correlation coefficient between the change in employment and in unemployment would be -1.0, whereas in actual fact it amounts to a mere -0.4. Ignoring the case where both remain constant, this means that the following relationships between the two values could, in principle, occur:

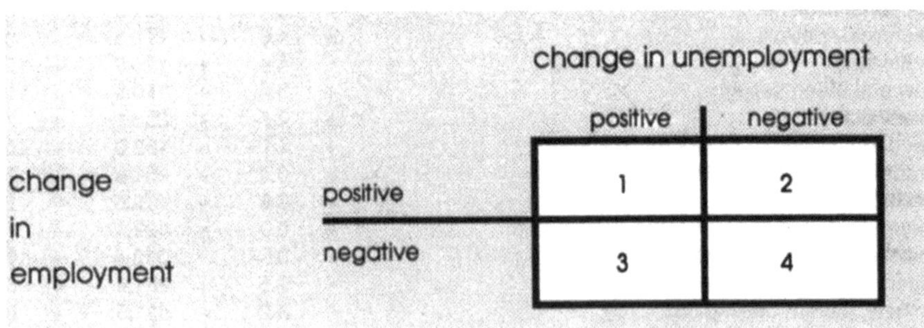

Figure 7.1: The Relationship between the Change of Employment and that in Unemployment

In case 1 employment increases with a simultaneous increase in unemployment, due to a more than proportional increase in the supply of labor. In case 2 employment increases while unemployment falls, i.e. there is a pronounced flow from unemployment into employment. In case 3 employment decreases whereas unemployment rises, the standard case so to speak, where those previously employed enter registered unemployment. The situation in case 4 is of decreasing employment and unemployment implying that the supply of labor has fallen to a greater extent than the decrease in employment. It is only in cases 2 and 3 that the situation is the one normally associated with the operation of the labor market in terms of the relationship between unemployment an employment.

Figure 7.1 serves as a further illustration of the dynamics of movements on the labor market. The economically active population (those in employment and registered unemployment) are far from being a constant 'block' of which some merely change their status between the two categories. The composition of this group is changing constantly with (younger) workers entering and (usually older) workers leaving. This is why situations such as cases 1 and 4 can occur.

Of course if the changes are analyzed at the level of individual occupational groups, the additional possibility exists of mobility between occupational groups. Workers remain in their occupational group as long as they are registered unemployed, as the unemployed are classified according to the occupation in which they were last employed. Hence it cannot be definitively concluded from an increase in employment in one occupational group and a fall in unemployment in that group that those previously unemployed and classified in this group have now entered employment in the same category. It may well be that they have found employment in another occupational group (classification is then according to the new type of employment). The increase in occupational group employment can thus be due to the integration of 'outsiders' into an occupation so that the occupational group's level of unemployment may remain constant or even increase.

Looking now at actual developments in the Federal Republic of Germany between 1980 and 1985, only cases 1 and 3 come into consideration since there was no reduction of unemployment in *any of the occupational groups* during this period. Occupations considered to be the 'jobs of the future' (e.g. engineers, industrial chemists) experienced employment growth, but at the same time the supply of labor in these groups also increased to such an extent that unemployment rose. The same is true for the expanding service occupations.

In those occupations where employment fell the supply of labor rose only in the following: painters, artists, carpenters, domestic occupations, interior designers and fitters. In all the other groups the supply of labor fell and therefore matched (at least in direction) the change in demand. The increase in the supply of labor in the groups mentioned above, contrary to the movement of labor demand (i.e. employment) is probably at least partly due to the situation on the training and apprenticeship sector of the labor market. The scarcity of training opportunities narrows the available choice, so that a decision is made solely on the basis of the availability of training opportunities, for no matter which occupation - "any sort of training is better than nothing". Calculations based on employment statistics, wage and salary earners and registered unemployed subject to social security payments; the supply of labor consists of wage and salary earners subject to social security payments plus the registered unemployed.

Of particular relevance for evaluating the degree of occupational flexibility of the West German labor market is the overall orientation of workers to new skill requirements. If the changing requirements fail to induce a change in the decisions taken by workers on (further) education and (re)training or in the opportunities for such training provided by public educational institutions, employers and labor exchanges, economic growth may well be blocked by a lack of skilled workers and the consequent skill mismatch will result in a form of structural unemployment. If on the other hand occupational mobility is very high such forms of unemployment will be avoided. Given a higher degree of skill adjustment the supply of labor, i.e. the total of wage and salary earners and the unemployed, ought to increase in the 'industries of the future' and decrease in occupations which are less in demand. The changes in employment and the supply of labor ought then to be positively correlated.

The WZB/AMB carried out such an analysis of the flexibility of the West German labor market for 38 occupational groups. We reached the conclusion that between 1980 and 1985 occupational flexibility of the supply of labor related to the changes in the employment level in the different groups was very high. There is thus no evidence for a 'skill gap' in the more dynamic sectors of the West German economy: in these occupations the supply of labor increased to a greater extent than demand, and thus reacted very flexibly to changing skill requirements.

8. A Summary of Our Conclusions

The analysis presented here of the relationship between innovative activity and labor market dynamics using a pooled cross-sectional/longitudinal approach with data on 21 branches in West German manufacturing industry shows that in more innovative branches the risk of becoming unemployed is on average lower than in less innovative ones. Clearly the exchange with the external labor market is less pronounced in less innovative industries. These findings are in contrast to the conjecture of the "euro-sclerosis theoreticians" who claim that the West German economy is not flexible enough. This is supposedly due primarily to insufficient labor market mobility resulting from the restrictions imposed by labor law and the welfare state. Our results complement the analysis conducted by Franz (1987) which concluded that labor market structures offer a poor explanation of persistently high unemployment in the Federal Republic. It is clear that precisely those branches which, because of their high level of innovation are to be seen as the motor of economic development, manage quite well without external labor market mobility.

There is also an input of new skills and qualifications into the production system resulting from the entry of younger workers onto the labor market. One might therefore expect the average age of workers in more innovative branches to be lower, and for the flow out of the active labor force and particularly into retirement to be especially high. However, these assumptions were not confirmed by the evaluation of the Mikrozensus conducted for the WZB/AMB by the Federal Statistical Office. The structure of enterprise size is of greater importance for early retirement than innovative activity.

The West German labor market is also extremely elastic with regards to skill adjustment. In those occupational groups in which employment increased significantly the supply of labor also increased, whereas it decreased in those occupations where employment fell. It can therefore not be maintained that West Germany is suffering from a "qualification gap" at the macro level.

As far as skills and qualifications are concerned, the elasticity of the West German labor market seems to be high and our findings indicate that a rapid rate of innovation can be achieved not only through external but also through internal flexibility.

PART III

MACRO ECONOMICS

Intersectoral Effects of the Use of Industrial Robots and CNC-Machine Tools - An Empirical Input-Output Analysis

Dietmar Edler, Renate Filip-Köhn, Frieder Meyer-Krahmer, Reiner Stäglin, Hans Wessels

1. Introduction

Input-output analysis is a tried and tested method of empirical economic research which has been applied to a wide range of questions and problems (Stäglin 1982), the advantages of which can also be used to good effect for the analysis of the labor market effects of modern technology. It provides a consistent analytical and data structure which can be disaggregated as required, enabling results to be gained at sectoral or macro level while maintaining the consistency of the two sets of results. Input-output analysis represents a potential "bridge" between detailed case studies, which usually cannot be generalized and, due to their partial nature, neglect important causal relationships within the economy as a whole, and macroeconomic analyses which do not take account of structural components and specific details. It is thus an important instrument for modeling the "micro-macro-bridge", one of the central research aims of the Meta Study.

The data entered into the model on new technology and its diffusion are based on the results of surveys conducted within the framework of the Meta Study (in particular the survey of technology users conducted by the 'Institut für Stadtforschung und Strukturpolitik Berlin', IfS, and the DIW's surveys of selected producers and users of new technology) together with expert interviews and the evaluation of the available specialist literature. Further sources of information were the statistics on industrial robots installed in the Federal Republic of Germany published by the Fraunhofer Institute for Production Technology and Automation (IPA) in Stuttgart, and branch-specific data from the 'Verein Deutscher Maschinen- und Anlaganbau' (VDMA) in Frankfurt.

The results presented here indicate that, even on the basis of such a wide spectrum of data, technology-specific microeconomic results can be transformed into macroeconomic results using input-output analysis. At the same time, it was only in the course of our study that the large volume of data required and the need to tailor the data specifically to the needs of the input-output analysis became fully clear. The conclusion to be drawn from this is that in future studies of this type, the surveys on whose data they are based ought to be more explicitly oriented to the input-output approach.

Our study was concerned with selected modern computer-aided assembly and production technologies. Due to the selection of the technologies studied and the type of input-output analysis used, the results will be presented in two sections. Section 2 describes the employment effects of the use of welding and assembly robots, CNC lathes and milling machines with the aid of a static input-output model. In section 3 the effects of the diffusion of industrial robots (differentiated into 11 operational areas/functions) on employment and occupational structure are described using a dynamic input-output model. In those areas in which the two concepts overlap, particularly in the case of welding and assembly robots, the studies have been conceptually linked.

2. The Effects of the Use of Industrial Robots and CNC Machine Tools - Calculations Using a Static Input-output Analysis

2.1 Employment Effects - Users and Producers

The employment effects evaluated in this section are restricted to those resulting from the use of welding and assembly robots, CNC lathes and milling machines in the FRG, including the effects of exported machines of this type. The input-output approach also allows the effects of the production and use of such machines on upstream branches and firms to be taken into account (indirect effects). First, the effects of the use of industrial robots and CNC machines in place of conventional equipment were studied. Second, the economic situation in 1980 was taken as a reference point in order to be able to follow the extent of the effects at different stages of the diffusion process and to isolate the influences exerted by technology from those of other factors. 1980 was chosen because at that time the level of diffusion was very low. The extent of the diffusion process was then studied at different points in time. The effects of the selected technologies were studied on the assumption that both production and use in 1980 and 1986 (and at the saturation point) had taken place in 1980, i.e. under conditions which were the same as those prevailing in 1980. A further assumption was that the diffusion of the new technology had no effects on exports.

Information on the changes in demand in user enterprises as a result of the use of computer-aided technology cannot be determined on the basis of enterprise surveys. Instead elasticities of demand with respect to sales prices were calculated and discussed.

Tables 2.1 and 2.2 show that there were considerable differences in the level of diffusion reached by the four selected computer-aided technologies in 1980. Neither type of industrial robot was used to any great extent at all (at the end of 1979 there were only six assembly robots in use in the Federal Republic) whereas at this stage the diffusion of CNC machine tools was much greater. At this point the positive direct employment effects of producing industrial robots outweighed the indirect negative effects in user firms. The employment effects are measured against the use of "traditional technology", i.e. in this case hand welding or multi-spot welding, traditional assembly techniques or the use of non-computerized machine tools.

Table 2.1: The Employment Effects of the Use and Production of Industrial Robots and CNC Machine Tools in Place of Conventional Technology (number of persons)
- summarized results of an empirical model calculation -

	caused by application in				
	mechanical enginneering	motor vehicles	electrical engineering	other branches	total
1980					
welding robots	0	+234	0	-1	+233
assembly robots	0	+5	+3	0	+8
CNC lathes	-7,386	-5,308	-1,704	-8,393	-22,791
CNC milling machines	-3,731	-776	-1,032	-2,165	-7,704
1986*					
welding robots	-749	-4,498	-308	-596	-6,151
assembly robots	-40	-169	-505	-174	-888
CNC lathes	-12,132	-6,157	-2,699	-11,442	-32,430
CNC milling machines	-11,978	-2,048	-2,897	-14,158	-31,081
on reaching saturation point*					
welding robots	-2,279	-11,145	-1,142	-2,041	-16,607
assembly robots	-2,979	-5,743	-14,975	-8,502	-32,199
CNC lathes	-27,149	-15,128	-8,641	-35,565	-86,483
CNC milling machines	-22,513	-4,264	-8,240	-31,193	-66,210

* calculated on the basis of the economic structure of 1980

Sources: enterprise surveys and expert interviews, evaluation of literatur, input-output analysis by the DIW, own calculations.

By 1980 the use of CNC lathes had led to a loss of 23 000 jobs, whereas a slight positive effect was registered for industrial robots with their low level of use and relatively high investment costs. At the 1986 level of diffusion the use of CNC lathes and milling machines would have resulted in a loss of 31 000 and 32 000 jobs respectively. The use of both types of industrial robots would have cost 6 000 and 1 000 jobs respectively. At the saturation point for these types of technology the negative employment effects would have amounted to 86 000/66 000 for the two types of CNC machine tools and 17 000/32 000 for the industrial robots. In relation to the volume of their application the negative employment effects of welding robots are not particularly high since - especially in the automobile industry - spot-welding robots usually replace multi-spot welding machines, with no employment effects in user firms. There was a more rapid increase in the diffusion of CNC lathes than of CNC milling machines between 1980 and 1986. A higher level of diffusion was thus assumed for the former for 1986, a fact which is reflected in the difference in employment effects.

In all the cases studied the negative, indirect employment effects in user sectors outweighed the positive effects in the producer sector. The slight increase in inputs required by the use of industrial robots led to comparatively minor positive employment effects in upstream branches (suppliers). The reverse is true of CNC machines, where the slight reduction in the use of inputs had a negative employment impact. The production of these new technologies induced direct and indirect (i.e. upstream) positive employment effects, due almost exclusively to an increase in output.

In the sectors listed in table 2.1 the employment effects are due primarily to the application of their products in the automobile industry (welding robots) and in the electrical engineering industries (assembly robots). Both types of CNC machine tools were used particularly intensively in the mechanical engineering industry.

The changes in employment due to the use of industrial robots and CNC machine tools are in most cases compatible with the changes in the skill structure of employment in the relevant branches (cf. chapter III.3 in Meyer-Krahmer 1989). A comparison of the occupation by industry matrices for 1976 and 1984 indicate job losses among welders in the automobile industry. These losses are due to structural and productivity-linked changes resulting from the use of welding robots and (earlier) of multi-spot welding machines. The number of fitters has also fallen, which must be seen at least partially in the light of the introduction of assembly robots. The drop in metal-machining operations in the automobile industry can be explained by the introduction of CNC machine tools.

Table 2.2: Demand Effects and Elasticities of Demand with Respect to Sales Prices which would Compensate for the Employment Effects Induced by the Use of Industrial Robots and CNC Machine Tools
- summarized results of an empirical model calculation -

	caused by application in				
	mechanical engineering	motor vehicles	electrical engineering	other branches**	total***
compensatory demand effects in per thousand					
1980					
welding robots	0.0	-0.2	0.0	0.0	0.0
assembly robots	0.0	0.0	0.0	0.0	0.0
CNC lathes	3.7	3.2	0.9	0.6	1.2
CNC milling machines	1.9	0.5	0.6	0.1	0.4
1986*					
welding robots	0.4	2.7	0.2	0.0	0.3
assembly robots	0.0	0.1	0.3	0.0	0.0
CNC lathes	6.2	3.7	1.5	0.8	1.6
CNC milling machines	6.1	1.2	1.6	1.0	1.6
on reaching saturation point*					
welding robots	1.2	6.6	0.6	0.1	0.9
assembly robots	1.5	3.4	8.2	0.6	1.6
CNC lathes	13.8	9.0	4.7	2.5	4.4
CNC milling machines	11.4	2.5	4.5	2.2	3.3
compensatory elasticities of demand with respect to sales prices					
1980					
welding robots	.	+7.2	.	-1.0	+6.1
assembly robots	.	+4.0	+1.0	.	+1.5
CNC lathes	-1.7	-1.7	-1.8	-2.1	-1.9
CNC milling machines	-1.4	-1.5	-1.7	-1.8	-1.5
1986*					
welding robots	-1.6	-1.6	-1.8	-2.0	-1.7
assembly robots	-1.5	-1.5	-1.9	-2.3	-1.9
CNC lathes	-1.6	-1.7	-1.8	-2.0	-1.8
CNC milling machines	-1.6	-1.7	-1.8	-1.9	-1.8
on reaching saturation point*					
welding robots	-1.7	-1.8	-1.9	-2.1	-1.9
assembly robots	-1.2	-1.2	-1.2	-1.2	-1.2
CNC lathes	-1.7	-1.8	-1.9	-2.1	-1.9
CNC milling machines	-1.6	-1.7	-1.7	-1.9	-1.7

* calculated on the basis of the economic structure of 1980
** manufacturing industry excluding the mechanical engineering, motor vehicles and electrical engineering industries
*** all manufacturing industries

sources: enterprise surveys and expert interviews, evaluation of literature, input-output analysis by the DIW, own calculations

2.2 Possible Demand Changes and their Positive Employment Effects

The demand effects calculated in our study indicate the extent to which output in the different sectors would have to increase in order to compensate the negative employment effects which arise from the use of the selected technologies. The effects are at their highest once the saturation point has been reached. In order to compensate for the negative effects of the use of welding robots in the automobile industry, for example, output in this sector would have to increase by 0.66%. For the use of assembly robots in the electrical engineering industry, the corresponding figure is 0.82%. To compensate for the negative effects of the use of CNC lathes (milling machines) production would have to increase by 1.38% (1.14%).

As can be seen from table 2.2, with the exception of the installment period for industrial robots, the elasticities of demand with respect to sales prices which would be required to compensate the labor-displacing effects of the use of new technology lie between -1.2 and -2.3. As far as the sales prices are concerned it is assumed that the changes in costs which arise from the use of industrial robots are passed on in full. Cost reductions stem from reduced labor costs in user firms, face cost increases from a higher level of depreciation (because of the increase in investment). For intermediate inputs, slight cost reductions were assumed for industrial robots, and minor increases for CNC machine tools.

Once the saturation point has been reached, the elasticities of demand with respect to sales prices which would be required to compensate for the negative employment effects of new technology lie between -1.2 and -2.1 for all sectors and types of computer-aided technology in our study. Again it is assumed that cost savings are fully reflected in price reductions. An expected elasticity of more than -1 would appear to be problematic, in some cases. Provisional results from the DIW's FIND model indicate that overall the elasticities may be lower than our calculations suggest. Even if it is taken into account that the level of demand is also positively influenced by increases in product quality, the diversification of the product range and more punctual delivery as a result of the use of new technologies, it cannot be expected that these factors will have a sufficiently strong impact to compensate for the displacement of labor as a result of the introduction of new technology through an induced increase in demand.

2.3 The Employment Effects of CNC Machine Tools and Industrial Robots at the Saturation Point

In 1986 welding and assembly robots accounted for 58% of all industrial robots in the FRG (in terms of numbers of units). However, once the saturation point has been reached this will fall to 51%. Even allowing for the special case of spot welding robots which replace multi-spot welding machines with negligible effects on employment, the use of all forms of industrial robots could lead to a loss of approx. 17 000 (1986) and over 100 000 (saturation point) jobs, not all of which will be compensated by the increased demand of the users of industrial robots. This holds on the assumption that the employment effects of track welding and assembly robots are similar to those of industrial robots as a whole (with the exception of spot welding robots).

In value terms, domestic production of CNC lathes and milling machines employed in the FRG accounted for about one third of all CNC machine tools in the metal-working industries (1986). If the employment effects of CNC lathes and milling machines are representative for those of CNC machines as a whole, this would imply a labor displacement of 190 000 (1986) and 450 000 (at saturation point). Here too these negative employment effects will not be fully compensated by the positive demand effects of falling prices.

3. The Effects of the Use of Industrial Robots to 1995 - Model Calculations using a Dynamic Input-output Approach

In view of the fact that the introduction of a new technology into, and its diffusion within the existing technological structure of an economy represents a classic example of a *dynamic* process, a dynamization of the input-output analysis would represent a significant methodological advance. Ideally such a model would provide a consistent methodological framework over time in which the modeling of the diffusion process of a new technology could be embedded. However, theoretical and empirical research still has a long way to go before it reaches this goal, because it is not yet possible for a number of important causal relationships to be incorporated into such a model.

This study represents the first time that a dynamic input-output model for the FRG has been implemented which enables technology-specific studies of this type to be made. Its basic methodological structure is the same as that developed at the Institute for Economic Analysis, New York University, for the US (Leontief/Duchin 1986, Duchin/Szyld 1984). The most important feature of the change-over from the static to the dynamic input-output model is the endogenous explanation of the process of investment together with its effects on the productive capacity of an economy. In the dynamic input-output model the investment process is modeled using a sectoral accelerator approach.

On the basis of exogenously determined values for final demand the development of investment, output and employment, differentiated by sector is determined endogenously within the model simulation. The simulated trends of investment, output and employment for the different sectors are consistent both with one another and with the resultant macroeconomic trends. They allow for the changes in the input of intermediate goods, capital and labor in the production process over time resulting from structural and technological change. Also included are the direct and indirect effects resulting from this on the process of investment and on employment and output trends. At the present stage of the model's development, it is not yet possible to take into account the causal relationships which are mediated by price, income and redistributive effects (cf. chapter III.2 in Meyer-Krahmer 1989). Nevertheless, in the final part of our study, an attempt was made roughly to estimate the importance of the causal relationships not explicitly accounted for in the model, and thus to assess the resultant impact on the employment trends.

In order to be in a position to analyze the effects of the diffusion of industrial robots until 1995, it is necessary to simulate economic development to 1995 without the diffusion of this new technology (reference scenario). The effects of the introduction of industrial robots are then measured by the difference between the two simulations. In the dynamic model the simulation requires the exogenous input of values for final demand in the different sectors until 1995. For this purpose a prognosis - made by the DIW in collaboration with experts from a wide range of firms - for the trend development of private consumption and exports was used. The most important values for economic development were the same in our reference scenario as in the DIW's prognosis (cf. Blazejczak in this volume). As no information is available on the future development of the parameter matrices, they remain constant from 1984 onwards in the reference scenario. This means that as far as the technological structure of the economy is concerned the reference scenario is "frozen" at the situation prevailing in 1984. The structure of gross output, investment and employment, however, continues to change over time due to the uneven growth in final demand in different sectors. Thus the employment effects of the diffusion of industrial robots is measured against an economy in which after 1984 no further technological changes (except those concerning industrial robots) takes place.

Our point of departure is the assumption that the goods embodying the new technology are produced in an additional sector separate from the existing structure of sectors. The cost structure of this new-technology production sector is determined by the extent to which it purchases intermediate and capital goods from other sectors and the amount of labor of different categories it employs. The new technology leads to changes in the production process in user branches, the extent of the changes depending on the degree of diffusion in the respective branch. Again depending on the use of the new technology over time in a given branch changes will be induced in the quantity and quality of the intermediate goods and labor employed in production. The level and structure of the capital stock is also subject to change in the course of the diffusion process due both to investment in new technology and the absence of reinvestment in the old technology which has been replaced.

3.1 The Diffusion of Industrial Robots

The stocks of industrial robots simulated in the dynamic input-output model will be briefly described, as they provide a means of judging the value of the model of the diffusion process of the new technology over time. According to the results of the simulation, the stock of industrial robots is approx. 7 000 in 1985, climbing to 24 000 units in 1990 and 47 500 in 1995.

As far as can be judged from the data now available the forecasts for the stocks in 1990 and 1995 appear realistic. They fall between the values forecast by Hansmann/Roggon as a minimum and Schünemann/Bruns as a maximum value (Hansmann/Roggon 1984 and Schünemann/Bruns 1986).

Table 3.1: A Comparison of Forecasts of the Diffusion of Industrial Robots in the Federal Republic of Germany

	stock of industrial robots	
	1990	1995
Hansmann, Roggon	20,400	ca. 35,000
Schünemann, Bruns	29,914	ca. 50,000
Volkholz	39,160	-
own prognosis	24,000	47,500

sources: Hansmann/Roggon 1984; Schünemann/Bruns 1986; Volkholz 1982; own calculations

Based on the actual trend up to 1987 it can be said with considerable certainty that the prognosis made by Hansmann/Roggon for 1990 is too low while that of Volkholz is definitely too high.

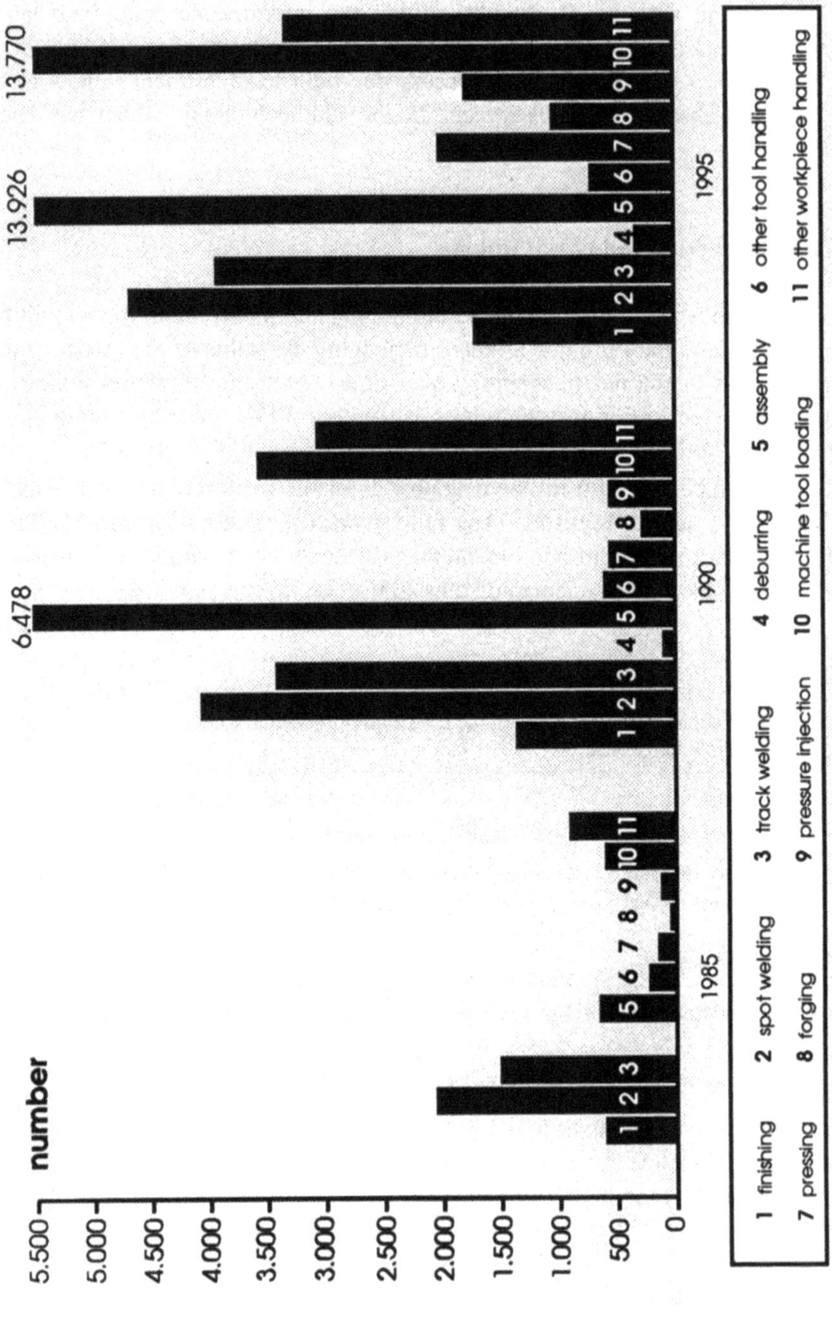

Figure 3.1: Simulated Stock of Robots (1985, 1990 and 1995)
- by application -

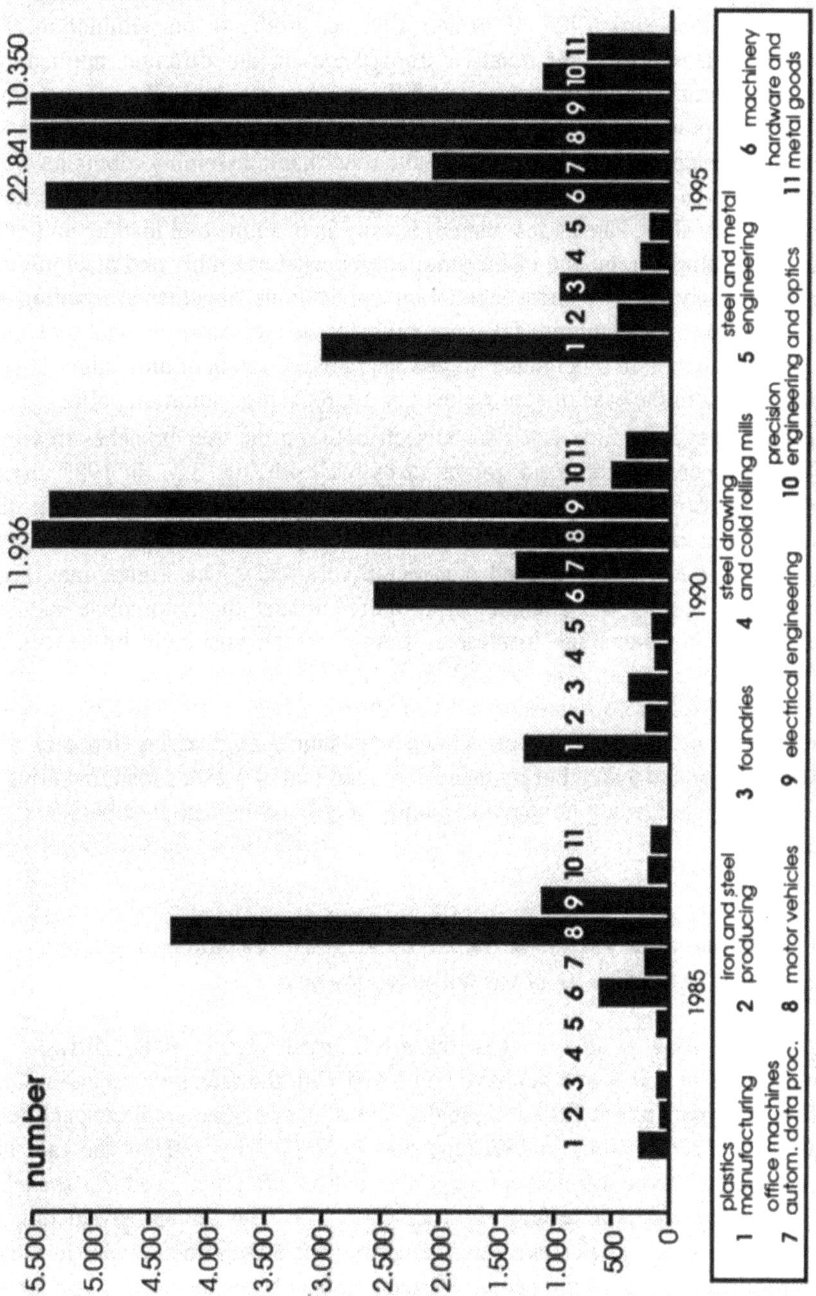

Figure 3.2: Simulated Stock of Robots (1985, 1990 and 1995)
- by user branch -

Fig. 3.1 provides an overview of the forecast stock of industrial robots by application for the years 1985, 1990 and 1995 according to our simulation. The figure shows clearly how the relative importance of the different applications changes considerably in the course of the diffusion process. In 1985 spot and track welding clearly predominate, whereas in assembly, for example, the use of industrial robots is very modest indeed. By 1990, on the other hand, assembly robots have the highest installation figures, followed by spot welding whose rate of growth, however, is relatively slow. The largest increases were in machine tool loading and other workpiece handling. At the end of the simulation period assembly and machine-tool loading are clearly the two most significant applications, together accounting for 60% of total robot installations. They are followed at a distance by spot and track welding robots, which in 1995 make up less than 20%. Growth of unit values is very slow and, at least in the case of spot welders, is approaching saturation point.

Clear changes over time can also be seen between the user branches in which the diffusion process of industrial robots takes place (cf. fig. 3.2). In 1985 around 60% of all industrial robots are installed in the automobile industry. While this remains the most important branch throughout the entire simulation period, by 1995 its predominance has been reduced considerably to 48%. This figure may seem somewhat high, but in the dynamic input-output model the automobile industry experiences an above-average increase in output, which positively influences the rate of robot installation. The second most important user branch is electrical engineering which increases its share of the total from 16% in 1985 to 22% in 1995. The proportion of industrial robots in the mechanical engineering industry also increases (from 9% to 11%), but by rather less than that of plastics manufacturing in which, by 1995 according to our simulation, 6.4% of industrial robots will be installed.

3.2 The Employment Effects of Industrial Robots in Producer and User Branches in the Course of the Diffusion Process

Taking the sum total in all branches, the employment effects of the diffusion of industrial robots are clear and negative compared with the reference scenario. Only in the first two years of the diffusion process can minor positive employment effects be registered; 1 200 persons in 1980 and 500 in 1981. This reflects the fact that during the course of the diffusion process the positive effects in producer branches occur before the negative effects in user branches. The annual production of industrial robots leads to positive direct employment effects in the year in which they are produced, whereas the negative effects in user branches usually last for the entire working life of the robots. The dynamic "tug-of-war" between these two processes for each investment year produces overall negative employment effects from 1983 onward, although these effects are very modest in the first few years. As the diffusion process gathers pace they increase rapidly to -48 000 in 1990 and almost -110 000 in 1995. The employment trend compared to the reference scenario is shown in fig. 3.3. The upper diagram illustrates the contrast between the employment trend with and without industrial robots, while the lower diagram shows the annual change in employment compared with the previous year; it has been included merely to illustrate the structure of the employment effects simulated by the dynamic input-output model over time. When considering the employment figures shown in fig. 3.3 it should be noted that at the present stage of development of the dynamic input-output model not all of the compensatory effects induced by a new technology can be incorporated into the model. It will therefore be necessary in the following to test for the extent to which certain plausible assumptions about compensatory effects (in particular demand effects in user branches) will affect the overall result.

It is interesting to see the degree to which the overall employment effects are due to the different causal relationships within the input-output model. They can be divided into four categories as follows. Direct producer effects is the term applied to those direct employment effects which occur in the production of industrial robots in the "robot sector". Indirect producer effects encompass the effects set in train in other sectors of the economy by the production of industrial robots via the exchange of intermediate and investment goods. Indirect user effects account for changes in the purchase of intermediate goods and the displacement effects of investment in old technology. Direct user effects are defined as the employment effects arising directly from the introduction of industrial robots.

Fig. 3.4 shows the relative importance of the indirect and direct effects. The negative consequences of the user effects gain rapidly in importance and dominate the picture from 1983 onwards, increasing monotonically to approx. 60 000 in 1990 and 130 000 in 1995, the last year of the simulation. The producer effects also increase, but with fluctuations due to changes in production and in the demand for investment e.g. in the robot sector, from 2 000 persons in 1980 to approx. 20 000 at the end of the simulation period.

In order to illustrate the extent of the effects and the way they change in the course of the diffusion process the positive and negative employment effects have been added together to produce a gross effect. In 1980 direct user effects are assumed not to occur as this is merely an introductory period: the direct producer effects account for 40%, the indirect producer effects for about a third and the indirect user effects about a quarter of the gross value.

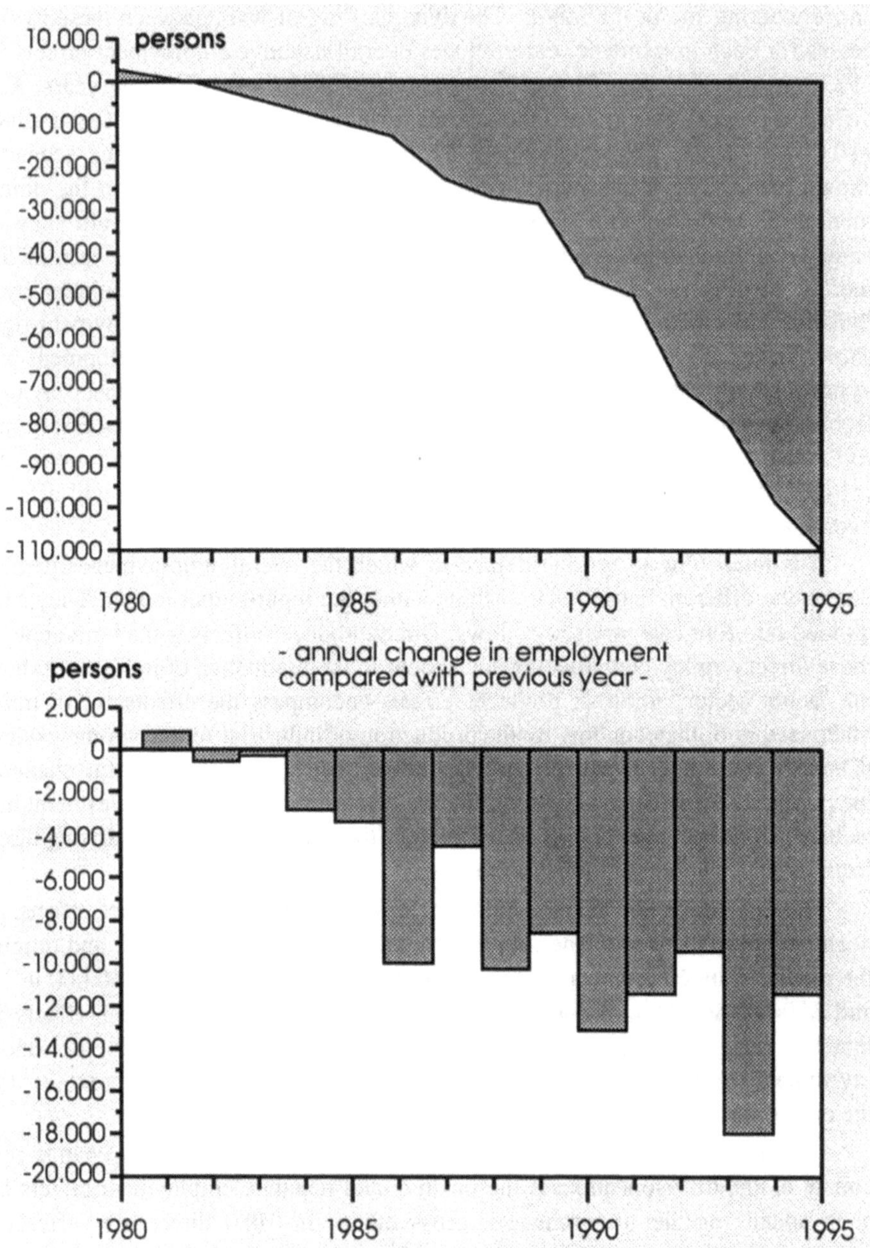

Figure 3.3: Change in Total Employment Due to the Diffusion of Industrial Robots

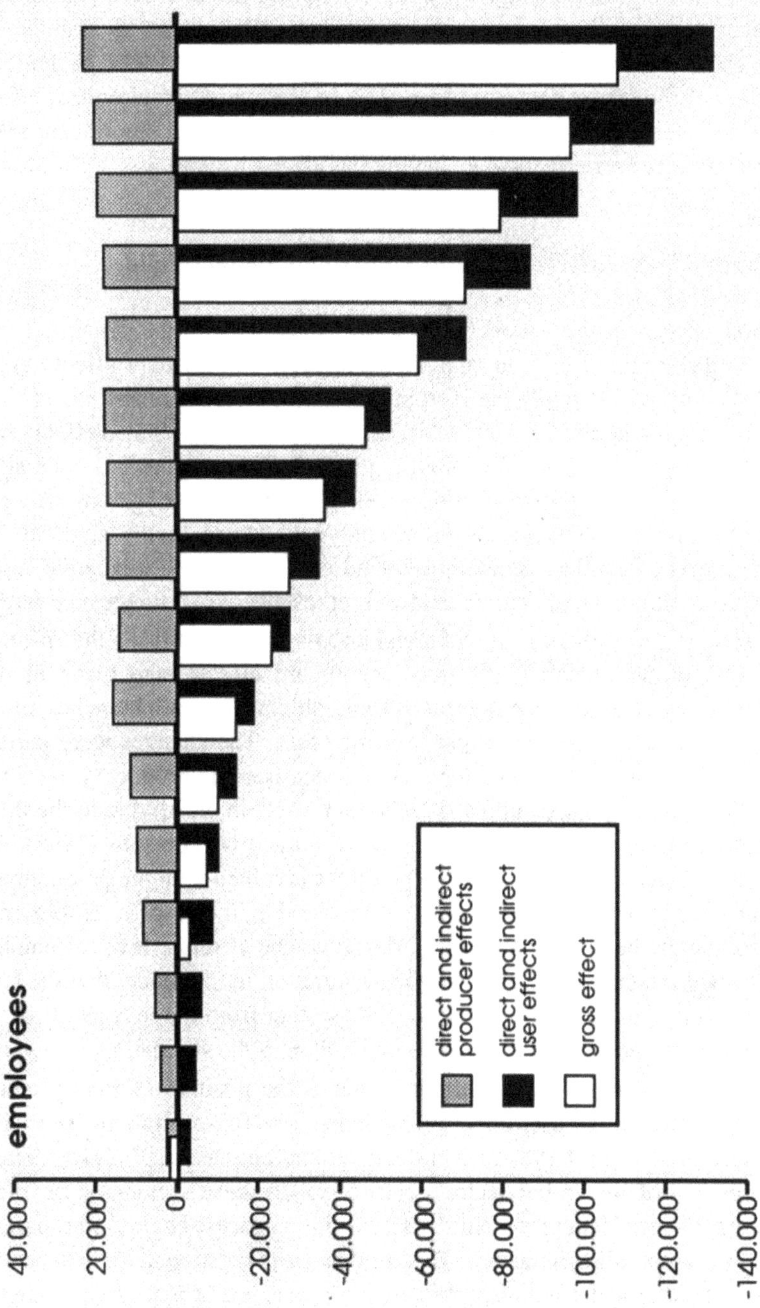

Figure 3.4: Employment Effects of the Diffusion of Industrial Robots
- by component category -

These relative weighting change very rapidly during the diffusion process. By 1985 the direct and indirect producer effects taken together make up slightly less than a third, while the direct user effects have increased to almost 60% of the total. In the final year of the simulation more than 80% of the overall employment effects are due to direct user effects. Direct and indirect producer effects account for about 7% each, with indirect user effects (4%) having the lowest weighting.

3.3 The Employment Effects by Sector and Occupational Group

The simulation of the diffusion process of industrial robots with the dynamic input-output model allows us not only to analyze the overall employment effects of the new technology, but also to study the sectoral and occupational dimensions of these effects. The model guarantees absolute consistency of the disaggregated effects with the overall result at all stages of the dynamic process. This emphasizes once again the importance of the input-output method as an instrument ensuring consistency in overcoming the distance between different analytical levels in the study of the employment effects of modern technology. This distinguishes the dynamic input-output model from partial analytical models and represents one of its great strengths.

The effects of the diffusion of industrial robots vary greatly in the different sectors. As was to be expected, the most significant effects take place in user branches, but, indirect employment effects were observed in all branches of the economy, although they were very slight in some cases. The changes were particularly pronounced in those sectors in which the indirect user and producer effects pull in the same direction. A good example is "other services" in which due to the diffusion of industrial robots employment increases by some 1 000 persons (1995). The largest positive effects are concentrated in the sector involved with the production of the new technology, which experiences a considerable increase in employment during the course of the diffusion process. As far as the absolute level of employment is concerned it is necessary to recall the assumption in the model that the labor coefficients remain constant at the level of 1984, so that from the mid 1980s on the actual employment trend would certainly be lower than in the simulation.

It is interesting to note that in certain branches the positive indirect producer effects initially outweigh the user effects. This is the case for example in "precision-engineering and optics" until 1984 and in the electrical engineering industry, one of the major suppliers of the "robot sector", until 1983. In these sectors, as in "office machines and automatic data processing", the diffusion process begins later than, for example, in the automobile industry. This is due to the differences between the various types of robot and their phased diffusion processes: it is in the automobile industry that the majority of spot and track welding robots are installed, whose diffusion process begins very early. The automobile industry is also the branch with the largest negative employment effects over the entire simulation period, accounting for 60% of all job losses in the final year of the simulation. The values in this branch are probably rather too high as it was assumed that spot welding robots would replace hand spot welding machines, whereas in actual fact the substitution partially affected multi-spot welding equipment, with only a slight labor-displacing effect (cf section 2). The automobile industry is followed at a considerable distance by the electrical and mechanical engineering industries. However, if the "robot sector", which in the model was considered as a separate sector, is included in the mechanical engineering industry as in the official statistics, it would experience an overall positive employment effect until 1990, and even in 1995 the negative effects would, at 5 000, be almost negligible.

The analysis of the occupational dimension of the employment effects also reveals a wide spectrum of effects. Due to indirect producer and user effects, most occupational groups experience slight improvements in their (quantitative) employment situation. For example the number of office workers throughout the economy increases by 1 000 (1995), although such workers are not directly affected by the introduction of industrial robots. In absolute terms fitters and electricians benefit most from the new technology with 10 000 and 9 000 additional jobs in 1995. This is a result not only of producer effects but also the increase in maintenance work in user enterprises. Our expert interviews indicated that those who take on the new jobs will require a considerably higher level of skills and qualifications than is usual in these occupational groups at present. However, the relative increase in these groups, i.e. compared with the absolute number employed in such jobs in the reference scenario are, at 1% and 1.5% respectively, relatively small. The greatest relative increase is for data-processing specialists at 1.7%, with clear employment increases for highly qualified occupational groups of engineers and technicians, particularly in electrical and mechanical engineering.

In the case of those groups negatively affected by the diffusion of industrial robots allowance must be made for the fact that in some departments of user enterprises it is often difficult to ascribe negative employment effects to a particular occupational group with any degree of certainty. This uncertainty should be borne in mind when evaluating the results presented here. The greatest job losses are suffered by the "welders" with around 30 000 less jobs by 1995. If, however, it is assumed that in the automobile industry spot welding robots replace multi-spot welders rather than hand welding machines, the employment reduction in this occupational group is only half as great. The welders are followed by the following three groups: "metal processors with unspecified activity", "electrical and other assemblers", "laborers without special job requirement", all of which are characterized by a low skill level. However, skilled workers are also negatively affected: an example is "machine tool operators", although this group suffers only a loss of only 4% of its employment in 1995. The most serious relative effects (i.e. measured against the group-specific level of employment in the reference scenario) are experienced by "machine operators with unspecified activity" with losses of 25% in 1995, "welders" and also "metal processors with unspecified activity", the "laborers without special job requirement" and "electrical and other assemblers".

As a general conclusion from our studies in this area it can be seen that even one isolated form of new technology induces large-scale shifts in the sectoral and particularly in the group-specific composition of the future structure of employment. In view of the large number of new technologies, some of which have not yet begun their process of diffusion within the economy, it is clear that considerable efforts will have to be made in the field of labor-market policy in order to overcome the problems arising from such structural changes in the labor market.

3.4 Additional Compensatory Effects and their Possible Impact on Employment

The model calculation described in the previous sections include all direct and indirect effects on output, investment and employment induced by changes in the input of intermediate goods, capital goods and labor. The causal relationships which are mediated by price, income and redistribution effects have so far not been taken into consideration. Since at the current state of empirical and theoretical knowledge it is not possible to endogenize these effects within the model, a simulation calculation was made in order to try and obtain a rough idea of their order of magnitude.

The introduction of a new technology leads to changes in the cost structure of the production process in user branches. In the case of industrial robots the cost shares of intermediate and capital goods increase slightly, while labor costs, due to the displacement effects of the new technology, sink to a greater extent, so that overall costs of production are lower than in the reference scenario. The cost savings shown in table 3.2 can only be considered estimates since, although they are based on gross income differentiated by sector, the average income of the specific groups of workers displaced from the production process by the new technology remains unknown.

Table 3.2: Changes in Cost Structure due to the Diffusion of Industrial Robots in User Branches Compared with the Reference Scenario
- changes in % of the value of output -

	1985	1990	1995
08 plastics manufacturing	-0.056	-0.209	-0.524
13 iron and steel producing	-0.008	-0.045	-0.126
15 foundries	-0.062	-0.251	-0.699
16 steel drawing and cold rolling mills	-0.006	-0.028	-0.072
17 steel and metal engineering	-0.022	-0.066	-0.081
18 mechanical engineering	-0.036	-0.158	-0.297
19 office machines, automatic data processing	-0.095	-0.333	-0.492
20 motor vehicles	-0.235	-0.663	-1.178
23 electrical engineering	-0.058	-0.240	-0.458
24 precision engineering and optics	-0.035	-0.174	-0.352
25 hardware and metal goods	-0.021	-0.080	-0.156

source: input-output calculations by the DIW

Because empirically tested, sector-specific values for these elasticities are not available it was necessary to make two simplifying assumptions in order to be able to estimate the possible consequences for changes in prices and demand if these cost reductions were to be passed on in the form of lower prices. The first is that the cost reductions in the different user branches are passed on in full, leading to corresponding reductions in the prices of goods produced. The second is that these price reductions lead to a corresponding increase in the demand for these goods, i.e. their elasticity equals unity. In our view this represents the upper limit of the compensatory price and demand effects which may be expected, as the actual elasticities, both of prices in respect of costs and of demand with respect to prices are likely to be less than 1 in the majority of branches. Alternatively, if the cost saving are not passed on to their full extent, firms in the user branches benefit from higher incomes, which will then be divided in some way between wages and profits. Depending on the way this distribution process takes place and on the different income elasticities of demand of those drawing the two forms of income, a further round of demand effects will be induced in other sectors. Because at the present state of research it is not yet possible to model the distribution or the resulting demand effects, no attempt was made to analyze possible second-order compensatory effects of this type.

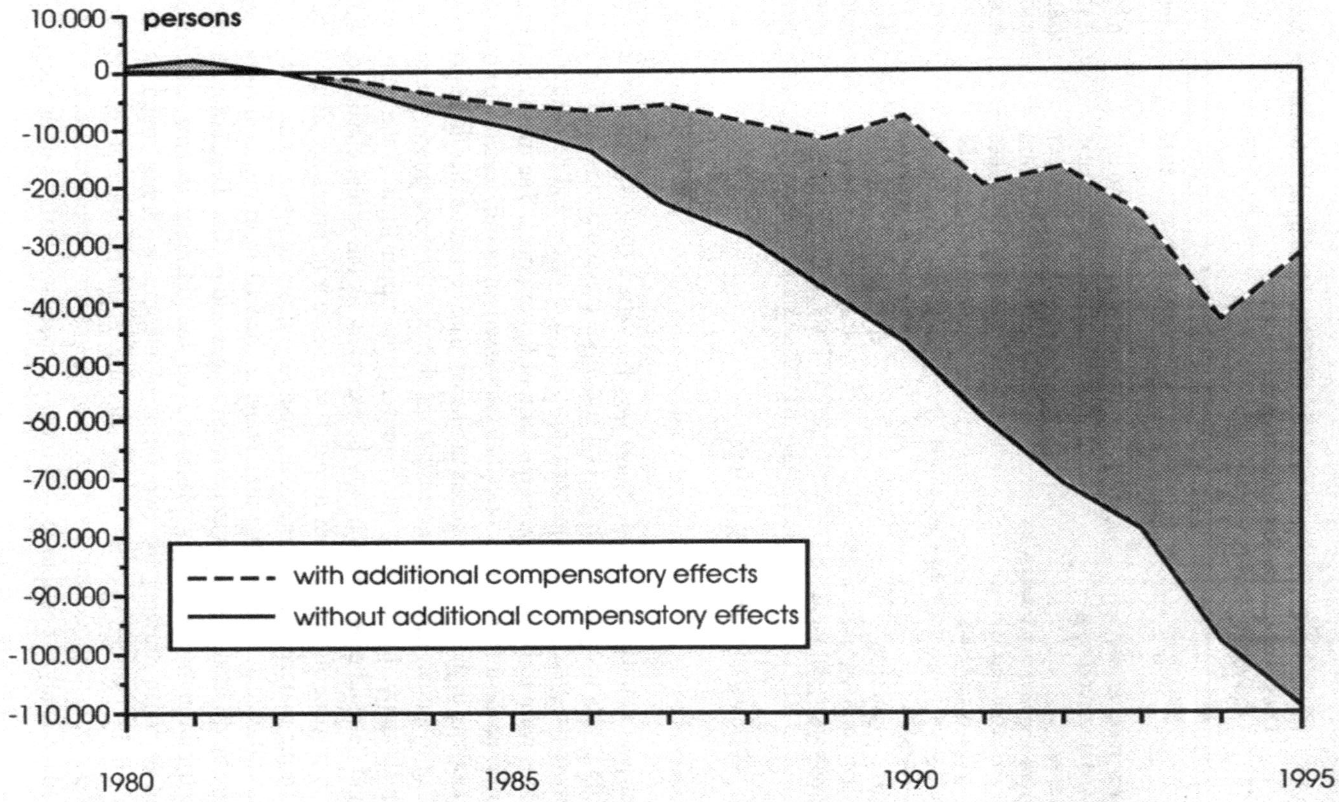

Figure 3.5: Employment Effects of the Diffusion of Industrial Robots
- with and without additional compensatory effects -

Table 3.3: Employment Effects of the Diffusion of Industrial Robots with Additional Compensatory Effects
- Change in per cent of Employment in Baseline Projection -

Change by sectors

		1985	1990	1995
08	plastics manufacturing	-0.15	-0.54	-1.41
13	iron and steel producing	+0.04	+0.18	+0.26
15	foundries	-0.01	+0.06	-0.41
16	steel drawing and cold rolling mills	+0.07	+0.31	+0.49
17	steel and metal engineering	-0.05	-0.01	+0.31
18	machinery	-0.18	-0.10	-0.13
19	office machines, automatic data proc.	-0.32	-1.03	-1.58
20	motor vehicles	-0.75	-1.78	-3.20
23	electrical engineering	-0.07	-0.22	-0.68
24	precision eng. and optics	-0.01	-0.12	-0.40
25	hardware and metal goods	-0.02	+0.01	-0.08

Change by occupational categories

		1985	1990	1995
31	plastics product makers	-0.11	-0.41	-1.26
35	machine tool operators	-0.03	-0.55	-3.22
37	welders	-7.89	-17.41	-18.90
38	metal processers n.e.c.*	-0.35	-2.23	-7.84
41	mechanics	+0.21	+0.87	+1.54
45	electricians	+0.25	+0.94	+1.72
47	electrical and other assemblers	-0.10	-1.95	-4.73
56	painters	-0.81	-1.47	-1.55
103	data-processing experts	+0.43	+1.21	+1.79
107	mechanical engineers and -technicians	+0.22	+0.77	+1.18
108	electrical engineers and -technicians	+0.18	+0.65	+0.95
114	packers and wrappers	-0.32	-1.47	-1.40
116	labourers without special job requir.	-0.46	-2.42	-7.78
120	machine operators n.e.c.*	-2.24	-10.41	-25.06

*n.e.c.: not elsewhere classified

Source: Input-Output-Analysis by DIW.

As was to be expected, the simulation calculation with the dynamic input-output model on the above assumptions concerning price and demand elasticities produces a considerable reduction in the negative employment effects resulting from the diffusion of industrial robots. Fig. 3.5 illustrates the course of the employment trend with and without the compensatory effects described above.

Even when these compensatory effects (which are likely to be overestimates) are taken into consideration the overall effect of the introduction and diffusion of industrial robots on employment remains negative, amounting to 35 000 persons in 1995. Even allowing for the exclusion of certain effects, such as on product quality, and the flexibilization of the production process, which are difficult to quantify using the present method, the overall employment effects of the diffusion of industrial robots would seem to be negative. The probable extent of these effects will lie somewhere in the shaded area of fig. 3.5, i.e. between the lines indicating employment with and without additional compensatory effects. Whatever the precise quantitative effects, one thing is certain: the diffusion of industrial robots will result in changes in the sectoral and occupational structure of employment. Table 3.3 shows the changes in employment in user branches and for selected occupational groups, including the compensatory effects compared with the reference scenario.

A comparison with the calculations without additional compensatory effects shows that some of the user branches with only a low level of diffusion of industrial robots enjoy a growth in employment, albeit a small one. In such branches the indirect producer effects predominate over the direct user effects throughout the simulation period. The largest fall in employment, both relatively and absolutely, will, however, still occur in the automobile industry.

The changes in occupational structure, on the other hand, are much more significant. Even when the additional compensatory effects are taken into account some occupational groups are subject to a considerable impact from the diffusion of industrial robots. This applies in particular to machine operators, welders and unskilled workers especially in the metal industries. Thus the inclusion of compensatory effects does not lead us to retract the claim made above that considerable changes in occupational structure will occur in the future as the result of just this one form of modern technology.

TANDEM: Simulations within the Innovation-Growth-Employment-Circuit

Werner Frühstück and Michael Wagner

The introduction of new technology is closely linked to a broad range of innovative activities, which represent one of the major sources of economic growth for the private sector in the Federal Republic of Germany. Competition, both at home and from abroad, stimulates firms to pick up on innovative impulses and to pass them on to other firms. The result is an "endogenous" circuit which guarantees a high level of innovative activity in the West German economy, even when there is a dearth of powerful "exogenous" stimuli such as radical new basic inventions.

Within this structure of endogenous innovation dynamics the labor market occupies a special place, since the implementation of new technology requires the ability and willingness to restructure job profiles at all levels, be they directing or operative activities. The increase in output which this implies is matched by improved skills and qualifications and growth in real wages. At the same time the demand for units of labor time is reduced, leading to unemployment in some sectors. Under such conditions even the high degree of occupational mobility of wage and salary earners in the Federal Republic is insufficient to prevent unemployment. A compensation of this trend can only occur through an acceleration of GDP growth or a redistribution of the volume of labor-time between different groups of workers.

The data bases described in this volume, some of which have been developed for the first time within the framework of the Meta Study, and the wide range of available econometric models offer the opportunity to make quantitative estimates of the dynamics of the innovation-circuit, and of the labor market effects connected to it. This is the aim of Tandem as a simulation model; to provide insights into the working properties of this circuit under changing external conditions. The Tandem model described in this paper is taken from a more comprehensive research context, the aim of which is to develop a quantitative framework in order to facilitate heuristic simulations of innovation dynamics in the Federal Republic of Germany.

1. Functional Relationships

The structure of Tandem is modular: each area of the innovation-growth-employment-circuit is modeled independently. The specifications required for this are based, both, on an explicit formal model (Hanappi/Wagner 1989), and on a number of assumptions and theoretical considerations which can be illustrated in the form of six hypotheses. The hypotheses relate to the selection of the central variables, the functional relationships between the variables and the direction and strength of the causal chains.

This section provides a summary of the theoretical considerations and empirical evidence for each of the six hypotheses. These are then supplemented by our results from the empirical specification of Tandem.

Hypothesis 1: Innovative Activity

The innovative activity necessary for the introduction of new technology is one of the stable characteristics of firms and branches which are guided by the same market signals as govern decisions on investment (such as relative factor prices, return on capital and sales expectations).

The hypothesis of the relative stability of branch-specific innovative activity is based on the view that firms invest in activities aimed at innovation (particularly R&D) as a standard instrument to secure the economic success of the firm. They therefore try to ensure that they do not lose out in the struggle with their competitors for an advantageous place within the framework of monopolistic competition (Dasgupta/Stiglitz 1988). In addition, the required complementarity between technical and organizational adjustment produces a tendency to inertia regarding innovative behavior (Cyert/Mowery 1987). This has been confirmed by empirical studies (Maresse/Siu 1984) which have shown, for example, that R&D activity is considerably more stable than investment.

In Tandem the stability of innovative behavior is represented by an autoregressive component in the "innovation function". This specification has proved empirically reliable. The sectors "electrical engineering" and "pharmaceuticals" in particular are characterized by a pronounced endogenous dynamic in this area.

The view that innovative activity is influenced by the same factors that affect investment behavior has been suggested by Schmookler (Schmookler 1962). It also appears reasonable from a theoretical perspective insofar as know-how is considered to be a sort of capital *sui generis*, a view which, however, will be opposed by economists who regard the aggregation even of real capital with scepticism.

In Tandem this hypothesis enables us to take investment as a proxy variable for the values of the costs of and returns to innovative activity. It appears that sectors such as the engineering or automobile industries exhibit a relatively high short-term sensitivity to changes in these cost and return variables.

The econometric estimates for Tandem (conducted with the help of Ifo's innovation time-series and with R&D data prepared by the DIW) do not seem to contradict the central proposition of Hypothesis 1. In this sense it provides a useful approach for specifying Tandem in a way which is subject to empirical falsification.

Hypothesis 2: Productivity and Demand Effects

Innovative activity leads to an increase in factor productivity (especially that of labor), while, within the framework of quality competition it also reduces the price elasticity of demand for goods sold on those markets in which the innovative firm offers its products.

Innovative activity arises as part of the normal processes of market competition. Its aim is to reduce costs and increase revenues. This has been confirmed by the IFO Innovation and Business-cycle Test for all branches of manufacturing industry in the Federal Republic of Germany (IFO 1988).

In order to measure the effects of innovations on production it is necessary to construct a stock value corresponding to innovative activity. A useful approach is the concept of a "stock of innovative knowledge" which is derived from the cumulative aggregation of current expenditure on innovation (making an allowance for depreciation). In this way "innovative know-how" can be incorporated into the traditional theory of the firm, thus facilitating the specification of econometric equations.

This is in fact the method adopted by Tandem. Current expenditure on innovation is aggregated to a "know-how stock", changes in which influence the "optimal" production technique at a given point in time.

The economic effects of new technology are represented in Tandem by the relation between labor input and the volume of output, i.e. the level of the innovation-capital stock systematically influences the level of labor productivity. This can result either from organizational/technical adjustment at enterprise level or from revenue increases determined by the market. This method takes both product and process innovation into account.

Empirically this approach proved to be very satisfactory for the West German case. Econometric estimates indicate a significant relationship between the innovation-capital stock and labor productivity (controlling for relative factor prices and differences in the degree of capacity utilization in each industry). This result is particularly evident for manufacturing industry, but can also be found in the service sector, especially in retail and wholesale trade, communications (Federal Post Office) and "other services". The trend was less clear for the banking and insurance sector, an outcome which corresponds to the detailed branch-level study conducted by Infratest (1989).

Although our approach stood up well to empirical estimates, some critical remarks need to be made and borne in mind. They relate primarily to the following points: the extent to which innovative know-how can be adopted by specific firms or branches; the rate of "depreciation" of stocks of knowledge; the choice of a price index to measure real expenditure on R&D; the simultaneity problem of market structure (price elasticities) and innovative activity.

As far as the adoption of innovative knowledge is concerned, the point has been made by classical economists that pioneers must expect to see their surplus-profits decline over time because of competition with imitators. For this reason there is a considerable difference between the social and private benefits of innovation (Arrow 1962). In addition recent studies have shown marked differences in the extent to which applied R&D can be adopted in different branches (Levin et al. 1984).

The insufficient "appropriability" of innovative knowledge influences the value (and so implicitly also the "depreciation") of the innovation-capital stock. Depending on the estimates for the extent of such imitation and replacement effects, various studies have arrived at widely differing values for the implicit annual "rate of obsolescence" for stocks of knowledge. They range from about 25% (Schankermann/ Nadiri 1984), 15% (Frascati Manual of the OECD), under 10% (Griliches 1980) to 5% (Schadt 1978) and lower (Mansfield 1968). For empirical estimates and simulations with Tandem implicit obsolescence rates of between 0% and 20% were experimented with. The use of the lower rates produced the more plausible results.

Similarly the problem of accounting for changes in the value of stocks of knowledge proves to be thorny (Mansfield 1984). For Tandem the GDP deflator is used to obtain real values for innovation expenditure. This makes the results comparable with those of the DIW (1988).

Finally there is the simultaneity problem of "market structure" and "innovation". The most recent studies for the Federal Republic of Germany (ISG 1988) confirm international evidence that there is a positive correlation between market concentration and innovative activity. In theoretical works (Dasgupta/Stiglitz) this relationship is considered as characteristic of a simultaneous process. In Tandem, on the other hand, it is implicitly assumed that innovative activity is the driving force behind the reductions in the price elasticity of demand for the goods of innovative firms.

Without losing sight of the problems of empirical measurement and theoretical conceptualization (not least the question whether "knowledge" can be aggregated at all into a stock of capital), we consider the approach selected for use with Tandem to be an acceptable compromise: the perspective described in Hypothesis 2 has proved effective.

Hypothesis 3: The Diffusion of the Effects of New Technology

The stocks of knowledge which are accumulated during the development of new technology exert a direct influence on the firms in which the innovative activity was undertaken. At the same time such activity also influences the production technology of firms which purchase investment or intermediary goods from innovative firms and sectors. Thus the diffusion of the effects of new technology is determined by the specific dynamic of two factors: accumulation and absorption.

The fundamental idea behind Hypothesis 3 has been expressed in the view held by classical political economy that the division of labor allows greater specialization which in turn increases the productivity of individual activities and the economy as a whole. This idea has been taken up by the "new growth theories" (Roemer 1987). It is argued that the differentiation of input factors (product innovation) results in increased economies of scale in producer firms (process innovation) leading to an overall increase in GDP.

The significance of the transfer of innovative impulses via the exchange of goods between sectors has been empirically confirmed for the Federal Republic of Germany by the studies conducted by IFO (1988) and by the DIW (1988). They indicate that there is a considerable overlap between the "accumulating" (contributing) and the "absorbing" (recipient) sectors, although they are not fully congruent. The engineering industry, for example, contributes much more than it absorbs, whereas large sectors in the service sector (e.g. banking) absorb considerable amounts of innovative activity, without passing on innovations of their own to anything like the same extent.

In Tandem (version 2) this feedback mechanism between accumulation and absorption is included in the flow of investment goods. The greater the innovative activity in the sectors producing investment goods, the greater the accumulated innovation-capital stock. However this stock can only be absorbed if firms in other sectors carry out investment. Thus a necessary condition for the diffusion of the effects of innovative activity is a sufficiently high level of fixed capital formation.

In this respect the dynamic which has been described on the basis of the investment-good-flow matrix matches the conception of "vintage" models, in which marginal capital productivity is always higher than the average value (Jorgensen

1988). However, such "ageing" effects are due not only to technical progress which is embodied in and transmitted by investment goods: productivity growth also results from the external effects arising from the development of human capital (Prescott/Boyd 1987).

These synergy effects are taken up by Tandem in the form of "autonomous" components in the structure of the innovation-capital stock. Increases in the stock of knowledge are considered to be "autonomous" if they can be explained neither by the causal chain "innovation-investment" nor by the exchange of investment goods. Empirically, there is a close relationship between the two forms of innovation-induced effects (exogenous and endogenous), as has been emphasized by Metcalfe (1987). Working along these lines Hypothesis 3 provides a satisfactory approach for constructing an empirically-based model of the diffusion of innovation.

Hypothesis 4: Direct Labor Market Effects

For a given level of output the use of new technologies reduces the required input of labor (working hours). The hypothesis of the rationalization effect of new technology, which has often been empirically tested, is frequently the point of departure for controversies over economic policy. However, it has been established that at both the micro and macro levels innovative activity tends to be labor saving. Kugler/Müller/Sheldon (1989) at the macro level and Flaig/Stadler (1989) from a micro perspective have again shown this to be the case in the Federal Republic. However, the extent of this labor-saving bias varies considerably between the different branches: the latest studies from Infratest (1988), for example, have indicated only limited effects in the banking sector. Cyert/Mowery (1987) have found a high degree of sectoral differentiation of the degree of the labor saving bias in the USA.

The sectoral estimates of the "productivity function" (labor productivity) for Tandem is in accordance with the findings of these studies. Large rationalization effects are particularly characteristic of manufacturing industry (e.g. mechanical and electrical engineering), but even in wholesale and retail trade innovations lead to a considerable reduction in the demand for working hours for a given level of output.

The extent to which direct rationalization effects can be compensated by an increase in employment opportunities due to an expansion of demand (because of the price reductions made possible by innovation) cannot be analyzed within the framework of the Tandem model. Leaving the problems of measurement aside (Oppenländer 1987) it can be assumed on the basis of model calculations that the direct rationalization effects are only seldom fully compensated by corresponding demand effects (DIW 1988). At the same time, there may be a high degree of compensation at the macro level, as has been shown by Cyert/Mowery (1987) for the USA.

Hypothesis 5: Macro-economic Adjustment Processes

Innovative activity results in changes in the techniques of production which in turn induce supply-side shocks which, within the interactive framework provided by the market economy, bring about the following macroeconomic adjustment processes: increased investment, higher real wages, improved terms of trade, export growth, accelerated growth of GDP. These adjustment processes can be identified in numerous macroeconometric models (at least those which are based on an economy with a similar institutional and economic structure as the Federal Republic of Germany). To take an example, they emerge as very robust results from the DIW's Langfrist Model.

Nevertheless, the precise quantitative dimensions of these adjustment effects vary according to the strategies pursued by influential groups and institutions within the economy. The extent of the increase in money-wages associated with the increase in labor productivity, for example, depends on the agreements reached between unions and employers. This in turn (assuming no change in exchange-rate policy by the central bank) influences the terms of trade, on which the profitability of the export trade depends. Such strategic constellations influence the extent to which productivity increases are transformed into positive income or employment effects (DIW 1988). All the same the direction of the adjustment processes remains unaffected by such strategic considerations, so that Hypothesis 5 proved most suitable as part of the modeling approach for Tandem.

Hypothesis 6: Exogenous Shocks - Endogenous Circuits

Innovative activity and its macroeconomic and labor-market effects are partially determined by "exogenous" shocks (scientific progress, inventions with wide-ranging effects) and partially due to the endogenous functional relationship of the circuit "innovation-technology-macroeconomy" shaped by market forces.

In most of the analytical models dealing with the economic effects of technical progress, innovative activity is treated as an exogenous variable. Tandem, by contrast, to take up the call for an "endogenization of innovation" (Boskin 1986, Blattner 1987). In particular a nexus between investment and innovation has been established, a link which is in accordance with the mainstream view on the microeconomics of innovative activity within individual firms.

At the same time, not all innovative activities are derived from an economic circuit in the narrow sense of the term. New technologies are often the fruit of research which is financed out of public funds, whereby the technological possibilities which such research opens, offer incentives for firms in the private sector to attempt to acquire new technologies (Jaffee 1986). Such activity by public bodies is to be considered an exogenous variable relative to the dynamics based on competition driven by profit maximization.

Similar considerations apply to the profile of skills and qualifications of wage and salary earners. Social expenditure on education is relatively independent of short-term profit-and-loss calculations. Nevertheless it influences the extent to which the accumulated know-how in the private sector can be used. The more highly qualified the labor force, the more revenue can be generated by the use of innovative potential. It proved possible and beneficial to incorporate such "exogenous" determinants of innovation into the framework of Tandem. It should, however, be noted that such determinants affect the speed but not the direction of the endogenous functional relationship.

2. The Two Versions of Tandem

Tandem's basic structure consists of three modules which, with the use of suitable interfaces, are bound together into one unified simulation model. They are: innovative activity (module 1), production technology (module 2) and macroeconomic functional relations (module 3). The interfaces are provided by functional relations between the central variables of each module.

In order to model the central relationships between investment, innovation and productivity two versions of Tandem were constructed. The differences between the versions lie in the specification of certain equations, the data bases used and the number of industries considered.

Tandem 1: The Dynamic of Innovation in Manufacturing Industry

Tandem 1 was specified to suit the study of the innovation dynamic in manufacturing industry. The most important functional relationships of this version are illustrated in the block diagram (Fig. 2.1). Investment (both current and lagged) is the central determinant of innovation at the enterprise level. The expenditure on innovation, in turn, (together with "autonomous technical progress") furthers the accumulation of technical knowledge which, *ceteris paribus*, is responsible for productivity growth.

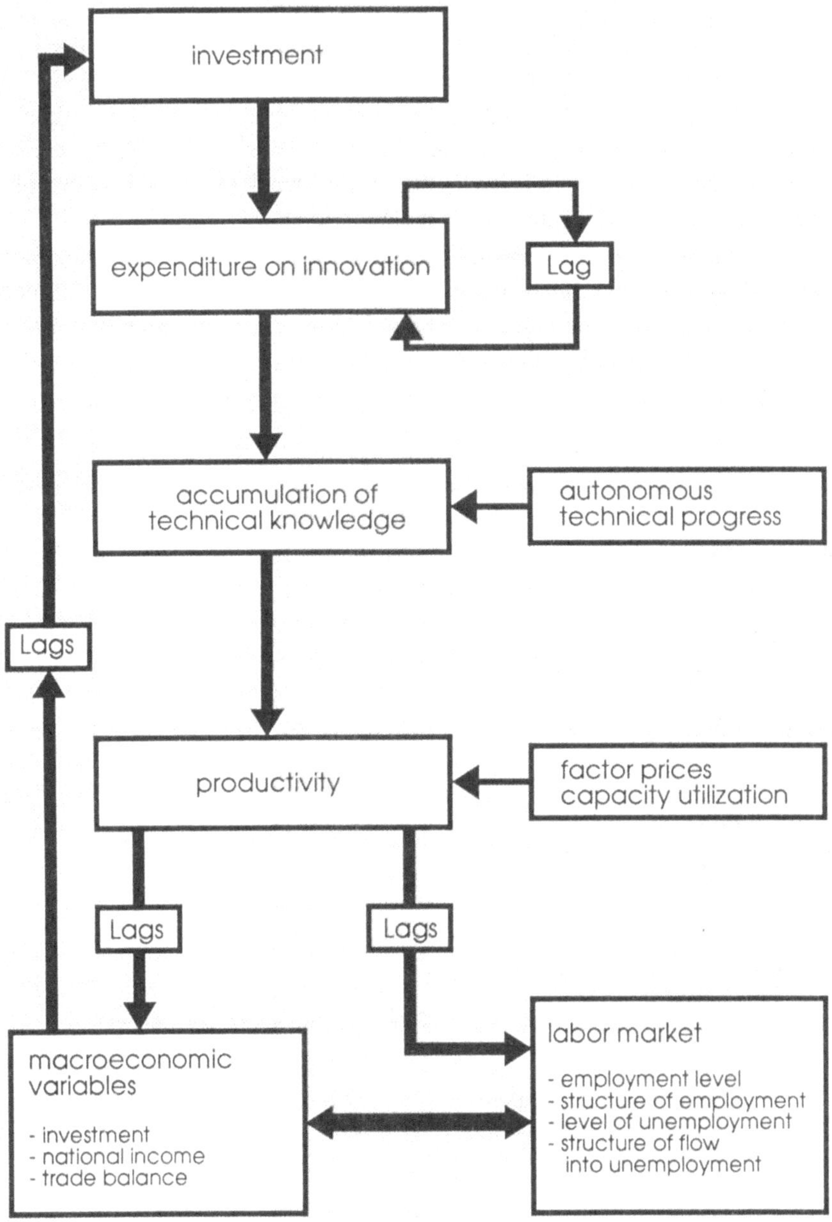

Figure 2.1: Flow Diagram for Tandem 1

Figure 2.1: Flow Diagram for Tandem 1

Industry-specific changes in production technology induce changes in labor productivity at the macro level which in turn (and with a time lag) induce macro-economic adjustment processes. In the model the adjustment processes of the (aggregated) macro variables such as investment, national income, employment, unemployment, and the trade balance are explicitly formulated. The main circuit is closed when the change in productivity exerts a feedback effect on investment. Thus investment plays the major role in the simulation model.

The equations were empirically estimated for 30 sectors of manufacturing industry. Data from the Ifo-Institute in Munic served as the basis for deriving innovation indicators in the model. These indicators are more broadly defined than the data on R&D prepared by the DIW.

Table 2.1: The System of Equations in Tandem 1

innovation function

(2.1) $g(N_{S,t}) = a_{S,1} \cdot g(N_{S,t-1}) + a_{S,2} \cdot g(I_{S,t})$

technical knowledge

(2.2) $g(T_{S,t}) = N_{S,t}/T_{S,t-1} - ob + ta$

productivity function

(2.3) $g(P_{S,t}) = a_{S,3} \cdot g(T_{S,t})$

investment function

(2.4) $g(I_t) = a_4 \cdot g(P_t) + ... + a_7 \cdot g(P_{t-3})$

Key:

g(.)	rate of change of a variable
t	time index
S	sector index (variables without a sector index are aggregate values)
I	investment
N	innovation indicator
T	stock of technical knowledge
ob	rate of obsolescence of technical knowledge
ta	rate of autonomous technical progress
P	labor productivity
$a_{S,i}$	sectoral elasticities
a_i	aggregate elasticities

Tandem 2: Accumulation and Absorption of Technical Knowledge

The model specification of Tandem 2 is more complex than that of Tandem 1. It also conceptualizes the production and diffusion of technical progress in a different form. The engineering (sector 19, cf. table 2.3), automobile (sector 21) and electrical engineering (sector 24) industries are the central areas of accumulation of technical knowledge. (Together they account for about half the total expenditure on R&D in the Federal Republic.) The diffusion of technical knowledge is via investment: each sector absorbs (to an extent measured by a sector-specific weighting) know-how via the capital goods which they "import" from the accumulating sectors. The sector-specific weights are determined by the sectoral level of investment relative to the gross level of output.

Fig. 2.2 illustrates the way in which the central functional relationship was modeled in Tandem 2. Again investment is the central determinant of innovative activity, whereby in Tandem 2 the latter is defined according to expenditure on research and development as measured by the data prepared by the DIW. The expenditure on R&D in turn determines the accumulation of technical knowledge in a given sector. The absorbed technical progress then determines changes in labor productivity in each sector. The macroeconomic effects, the changes on the labor market and the feedback effects to investment in the different sectors are modeled in a similar way to Tandem 1.

Version 2 of Tandem is based on estimates for all sectors of the economy with the exception of the "public sector", "renting residential accommodation", and "non-economically active private households and organizations". This amounts to 48 sectors in all. This expansion of the perspective compared with Tandem 1 helped to include agriculture and forestry, energy and mining, construction, wholesale and retail trade, transport and services. Data on R&D expenditure (prepared by the DIW) were used as innovation indicators: they are more narrowly defined than Ifo's data on innovation expenditure.

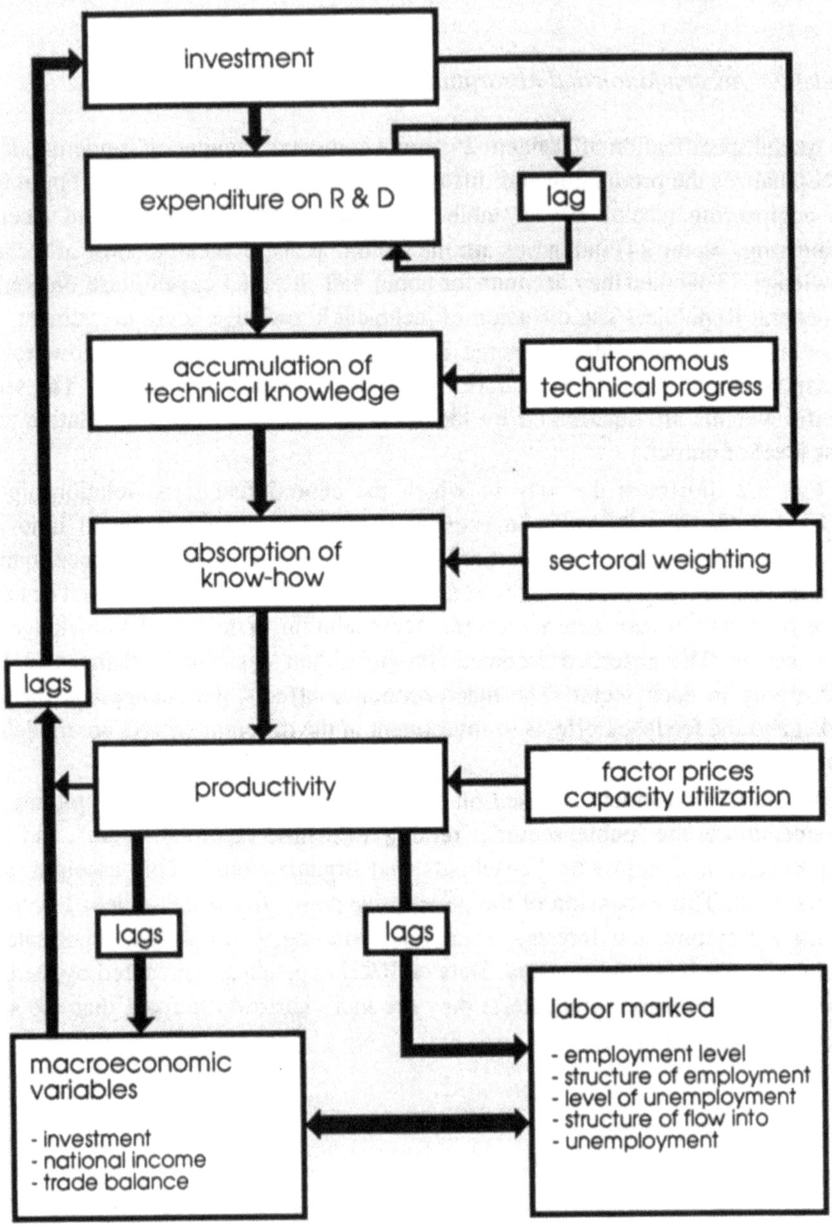

Figure 2.2: Flow Diagram for Tandem 2

Table 2.2: The System of Equations in Tandem 2

innovation function

(9.5) $g(N_{S,t}) = b_{S,1}.g(N_{S,t-1}) + b_{S,2}.g(I_{S,t})$

technical knowledge: accumulation function

(9.6) $g(T_{S,t}) = N_{S,t}/T_{S,t-1} - ob + ta$

technical knowledge: absorption function

(9.7) $HW_{j,t} = (I_{j,t}/B_{j,t}).(\Sigma_S c_{jS}.T_{S,t}/B_{S,t})$

productivity function

(9.8) $g(P_{S,t}) = b_{S,3}.g(HW_{S,t})$

investment function

(9.9) $g(I_t) = b_4.g(P_t)+...+b_7.g(P_{t-3})$

Key:

g(.)	rate of change of a variable
t	time index
S	sector index (variable without a sector index are aggregate values)
N	innovation indicator
ob	rate of obsolescence of technical knowledge
ta	rate of autonomous technical progress
$HW_{j,t}$	absorbed technical knowledge in sector j
$I_{i,t}$	investment in secor j
$B_{j,t}$	gross output in sector j
$T_{S,t}$	accumulated technical knowledge in sector S
c_{jS}	share of the investment that sector j purchases from sector S
P	labor productivity
$b_{S,i}$	sectoral elasticities
b_i	aggregate elasticities

Table 2.3: Sectoral Classification of Tandem 1 and 2

Sector		Tandem 1	Tandem 2
1	agriculture, forestry, fishing		x
energy and mining			
2	electricity supply (incl. heat and power stations)		x
3	gas supply		x
4	water supply		x
5	coal mining, coking plants		x
6	other mining		x
chemical industries			
7	chemical industry and production and processing of fossile materials	x	x
8	crude oil processing	x	x
9	production of plastic goods	x	x
10	rubber processing	x	x
11	quarrying, processing of stone	x	x
12	ceramics	x	x
13	glass production and processing	x	x
iron and non-ferrous metal production			
14	iron producing industry	x	x
15	non-ferrous metal production	x	x
16	foundries	x	x
17	drawing and rolling mills	x	x
18	steel construction	x	x
engineering and automobile industries			
19	engineering	x	x
20	production of office machinery	x	x
21	automotive production	x	x
22	ship building	x	x
23	aeronautical/aerospace	x	x
electrical/electronic			
24	electrical engineering	x	x
25	fine mechanics, optics, watch production	x	x
26	production of sheet metal and metal goods	x	x
27	production of musical instruments, toys,	x	x

Continuation of Table 2.3 (Sectoral Classification of Tandem 1 and 2)

Sector		Tandem 1	Tandem 2
wood and textile products			
28	millwork and wood products	x	x
29	wood processing	x	x
30	pulp and paper production	x	x
31	paper and allied products	x	x
32	printing and copying	x	x
33	textile production	x	x
34	leather and leather products	x	x
35	clothing	x	x
food			
36	food and drink production	x	x
37	tobacco processing		x
construction			
38	construction		x
39	remodelling		x
trade			
40	wholesale trade, intermediaries		x
41	retail trade		x
transport and communications			
42	railways		x
43	shipping, canals, ports		x
44	road and other transport		x
45	communication (Fed. Post Office)		x
services			
46	financial institutions		x
47	insurance		x
48	accommodation rental		
49	other services		x
50	public sector		
51	non-economically active private households and organizations		

3. Innovation and Investment

One of the aims of Tandem is to endogenize the dynamics of innovation. This is accomplished by incorporating an "innovation function" into the model in which the branch-specific innovative activity is made a dependent variable of past innovation and the investment behavior of the industry. By interpreting investment as a proxy variable for the cost and return variables of innovation (Hypothesis 1) market trends and changes in relative prices are included as major determinants of innovative activity, alongside technical and organizational influences (autoregressive components).

The precise specification of the innovation function of Tandem is as follows:

$$(3.1)\quad g(N_{S,t}) = a_{S,1} \cdot g(N_{S,t-1}) + a_{S,2} \cdot g(I_{S,t})$$

where:

$g(N_{S,t})$ = rate of growth of innovation in sector S in period t
$g(N_{S,t-1})$ = rate of growth of innovation in sector S in period t-1
$g(I_{S,t})$ = rate of growth of investment in sector S in period t

The expressions $a_{s,1}$ and $a_{s,2}$ refer to the (sectoral) elasticities of innovation with respect to past innovation and current investment. The formal structure of the innovation function is identical for both versions of Tandem. However, they are subject to different interpretations in the two models.

In version 1 of Tandem the data base for the innovation indicator is provided by the annual figures obtained by the Ifo-Institute as part of their innovation test. It refers to real expenditure on innovation in 30 sectors of manufacturing industry in the Federal Republic of Germany for the years 1970 to 1986 (sectors 7 - 36 in table 2.3). For Tandem 2 R&D data were used, prepared for us by the DIW. They cover real expenditure on R&D, by sector for the Federal Republic between 1961 and 1983. In both versions the data used to measure investment were the annual statistics on real investment in plant and equipment by sector (excl. residential construction).

The empirical estimates of the innovation function show a significant positive relationship between investment and innovation for a majority of sectors. At the same time the investment elasticities of innovation exhibit considerable industry-specific fluctuations: the estimates range from very low values to values slightly above 1. The results for the elasticity of current innovations with respect to past innovations are within the same range.

Table 3.1: The Elasticities of the Innovation Function

Tandem 1
elasticities of innovation with respect to

	innovation	investment
chemical industry	0.59	0.45
	(2.70)	(1.84)
engineering	0.41	0.61
	(2.73)	(3.98)
automotive industry	0.56	0.47
	(4.29)	(3.41)
electrical engineering	0.72	0.31
	(4.31)	(1.66)

Source: own calculations (t statistics in parentheses)

Tandem 2
elasticities of innovation with respect to

	innovation	investment
chemical industry	0.90	0.11
	(41.27)	(5.16)
engineering	0.83	0.16
	(28.2)	(5.90)
automotive industry	0.75	0.24
	(11.74)	(4.14)
electrical engineering	0.80	0.21
	(17.27)	(4.45)

Source: own calculations (t statistics in parentheses)

The most important results are shown in table 3.1. In the core sectors the investment elasticities amount to: engineering 0.61, automobiles 0.47, chemical industry 0.45 and electrical engineering 0.31 in version 1, and 0.61 (engineering), 0.24 (automobiles), 0.11 (chemical industry), and 0.21 (electrical engineering) in version 2.

In general the sensitivity of innovation to that of the previous period is greater than to changes in investment behavior. This indicates that, in the Federal Republic at least, technical and organizational conditions exert a greater influence in the short term on innovation than market conditions and changes in relative prices.

4. Accumulation and Absorption of Technical Knowledge

One of the cornerstones of the Tandem model is that the level of available technical knowledge influences the production technology used by firms. This means that the level of labor productivity in a sector depends not only on current expenditure on innovation but also on the stock of technical know-how which has already been accumulated.

This view of the knowledge-innovation-productivity relation is based on several considerations. While expenditure on innovation can be empirically determined but it represents merely the input into innovative activity, from which an output of process and product innovations may or may not flow. As an (annual) flow value it represents the investment in the production of technical knowledge. The stock value, on the other hand, is dependent on a whole series of other factors, including the growth of technical knowledge by the public and private education system or the flow of information between buyers and sellers in the normal course of business.

In Tandem technical knowledge takes on the function of a "hinge" between expenditure on innovation and the trend of productivity growth. The two sides of the hinge are represented by the accumulation and absorption of technical knowledge.

This conceptual differentiation serves on the one hand to open a methodological "black box" by attempting to meet the demands for new approaches to the modeling of investment in innovative activity and the diffusion of technical knowledge in empirical research into innovation. At the same time this distinction makes it possible to use alternative model specifications.

Tandem 1

Tandem 1 is limited to expenditure on innovation in manufacturing industry. The basic idea is that in each of the 30 sectors technical knowledge is accumulated which then determines the changes in production technology in that sector.

The precise specification of the "technology function" is as follows:

(4.1) $T_{S,t} - T_{S,t-1} = N_{S,t} + (ta-ob).T_{S,t-1}$
resp.

(4.2) $g(T_{S,t}) = N_{S,t}/T_{S,t-1} - ob + ta$,

where:
$T_{S,t}$ = stock of knowledge in sector S in period t
$N_{S,t}$ = expenditure on innovation in sector S in period t
ta = rate of autonomous technical progress
ob = rate of obsolescence of technical knowledge
g(.) = growth rate of a variable

In equations 4.1 and 4.2 the stock of knowledge of a sector is determined by three elements: expenditure on innovation, "depreciation" and by an autonomous growth factor. The sectoral expenditure on innovation is determined endogenously by the innovation function. The non-induced part of technical progress is represented by an exogenous rate of growth. The rate of obsolescence represents that part of technical knowledge which, in the course of time, is replaced by new knowledge or becomes obsolete for other reasons.

According to equation 4.1 the change in technical knowledge is a net value which results from the difference between the "gross increase" (expenditure on innovation and autonomous changes) and "depreciation". The level of know-how determined in this way affects the production function.

Tandem 2

The modeling of the production and diffusion of technical knowledge is more complex in Tandem 2. A distinction is made between accumulation and absorption, as follows:

- the accumulation function determines the extent to which a sector develops know-how of modern technology;
- the absorption function indicates the extent to which a sector absorbs accumulated technical knowledge from other branches.

The formal specification of the accumulation function corresponds to the technology function of Tandem 1:

(4.3) $T_{S,t} - T_{S,t-1} = N_{S,t} + (ta-ob).T_{S,t-1}$
resp.

(4.4) $g(T_{S,t}) = N_{S,t}/T_{S,t-1} - ob + ta,$

where:

$T_{S,t}$ = stock of knowledge in sector S in period t
$N_{S,t}$ = expenditure on innovation in sector S in period t
ta = rate of autonomous technical progress
ob = rate of obsolescence of technical knowledge
g(.) = growth rate of a variable

The absorption function determines the extent to which each of the 48 sectors in the model "imports" know-how from the three strategic (and the remaining) sectors. It should be noted that all sectors also absorb technical knowledge which has been generated by the sector itself. The absorbed technical knowledge is responsible for changes in production technology.

The specification of the absorption function is as follows:

(4.5) $HW_{j,t} = (I_{j,t}/B_{j,t}).(\Sigma_S c_{jS}.T_{S,t}/B_{S,t}),$

where:

$HW_{j,t}$ = HW index of the absorbed technical knowledge in sector j
$I_{j,t}$ = investment in sector j
$B_{j,t}$ = gross output in sector j
$B_{S,t}$ = gross output in sector S
$T_{S,t}$ = accumulated technical knowledge in sector S
c_{jS} = share of investment purchased by sector j from sector S

The expressions $(I_{j,t}/B_{j,t})$ and $(\Sigma_S c_{jS}.T_{S,t}/B_{S,t})$ are index values with 1970 as the base year. Equation 4.5 indicates that two factors are decisive for the level of absorbed knowledge: the level of investment and that of accumulated knowledge. The higher the level of investment in a sector relative to its gross output, the more new technologies are absorbed. The higher the accumulated technical knowledge of sector S (relative to gross output), from which sector j purchases capital goods, the higher the absorption of new technology.

This distinction between accumulation and absorption lends much greater weight to the role of investment in determining technical progress in Tandem 2. On the one hand, investment is the primary explanatory variable of innovation in the innovation function (3.1), i.e. those variables which influence investment in equipment in the strategic sectors (relative prices, profit expectations, technical possibilities) also determine expenditure on innovation in these sectors. In the accumulation function this expenditure determines, making due allowance for obsolescence and autonomous technical progress, the level of technical knowledge. On the other hand, investment is also of decisive importance for absorption. The extent to which technical progress is transmitted to other sectors depends on the relationship of investment to gross output in the absorbing sector. The index of absorbed technical knowledge $(HW_{j,t})$ is thus sensitive to both the intensity of investment and to the technological content of investment in the absorbing sector.

5. The Effects on Productivity

Of all the motives for firms to engage in innovative activity the most important are, first, an increase in output and, secondly, an improvement in product quality. To this extent the adoption of technical and organizational knowledge is subject to private profit calculations at the micro level. Innovation is usually only undertaken when it leads to increased revenue for the firm through improved productivity or the opening up of new markets.

These are the considerations on which the modeling of the effects of new technology on productivity in Tandem is based. The technical and organizational knowledge employed in production is - alongside the factors labor and capital - an important determinant of the level of output. It is however not the current access to new knowledge about products and production processes which, *ceteris paribus*, determines the level of factor productivity and production, but the stock of knowledge accumulated in the past.

The "productivity function" of Tandem is as follows:

(5.1) $P_{S,t} = a_{S,3}.T_{S,t} + d_{S,1}.(U_{S,t} - W_{S,t}) + d_{S,2}.C_{S,t}$ für Tandem 1
resp.

(5.2) $P_{S,t} = b_{S,3}.HW_{S,t} + e_{S,1}.(U_{S,t} - W_{S,t}) + e_{S,2}.C_{S,t}$ für Tandem 2

where:
$P_{S,t}$ labor productivity of the sector S in the period t
T_S technical knowledge (accumulated expenditure on innovation)
$HW_{S,t}$ absorbed technical knowledge (HW indicator)
$U_{S,t}$ capital costs (user costs per hour)
$W_{S,t}$ wage costs (per hour)
$C_{S,t}$ degree of capacity utilization
$a_{S,3}$, $d_{S,1}$, $d_{S,2}$, $b_{S,3}$, $e_{S,1}$, $e_{S,2}$ ... elasticities

The differences between the specification of the productivity function in Tandem 1 and 2 lie in the choice of the innovation indicator. In accordance with the conception of Tandem 1 the accumulated expenditure on innovation (as defined in the previous section) was used as the indicator for the "productive services" of the know-how used in production. In version 2 it is not the accumulated stock of knowledge of a sector which serves as the explanatory variable for sectoral productivity (growth), but rather the technology absorbed through investment. The use of the HW innovation indicator in Tandem 2 offers the opportunity to simulate the effects of new technology on productivity for sectors outside manufacturing industry. Relative changes in the prices of labor and capital are included in estimating the equation in order to control for the influence of labor and capital markets on productivity growth.

Table 5.1 summarizes the estimation results obtained for Tandem in selected sectors. It shows a significant positive relationship across the board between the stock of knowledge of a sector and its level of labor productivity. In the specification of Tandem 1, the elasticity of productivity with respect to technology for the strategic sectors is 0.32 (chemical industry), 0.50 (engineering), 0.37 (automobile industry) and 0.72 (electrical engineering). The weighted average of the elasticities is about 0.40.

Table 5.1: Production Function

Tandem 1: Elasticities of innovation with respect to

	technichal knowledge	factor prices	capacity utilization
chemical industry	0.32	-0.88	0.27
	(3.25)	(-13.50)	(1.07)
engineering	0.50	-0.63	-0.24
	(4.64)	(-10.63)	(-0.91)
automotive industry	0.37	-0.63	0.12
	(13.50)	(-20.23)	(1.85)
electrical engineering	0.72	-0.65	-0.68
	(7.03)	(-6.18)	(-3.35)

Source. own calculations (t statistics in parantheses)

Tandem 2: Elasticities of innovation with respect to

	technichal knowledge	factor prices	capacity utilization
chemical industry	0.15	-0.89	0.79
	(1.62)	(-1.04)	(4.07)
engineering	0.25	-0.44	0.45
	(3.16)	(-4.70)	(2.76)
automotive industry	0.22	-0.52	0.57
	(4.14)	(-6.06)	(5.26)
electrical engineering	0.53	-0.68	-0.15
	(4.74)	(-4.68)	(-0.68)
clothing industry	0.24	-0.66	0.37
	(1.34)	(-3.69)	(1.01)
wholesale trade	0.33	-0.49	0.18
	(4.33)	(-7.55)	(1.16)
Federal Post Office	0.26	-1.12	0.33
	(3.56)	(-10.29)	(2.20)
financial institutions	0.04	-0.58	0.90
	(1.40)	(-4.41)	(4.00)
insurance	0.12	-0.82	0.77
	(1.40)	(-4.41)	(4.00)

Source. own calculations (t statistics in parantheses)

The estimates for Tandem 2 also include the results for the service sector. Here too in the overwhelming majority of cases technical progress and productivity are positively correlated. The elasticities are, for example, in wholesale trade 0.33, in communications 0.26 and other services 0.33 - they are thus all highly significant. The parameters for financial institutions and for insurance are, however, insignificant at the 10% level.

6. Patterns of Macroeconomic Adjustment

There are several macroecononometric models of the Federal Republic of Germany available with which these adjustment processes induced by "supply-side" changes in labor productivity can be studied. The most detailed of these analyses has been conducted by the DIW (1988) in which Jürgen Blazejczak has developed a range of scenarios within the framework of the DIW's Langfrist model. The insights which this model has provided for the purpose of a simulation model such as Tandem can be illustrated in the form of a small number of parameters.

These parameters describe the dynamic structure of the adjustment behavior of the DIW's long-term model when the values for labor productivity are increased. This enables the functional relationships within the macroeconomic structure of the Federal Republic of Germany to be described as a sort of black box in which changes in labor productivity lead to changes in specific macroeconomic variables. In the case of Tandem these are GDP, employment, unemployment exports, imports and investment.

Table 6.1 shows the multipliers in their reduced form according to the results of the DIW's Langfrist model. The most prominent feature is that even after 4 time-periods a supply-side productivity shock still has perceptible effects. Regardless of their dynamic specification, the multipliers are in accordance with the expected effects of a supply-side increase in labor productivity: investment is stimulated; nominal wages grow at a faster rate than consumer prices, i.e. real incomes rise; the terms of trade improve so that only limited export increases are possible; the trade surplus of the Federal Republic sinks. As far as the labor market is concerned, increased GDP growth is insufficient to compensate entirely for the fall in demand for working hours - i.e. unemployment increases.

Despite this, the extent of the increase in unemployment is far from being a constant. The more successful the attempts to transform innovative activity into positive results in international competition, the higher the increase in growth accompanying the rise in labor productivity. In order to account for this effect the DIW calculated a second set of multipliers which can be interpreted as being the reduced form "under improved conditions for foreign trade".

This method provides two versions of the simulation for use with Tandem. The "DIW-standard" version is based on multipliers which result from a long-term simulation under conditions approximating to those of the recent past. The "export offensive" version, on the other hand, assumes an increase in the share of world markets due to an increase in innovative activity. At first sight the differences in the multipliers may appear small (table 6.1) but in the long term the variations become quite significant.

Table 6.1: Makroeconomic Multipliers

- standard version -

effect of a 10% increase in labor productivity on macroeconomic variable in %

variable	effect in time period			
	t_0	t_1	t_2	t_3
GDP	0.8	1.4	1.2	0.8
imports	2.3	3.2	2.5	1.6
exports	0.0	-0.1	-0.1	0.0
investment	1.7	2.5	2.0	1.3
employment	-6.7	-2.8	-0.8	-0.1
unemployment	59.9	20.5	5.7	1.0

- export version-

effect of a 10% increase in labor productivity on macroeconomic variables in %

variable	effect in time period			
	t_0	t_1	t_2	t_3
GDP	2.9	1.7	1.2	0.7
imports	7.5	3.4	2.2	1.2
exports	6.5	0.2	0.1	0.0
investment	4.9	2.9	1.9	1.0
employment	-5.1	-1.6	-0.4	0.0
unemployment	30.1	11.8	2.5	0.3

Source: DIW Langfrist model: information provided by Jürgen Blazejczak

For the central circuit of Tandem the greatest importance is to be attached to those multipliers which capture the relationship between labor productivity and investment. This is because investment provides the pivot for the feedback effects onto innovative activity. The results show a monotonic positive relationship between innovation, labor productivity, investment and innovation for the Federal Republic of Germany.

7. Selected Simulations

The simulations using Tandem show the trends of innovation, labor productivity, GDP, employment and investment when the circuit represented by Tandem is allowed to work without further shocks over a long period. Such scenarios amount to thought experiment in order to analyze the logic of the functioning of interdependent systems. They are not intended to be used for projections or forecasts of expected economic developments. The comparison of the results of the simulations does, however, provide insights into the dynamic of innovation in the Federal Republic under the assumptions which were made when specifying the model.

The simulation period is 1970 to 1986 for standard simulations and 1970 to 1988 for problem-oriented simulations. Since the choice of the base year has no influence on long-term rates of change, so it proved unnecessary to conduct simulations for other periods of time.

The standard scenario and problem-oriented versions

The standard scenario of Tandem 1 offers the following view. The expenditure on innovation grows at a rate of 1.6% per annum. This stimulates an average growth rate of innovation of 0.7%. The increase in the stock of know-how which this brings about (4.8% p.a.) induces productivity growth of about 1%.

In Tandem the increase in labor productivity is not fully compensated by the increase in output, so that the level of employment sinks by 1% every year. Over the entire simulation period (1970-1986) unemployment increases by about 8.1%. In this circuit the adoption of new technology is responsible for an increase in the rated unemployment of 1.7% (over a period of 17 years).

In contrast to the standard scenario, the "export offensive" version is based on the assumption that one of the central compensation strategies against unemployment is the translation of innovative activity into an advantage in international competition in terms of product quality. If this is successfully accomplished (to an extent compatible with recent experience) the increase in unemployment due to productivity growth is cut by half. This version requires that exports increase by 0.8 percentage points faster than in the standard scenario. Under such conditions the number of persons unemployed increases by 241 000 (instead of 445 000) and the unemployment rate is raised by 1.1 (rather than 2.1) percentage points.

Econometric studies (DIW 1988) imply that this "export offensive" version of Tandem is empirically more plausible than the standard version, which served as a reference point for the various scenarios. The scenario of a technology-oriented export offensive exhibits certain other positive characteristics: in addition to the reduction in the increase in unemployment, the success in international quality competition leads to an increase in economic growth, increased investment and a faster rate of productivity growth.

A contrast to the "export offensive" scenario is provided by the "reduced innovation" simulation. This strategy is operationalized in the simulations by making two alterations to the specifications. First, the "autonomous" growth of the stock of innovative knowledge is reduced from 5% to 2.5% p.a., with the result that the innovation-capital stock depreciates at a net rate of 2.5% p.a.). Second, the elasticity of current innovation with respect to investment and the innovation of the previous time period is reduced by half.

As far as unemployment is concerned a defensive strategy of this type proves rather effective: under such conditions Tandem's circuit increases the unemployment rate by only 0.6 percentage points. However, wage and salary earners would have to pay a high price for such a strategy. Compared with the standard simulation, growth rates of GDP, investment and labor productivity are all cut by almost half. The restructuring process within the labor force towards better qualified (and so better paid) employment relations in manufacturing industry would come to a standstill.

Reduced Form

The simulations described above provide an illustration of the dynamics of the "innovation-technology-(productivity)-growth-cirsuit" for the Federal Republic of Germany.

The results can be obtained by using a reduced form of Tandem. Table 7.2 provides a summary of the direction of the influence which certain parameters exert on the development over time of specific variables (with the standard version as a reference scenario). The relationships can be summarized as follows:

- If innovative activity is translated into an offensive export strategy GDP, exports, imports and employment increase compared with the reference scenario, while labor productivity, unemployment and the rate of unemployment decrease.

Table 7.1: The Standard Scenario and Problem-Oriented Scenarios

average rates of growth in %

variable	standard scenario	export offensive	reduced innovation
GDP	0.4	0.6	0.2
exports	0.0	0.7	0.0
imports	0.9	1.4	0.5
investment	0.7	1.0	0.4
innovation	1.4	1.7	0.1
technical knowledge	4.9	4.9	2.7
labor productivity	1.0	1.0	0.5
employment	-1.1	-0.7	-0.5
unemployment	8.1	5.2	4.0
unemployment rate*	1.7	1.0	0.7

* difference compared with the base year in percentage points
Source: own calculations

The assumptions in the scenarios:

a) Standard scenario: The results of the standard scenario are based on empirically estimated coefficients of the equations in the simulation model. Assumptions were made only for the two exogenous parameters "autonomous technical progress" and the "rate of obsolescence". In this scenario a value of 5% was assumed for both parameters, so that their effects cancel out.

b) Export offensive: The difference between the export offensive and the standard scenarios lies in the macroeconomic patterns of adjustment which were obtained by simulations with the DIW's Langfrist model. The calculation of the macroeconomic multipliers in this scenario is made on the assumption that the FRG is able to increase its exports in the year of simulation by about 1%.

c) Reduced innovation: In this scenario the rate of obsolescence was increased from 5% to 12%, while the causal relationship between investment and innovation was reduced by 50%.

Table 7.2: Contrasting Effects

- causal relationships in Tandem 1 -

dependent variable (rates of growth)	improved position in international competition	accelerated obsolescence of technical knowledge	increased rationalization efforts
GDP	+	-	-
exports	+	(+)	(+)
imports	+	-	+
investment	+	-	+
innovation	(-)	(-)	(+)
labor productivity	-	-	+
employment	+	+	-
unemployment	-	-	+
enemployment rate*	-	-	+

- interaction relationships in Tandem 1 -

dependent variable (rates of growth)	of competition and obsolescence	of obsolescence and rationalization	of rationalization and competition
GDP	(-)	(-)	+
exports	(-)	(-)	+
imports	(-)	(-)	+
investment	(-)	(-)	+
innovation	(+)	(-)	(+)
labor productivity	(+)	(-)	(-)
employment	-	(+)	+
unemployment	+	(-)	-
unemployment rate*	+	(-)	-

NB. Signs in parentheses are for those causal relationships which are based on estimates which are not significant.

* difference compared with the base year in percentage points

- If producers employ innovative activity primarily for a more effective rationalization of production, the growth of GDP, investment, imports labor productivity is more rapid, while employment falls and unemployment rises. The skill structure of employment and the risk of getting unemployment are similar to those of the standard version.
- If the rate of obsolescence of innovative know-how in the FRG increases (e.g. because of an increase in innovative activity in the USA or Japan) the growth rate of GDP is lower as is that of investment, productivity, imports and exports. The employment situation improves, unemployment falls.

8. Outlook

This brief presentation of the Tandem model suffices to show that heuristic simulation models can help to get a glimpse into the dynamics of innovative activity in the FRG. There is ample room for improvement though:.

- The direct estimation of simultaneous equations (a model of the type presented in Hanappi/Wagner 1989) instead of the use of "reduced form models" and a necessarily arbitraty distinction between exogenous and endogenous variables as a basis for simulation.
- The introduction of "innovation variables" in disaggregated macroeconometric models (such as the DIW's FIND model) in order to be able to take account of changes in price elasticities of demand on both product and factor markets and changes in the optimal relative factor input simultaneously, rather than merely establishing a "real" nexus between innovation and the structure of production.
- The explicit recognition of the interdependent nature of innovative activity in different sectors and the dependence on the overall economic situation (such as relative prices and expectations of future sales), instead of the use of investment as a proxy variable.
- The endogenization of the depreciation of technical know-how depending on the level of current expenditure on innovation.

On the basis of the progress made within the Meta Study there is reason to believe that such improvements should be available in modelbuilding for the Federal Republic in the near future.

Perspectives for Macro-Economic Development at Different Rates of Innovative Activity

Jürgen Blazejczak

1. Introduction

The aim of the study presented here (see also Blazejczak 1989) was to determine, in the form of quantitative scenarios, the macroeconomic effects of an accelerated introduction of technological change on the perspectives for economic development in the Federal Republic of Germany until the end of the 1990s. The emphasis was on a quantitative evaluation not only of the labor-shedding, but also the compensation effects of technological change , and to determine the conditions under which more intensive innovative activity would have positive employment effects at the macro level.

The definition of innovation used was a very general one - changes in a highly complex production function, i.e. in n-dimensional space new production possibilities are set up (Siebert 1986). Thus, besides product and process innovation, the definition also includes organizational changes and new marketing strategies. This broad concept of innovation is operationalized by describing the effects of innovation on the main economic aggregates. In this study the emphasis is on the effects of innovation on the productivity of the factors of production, international competitiveness and the level of investment.

The explicit investigation of the relationship between expenditure on innovation and its success does not form part of this analysis. It will suffice to show that it is possible - through an accelerated introduction of innovations - to increase the tempo of productivity growth and to improve international competitiveness and, on the basis of other more detailed studies, some of which have also been conducted within the framework of the Meta Study, to determine the *extent* to which this is possible.

These primary effects of a forced rate of technological change are then entered into the DIW's econometric model for the Federal Republic of Germany (Blazejczak 1987) as exogenous inputs. This model provides a consistent framework for incorporating the results of case studies and structural analyses and 'translating' them into the categories of national accounting statistics.

Using the model it is possible to determine the quantitative effects of changes in these inputs on macro-economic variables which are endogenous to the model. Starting from a reference scenario (innovation increasing at the trend rate) different paths of economic development in the Federal Republic up to the year 2000 can be formulated under conditions of different innovation rates above that in the reference scenario. The uncertainties associated with such scenarios can be accounted for by sensitivity analysis: leading to upper, lower and medium scenarios.

In order to ascertain the effects of reduced innovative activity a further scenario was developed in which the rate of productivity growth slows and international competitiveness deteriorates. However, this scenario does not represent a development which is particularly likely to occur in the future, it has merely illustrative character.

2. The Extent of the Primary Effects of an Increase in Innovative Activity

2.1 The Attainable Increase in Productivity

Theoretical considerations suggest that increased innovative activity - such as greater R&D expenditure, modernization of plant, organizational improvements, imitation etc. - leads to accelerated productivity growth. The question as to the extent to which this can be achieved in the Federal Republic of Germany before the year 2000, on the other hand, is an empirical one, and as such empirical studies will be enlisted in an attempt to provide an answer.

Since the 1960s the trend rate of productivity growth has declined continuously. It fell from over 5% in the 1970s to 3.5% in the course of the 1980s. The trend to deteriorating productivity growth is a phenomenon which applies to almost all sectors of the economy and so cannot be put down to the structural effects of a shift in production to sectors with lower productivity.

The importance of innovative activity as a determinant of productivity growth has been confirmed by a considerable number of econometric studies, including comparisons between different countries and analyses based on panel data (Boyer 1988).

For the Federal Republic the relationship between R&D expenditure and productivity growth at the sectoral level has been analyzed within the framework of the Meta Study by Horn (1989) using the FIND-Model. Instead of the conventional procedure involving time trends, Horn used the R&D capital stock including the R&D

expenditure received from other sectors via purchases of investment goods as an explanatory variable for labor demand (Wessels 1989). This implies that all changes in labor demand which are not accounted for by the other determinants explicitly mentioned (output and relative prices) are attributed to R&D capital stock (Griliches 1980). As an answer to the question as to the increase in productivity growth which can be obtained by increased R&D expenditure, the values obtained by Horn thus ought to represent the upper limit.

The average for those sectors studied amounted to an elasticity of approx. 0.4 in the period from 1973 to 1980 and somewhat more than 0.5 between 1980 and 1983. In this period the R&D capital stock increased with an average annual rate of over 5%. Thus if the R&D capital stock were to be increased by a further 1 percentage point, Horn's results would imply that productivity growth would accelerate by around 0.5 percentage point in the sectors studied. The rate of increase in R&D expenditure which would be necessary to raise the growth rate of the R&D capital stock by 1 percentage point depends on the level of the R&D capital stock, and thus cannot be stated generally. However, it can be shown that the relation of R&D investment to the R&D capital stock would then have to increase by 1%.

Flaig and Stadtler (1987) studied the productivity effects of R&D expenditure based on panel data for 300 firms in the West German private sector. Their results provide no information on the net effect on productivity growth in the sectors of the economy as a whole, they are however consistent with the hypothesis that at branch level R&D capital stock, including 'imports' from other sectors, has a positive effect on labor productivity.

The quantitative importance of plant modernization for accelerating productivity growth can be measured using 'vintage models' in which technical progress is embodied in capital goods.

On the assumption of a constant increase of marginal labor productivity at the sectoral level, Görzig (1980) has shown that in such a model changes in the age structure of capital stock exert considerable influence on growth rates of labor productivity. In 1977 the increase in average labor productivity was approximately 1 percentage point lower than the value which would have resulted had the age structure of plant remained constant. In the long term, however, an acceleration of productivity growth could only be achieved through the continuous reduction in average plant age.

A further point is that, viewed from the aggregate level, productivity increases at the sectoral level may be compensated by shifts in the sectoral composition of production. Using a model which allows for increasing and decreasing rates of growth of marginal labor productivity of a sector as a result of changes in factor prices, Görzig simulated the effects of an accelerated increase in the level of investment in all sectors for the period 1981 to 1986 (Görzig et al. 1988). The results show that there is scope for an acceleration of productivity growth if in sectors with a high level of productivity older plant is replaced through additional investment and at the same time the productivity of new plant is increased.

Attempts have also been made to shed light on the relationship between the age structure of plant and productivity growth by summarizing the information on the age structure of plant in "newness" coefficients. In one such analysis Klodt (1984) found a significant positive correlation at the aggregate level, whereas at sectoral level the importance of this factor differed from sector to sector. The influence of the "newness" of plant on productivity has been analyzed by Horn (1989) using the factor-demand section of the FIND-model at a high level of sectoral disaggregation. The results do not indicate an unambiguous correlation between age structure and productivity growth in the various economic sectors.

Some evidence for the possible productivity growth which might be achieved by increased innovative activity is provided by a study of labor market perspectives to the year 2000 carried out by the IAB together with the Prognos AG (Rothkirsch/ Weidig 1985). While considerable reservations must be made in transferring the findings, the results of this study suggest that the additional productivity growth which can be achieved through accelerated innovative activity may lie between 0.5 percentage points and 1.5 percentage points p.a.

The comprehensive empirical study conducted into the reasons behind the slowdown of productivity growth in the Federal Republic by Klodt (1984) comes to the conclusion that one third of this can be attributed to a slowdown in technical progress. The main reasons for this are the slow rates of growth in R&D expenditure and the restricted diffusion of innovations embodied in capital goods due to the low rate of investment.

If the slowdown in annual productivity growth is put at 2 percentage points - from more than 5% in the 1960s to less than 3% since the mid 1970s - the conclusion that can be drawn from Klodt's analysis is that annual rates of productivity growth could have been approx 0.7% higher had innovative activity been maintained at a higher level.

Taking the results achieved by Horn described above, the annual increase in the R&D capital stock (which in his study serves as an indicator for technical progress) which would have been necessary to achieve such an acceleration in productivity growth would have to have been 1.5 percentage points higher than was in fact the case.

2.2 Required Additional Investment

One element in a strategy of increased innovative activity is the modernization of the capital stock. This section will inquire to what extent this is achieved through additional investment instead of normal replacement.

Additional investment is defined as that which is not determined by such motives as replacement, adjusting capacity to expected demand, or substitution due to shifts in relative prices but rather where the motive is to introduce new technology.

A necessary condition for increased innovative activity to result in additional investment is for technological change to be embodied in capital goods. Different innovations have widely differing degrees of embodiment; examples can be found where embodiment is nonexistent as well as those which make existing plant completely obsolete. Klodt (1984) comes to the conclusion that in the period studied (1960 to 1980) ".. in addition to non-embodied technical progress there had also been progress embodied in capital goods and that the latter was probably the more important for the development of labor productivity".

This does not help us to determine whether or not additional investment results from the innovation process. Based on the results from the Ifo-Institute's innovation test questionnaire Gerstenberger (1986) comes to the conclusion that there is nothing in net capital investment trends in manufacturing industry to indicate that the introduction of new technology is encouraging new investment. If industrial branches are studied separately it can be seen that it is only in innovative branches that real expenditure on investment in the mid 1980s was higher than at the beginning of the decade. This would lead us to suppose that new technology does indeed have a stimulative effect on investment. However, this result is not confirmed by statistical tests. The study of NC machine tools shows that since the end of the 1970s the sales of NC machine tools in West Germany expanded but only at the expense of sales of conventional machines. Overall Gerstenberger (1986) concludes that there is no empirical evidence of a stimulating effect on investment due to the increasing introduction of new technology in the Federal Republic of Germany in the first half of the 1980s. However, a different picture may emerge in future if new technology does not merely represent a more efficient replacement for existing technology.

The modernization of capital stock can also result in higher capital productivity which will have a dampening effect on investment activity. It is therefore necessary to obtain information on the effect of an acceleration of innovative activity on capital productivity.

Studies carried out by the DIW as part of their structural reports show that firms are clearly able to increase the productivity of their investment, e.g. when investment activity is at a low level. In the long term a trend towards increased capital productivity can be made out resulting from the increased importance of investment goods with a short productive life and a considerable capacity effect. They facilitate a rapid adjustment to changing market conditions brought about by product innovation. Such an innovation strategy is encouraged by technological developments which enable the capacity of existing plant to be considerably increased at little cost in terms of investment expenditure.

Soete and Freeman (1985), in a study of 6 OECD countries including the Federal Republic, have attempted to measure how the extent and the nature of future investment requirements are influenced by technological change. On the assumption that the technological trends for the period 1973 to 1981 continue unchanged they determine the growth rate of investment which would be necessary to produce the output at which employment would remain constant at the level of 1981. These calculations indicate that the Federal Republic would require annual economic growth rates of between 2.5% and 3.5% for the period 1981 to 2000. This in turn would require investment to increase at a rate of 5% p.a.

However, it is possible that the fall in capital productivity is actually less than assumed by Soete and Freeman (1985). If, for example, capital productivity were to fall by only 0.5% p.a., an increase in the capital stock of 2.5 - 3.5% would be sufficient to provide the productive capacity needed for a growth in output of two to three percent. From this it could even be concluded that the introduction of new technology had a dampening effect on investment.

Using the sectoral investment functions of the factor demand part of the FIND Model Horn (1989) has carried out a direct test of the influence of increased innovative activity on investment activity. In some important sectors he found a positive correlation between the R&D capital stock and the newness of the capital stock on the one hand, and investment on the other. However his empirical results do not allow us to make clear propositions as in some sectors a negative correlation was found between the technology indicators and the level of investment; often the estimated parameter values are not significantly different from zero.

On the basis of the available empirical evidence the expected effects of increased innovative activity on additional investment cannot be determined with any degree of precision. On the one hand it seems possible that the modernization of plant is achieved in the course of normal investment, especially when investment growth accompanying increased innovative activity is higher than under status quo conditions due to endogenous mechanisms, such as accelerator effects. On the other hand, technological developments are conceivable in which the modernization of the capital stock requires additional 'autonomous' investment. On available evidence it seems unlikely that long-term rates of investment growth can be achieved which exceed 3% per annum by any noticeable degree; this would certainly be considerably in excess of historical values.

2.3 Possible Improvements in International Competitiveness

If increased innovative activity leads to higher rates of productivity growth, reduced costs will, insofar as they are passed on in the form of lower prices, result in an improvement in price competitiveness. It would seem a reasonable assumption though that firms will use at least a part of their cost reductions to increase profits: such a pattern was identified by the DIW (1988) in structural report in the context of price advantages due to changes in exchange rates. This mechanism is endogenously modelled so that the following considerations can be limited to the effects of the improved quality of German products as a result of technological change.

It is generally accepted that a higher rate of innovative activity improves competitiveness, a view which, however, is not based on clear empirical evidence. That the export success of the Federal Republic is not restricted to the relatively narrow field of high-tech products but rather is based on the much wider field of 'technology-intensive' goods (DIW 1988) tends to support the hypothesis that there are considerable possibilities for greater success in foreign trade: increased innovative activity, including efforts to obtain a more highly qualified work force, organizational improvements as well as the development of new products etc., strengthens the foundations of the competitive position of this entire spectrum so that positive developments on the whole export front are to be expected.

A study of the effects of R&D expenditure on foreign trade conducted by Erber (1989) also indicates that there is considerable scope for expanding exports by raising the level of innovative activity. In the four sectors studied it is apparent that the knowledge accumulated through R&D has a strong positive influence on the level of real exports while the direction of its influence on real imports and the price indices of imports and exports varied from sector to sector.

Krupp (1986) on the other hand, in his analysis of the available empirical evidence on the question of the relationship between innovation and competitiveness - and in particular the studies in the structural reports of the DIW - found no clear evidence to confirm the hypothesis of a positive correlation. He also conducted regression analyses in which no stable correlation between the indicators for innovation and competitiveness could be established. Despite these results the author warns against simply accepting the opposite hypothesis, as this would "require a very much more exact analysis of the most important determinants of competitiveness". This, however, meets with numerous empirical difficulties such as the discrepancy between the functional delineation for indicators of competitiveness and the institutional delineation for indicators of innovative activity or the heterogeneity within the different branches.

In the macro-economic context there are factors which allow this uncertainty about the quantitative effects of increased innovative activity on competitiveness to be limited - at least as far as its 'ceiling' is concerned: the Federal Republic cannot indiscriminently increase its trade surpluses for long periods without increased likelihood of counter measures from its trading partners (DIW 1987b and DIW 1988). The trends which are already becoming apparent in industrial and trade policy provide an illustration of these barriers to the transformation of competitive advantages into a job-creating export expansion.

Where exactly the barriers to an expansion of Germany's export surpluses lie cannot be determined exactly, although historical developments can shed some light on the matter: they show that the Federal Republic has never succeeded in maintaining substantial improvements in its terms of trade over long periods. Also West Germany's share of world trade (both imports and exports) has changed relatively little, despite a considerable increase in the importance of foreign trade measured as a proportion of total output. In real terms the gap between the highest and lowest values for Germany's share of world exports amounts to 3 percentage points, and for imports the figure is a mere 1.5 points. In recent years the export share has reached values which in historical terms are high, so that a further expansion to any appreciable extent would appear to be unlikely.

3. Scenarios of Increased Innovative Activity

3.1 Exogenous Impulses of Increased Innovative Activity

The innovation scenario is based on the assumption that innovative activity is undertaken above and beyond the level assumed in the reference scenario such that at the macro level the average annual rate of productivity growth (by hour) is raised by 0.7 percentage points throughout the period 1987 to 2000. At the end of this period the level of productivity in the innovation scenario is almost 10% higher than in the reference scenario.

The acceleration of productivity growth is achieved via additional investment. As has been shown above, the relation between additional investment and the acceleration of productivity growth is subject to considerable uncertainty. In our study this uncertainty is taken into account by means of alternative simulations. In a medium simulation the acceleration of productivity growth is associated with additional autonomous investment of DM 10 000 Mill. per year (1980 prices).

The increased innovative activity will take effect sooner if the state paves the way with an innovation policy. No account has been taken of the effect of different forms of furthering innovation on the different spending departments which make up the state budget. As a simplification, subsidies from the state to private companies amounting to one third of additional autonomous investment expenditure were assumed. In the medium simulation these transfers (at current prices) amounted to on average DM 5 000 Mill.

The determination of impulses relating to foreign trade is particularly speculative, because they depend on the reactions of other countries. The assumption adopted was that of the reference scenario, i.e. that foreign countries do not react at all to the policy of forced innovation in the Federal Republic. Such an assumption can only be justified if Germany's export surplus does not increase substantially and for an extended period.

The structural determinants of real goods exports not explicitly modelled, in particular the quality of German products, are accounted for in the model by the elasticity of goods exports in relation to the volume of world trade. In the reference scenario this elasticity is taken to be 0.94. This means that the increase in German goods exports lags behind the growth of world exports, provided relative prices do not change in their favor. In the innovation scenario a higher elasticity is used in order to reflect increased competitiveness (in terms of quality). The medium simulation assumes a value of 1.1, i.e. an increase in world trade of 1% leads, ceteris paribus, to an increase in West German goods exports of 1.1%, and so to an increased share in world trade.

The improved quality of German products can also result in the substitution of domestic for imported products. In the model this is represented by a reduction in import elasticity. In the innovation scenario the partial elasticities of the import of goods (except crude oil, natural gas and petroleum products) were reduced by 0.1 points in respect of both domestic use and of exports.

The accelerated rate of productivity growth resulting from more intensive innovative activity reduces costs which may be passed on in the form of lower prices. Due to the simultaneous improvement in the competitive position in terms of quality, the relation of prices to costs for exported goods - to the direct benefit of profits - can be increased compared to the reference scenario. An increase in export-price elasticity with respect to unit costs from 0.5 (reference scenario) to 0.6 was assumed. While such a strategy leads to a deterioration in price competitiveness, this is compensated by the improvement in competitiveness in terms of quality.

If the quality competitiveness of German products on world markets improves, it is possible that importers will respond by cutting prices in order to defend market shares. It is assumed that the elasticity of the price index of imported goods (exceptions as above) in respect of the average price index for world trade (which accounts for structural influences on import prices not explicitly incorporated in the model) will fall by 0.05 compared with the reference scenario, to 0.45.

Given unchanged economic growth and constant average working time compared with the reference scenario, the acceleration of productivity growth induced by increased innovative activity would result, according to accounting identities, in a gross fall in employment of 2.3 million by the year 2000. However, compensation effects mean that total redundancies are less. In the medium simulation the compensation rate is 85%. The different effects are described in more detail below and are summarized in table 3.1.

3.2 Changes in the Components of Final Demand and the National Product

One major difference between the innovation and reference scenarios is that in the former the volume of investment increases at a faster rate than in the latter. One reason for this are the assumed changes in expectations and behavior patterns on the part of employers which, trusting in the success of a strategy based on innovation, are prepared to invest more than they have been in the same circumstances in the past. In addition, the relatively rapid success of the strategy, reflected in higher growth rates of domestic and foreign demand, leads to more optimistic expectations about future market developments, inducing further investment.

In the reference scenario the conditions for financing investment are relatively favorable even though the extremely favorable conditions of recent years will not continue. Equally, in the innovation scenario conditions for financing investment will be easier than in the 1970s: for the year 2000 the net self-financing quota in the medium simulation is, at 46%, higher than in 1980 (27%) and 1973 (33%).

In as far as shifts in relative prices are relevant - the empirical findings on this point are ambiguous - in the medium simulation their net effect is towards higher investment as the relative price of labor to capital increases more rapidly than in the reference scenario.

Table 3.1: Effects of Increased and Reduced Innovation

	average annual % change (2000/1987)			differences to the reference scenario in percentage points	
	scenario of reference	innovation scenario	scenario of reduced innovation	innovation scenario	scenario of reduced innovation
	GNP by type of expenditure at resp. current prices				
private consumption	4.0	4.6	3.0	0.6	-1.0
private fixed capital formation	4.7	5.3	3.7	0.7	-1.0
exports	5.5	6.1	3.4	0.5	-2.2
imports	6.0	6.5	4.6	0.5	-1.4
export surplus or deficit at current prices (DM billion)*	106.6	117.6	-11.3	11.0	-117.8
gross national product	4.2	4.7	3.0	0.6	-1.2
	GNP by type of expenditure at 1980 prices				
private consumption	1.9	2.8	0.7	0.8	-1.2
government consumption	1.6	1.6	1.6	0.0	0.0
private fixed capital formation	2.4	3.2	1.3	0.9	-1.1
- less residential construction	3.1	4.2	1.6	1.1	-1.5
public fixed capital formation	1.8	1.8	1.8	0.0	0.0
exports	3.6	4.1	1.7	0.5	-1.9
imports	3.7	4.4	2.0	0.7	-1.7
exports surplus/deficit (DM billion)*	47.6	30.6	21.9	-17.0	-25.7
gross national product	2.0	2.6	0.9	0.6	-1.0
	price level of national expenditure				
domestic expenditure	2.3	2.1	2.2	-0.2	0
exports	1.9	1.9	1.6	0.1	-0.3
imports	2.2	2.0	2.2	-0.2	0.1
terms of trade	-0.3	-0.1	-0.9	0.2	-0.6
gross national product	2.2	2.1	2.0	-0.1	-0.2
	income distribution and labor market				
productivity per hour worked	2.3	3.0	1.5	0.7	-0.8
employed labor force	0.2	0.1	-0.1	-0.1	-0.3
employed labor force(000's)*	26,490	27,750	25,530	-360	-960
unemployed (000's)*	960	1,190	1,580	230	620
hourly wage rate	4.6	5.4	3.6	0.8	-1.0
gross income from employment	4.3	4.9	3.3	0.6	-1.4
gross entrepreneurial and property income	3.5	4.3	1.9	0.7	-1.7
wages as % of GNP*	71.7	71.5	72.6	-0.3	0.8
dispos. income of priv. households	4.0	4.6	2.9	0.6	-1.1
	public-sector revenue and expenditure				
revenue	4.3	4.9	3.1	0.6	-1.2
expenditure	4.3	4.7	3.9	0.4	-0.4
financial balance (DM billion)*	-57.9	-29.0	-190.7	25.0	-132.4

* values in the final year of the respective time period

On the assumption that technical change has the effect of reducing not only the quantity of labor but also that of capital needed to produce a given level of output (i.e. capital productivity increases) the expansion (of productive capacity) motive for investment loses importance. On the other hand, a necessary condition of increased capital productivity is the modernization of the capital stock; i.e. the importance of the replacement motive increases.

In the medium simulation the volume of capital investment grows at an average annual rate of 4.2%. Judging from historical experience an acceleration of investment growth to above the rate of 4.8% - as assumed in the upper simulation - seems highly improbable. This is also clear from a glance at the figures for the investment share (investment by firms as a share of domestic product): in the year 2000 for the medium simulation the figure is 1 percentage point, for the upper simulation 2 percentage points higher than in the reference scenario.

Depending on the assumed elasticity of German goods exports in respect of the volume of world trade, the innovation scenarios predict an acceleration of export growth. In the medium simulation the annual average rate of export growth is 0.5 percentage points higher throughout the whole period than in the reference scenario. The upper and lower simulations allow for the possibility of a more and a less pronounced acceleration respectively. In the medium simulation the share of world trade accounted for by goods exports from the Federal Republic increased to 12.1% by the year 2000, whereas in the reference scenario the figure was only 11.4%.

In the reference scenario real economic growth of 1% p.a. was accompanied by an increase in imports of almost 1.9%: in the innovation scenario and on the assumptions described above, import elasticity was lower at 1.7.

By the year 2000 the constellation represented by the innovation scenario resulted in a lower trade surplus in real terms than the reference scenario. This is because not only exports but also investment and private consumption are positively influenced by the innovation strategy inducing additional demand for imports. Moreover, one of our assumptions was that firms attempt to increase their revenue from exports rather than extend market shares. This is reflected in the development of prices for exported goods which, despite the cost reductions in the innovation scenario, are no lower than in the reference scenario.

Compared with the reference scenario the price increases of import goods were dampened by the fact that importers make price concessions in order to defend market shares against the improved quality of German products. A further factor is that, under the conditions of the innovation scenario, the Federal Republic is even more successful at stabilizing prices. The changes in purchasing power parities lead to a slightly more pronounced tendency for the D-Mark to appreciate, although whether this trend will actually take effect depends on a wide range of factors whose evaluation is difficult in a long-term study such as this. If the D-Mark does indeed appreciate, the increase in import prices will be restrained, which would in turn lead to improved terms of trade for the Federal Republic compared with the reference scenario.

Under these conditions the revenue from the trade surplus is stabilized despite the fact that it has shrunk in volume terms. In the medium simulation of the innovation scenario the nominal value of the surplus is DM 11 000 Mill. higher than in the reference scenario in the year 2000. However, the surplus as a share of GDP in that year is the same for both scenarios at approx. 3.2%.

In the innovation scenario real incomes grow more rapidly than in the reference scenario, a trend which will be described below in more detail in the context of income distribution. The disposable income of private households deflated by the price increase of consumer products increases at a rate which is about 1 percentage point higher than in the reference scenario. The increase in real private consumption is accelerated to the same extent, if structural changes associated with the intensification of innovative activity do not change the propensity to save. Variables which determine the level of private consumption apart from the level of income (prices, interest rates, savings, liquid funds, capital gains, credit conditions) follow roughly the same course as in the reference scenario. Some influence on the propensity to save may be expected from a more rapid replacement of consumer durables by private households, as they are more rapidly devalued by the supply of more innovative products. As no evidence is available to corroborate this supposition, this effect is not considered here.

Under the conditions described above real GDP growth is accelerated considerably. The average of 2% p.a. in the reference scenario increases to 2.6% p.a. in the medium simulation of the innovation scenario. At the end of the simulation period this growth rate differential amounted to a difference of DM 170 000 Mill. in 1980 prices.

3.3 Costs, Prices and Income Distribution

Depending on the result of the processes of wage and price determination, an acceleration of productivity growth can result in changes in relative costs and in income distribution.

In the innovation scenario the increase in unit labor costs is higher than in the reference scenario (in the medium simulation by approx. 0.3 percentage points). As the wage rate - according to the definition used in the model - includes employees' and employers' contributions to social insurance, changes in unit labor costs also reflect a considerable proportion of changes in indirect wage costs.

The accelerated modernization of the capital stock leads to higher replacement costs as a result of a higher rate of scrapping. This means that, despite the more moderate rate of price increases, user costs of capital increase more rapidly than in the reference scenario. The total costs of capital depend not only on the user costs of capital but also on the level of the capital stock. While the total cost of capital is higher then in the reference scenario for the year 2000, unit capital costs (cost of capital against output) in the medium simulation climb more slowly than in the reference scenario.

Taken together the labor and capital costs borne by firms increase more slowly in the innovation than in the reference scenario. Price concessions by importers and the substitution of domestic for foreign goods have the effect of moderating the increase in unit import costs compared with the reference scenario. The burden of indirect taxes (in relation to output) is also less pronounced.

Taking all these factors together, it can be concluded that successful innovative activity on the part of enterprises enables them to reduce their costs considerably. To some degree these cost reductions are passed on in the form of reduced price increases: the rate of increase of input prices is 0.2 percentage points lower than in the reference scenario, and the dampening effect is somewhat greater still for private consumption goods.

In the case of technological change without social friction assumed in the model, recipients of wage and profit income benefit to practically the same extent from increased economic growth (see also Horn 1989). Gross income of wage and salary earners as a share of social product (wage share) is only slightly lower in the innovation scenario.

The reference scenario represents a status-quo projection. As far as the activities of the state are concerned this means that calculations are based on present behavior and on the existing system of institutional regulation. Changes which could be predicted with some degree of certainty - such as the tax reform which is due to come into force in 1990 - were also taken into account. The status-quo assumptions for the behavior of the state were maintained for the innovation scenario with the exception of transfer payments. However, in view of the changes in the position of the state budget in the innovation scenario, the state may see the way open for further tax reductions or changes in the contributions to social insurance. At the same time public measures to support the innovation strategy of private firms is desirable and necessary. Areas for state policy would include investment in environmental protection, the provision of an efficient infrastructure and a public spending policy more clearly oriented towards innovation. Such policy changes are, however, not considered here.

The results of income redistribution via the state, as reflected in the difference between gross and net incomes, remain relatively unchanged. The structure of public spending is not likely to experience great changes compared with the reference scenario and the share of transfer payments in the disposable income of private households will remain relatively constant. The effects on budget deficit, however, are much greater. Under the status-quo assumption regarding institutional regulation, the state budget benefits to the tune of DM 25 000 Mill. (medium simulation, year 2000) compared with the reference scenario.

The increase in revenue flowing into the public purse is closely linked to the acceleration of economic growth. The particularly large increase in tax revenue is a reflection of the fact that at higher levels of economic growth the tax share in national product also increases. In the medium simulation the tax revenue over and above that in the reference scenario amounts to almost DM 70 000 Mill. in the year 2000. The income from social insurance contributions grows at a rate slightly below that of GDP. Overall the rate of increase of total public revenue is equal to that of GDP.

The impacts on public spending are rather different. The most important component of public expenditure is public demand for goods and services including labor services. If, as assumed, it does not expand beyond the level in the reference scenario in real terms, the rate of increase of public expenditure will correspond to that of wage rates. This would result (medium simulation) in a 0.4 percentage point increase in average annual growth rates. Public demand as a share of GDP would therefore be lower than in the reference scenario: in the medium simulation by almost 1 percentage point.

Current transfers to private households, especially social security benefits, account for almost as great a share of public spending as public demand. In the innovation scenario the rate of increase of this component is 0.7 percentage points higher, a figure which conceals a differentiated development of the different forms of transfer payment which cannot be described in detail here.

3.4 Labor Market Effects

The constellation described above of changes in productivity and GDP growth implies changes in the level of employment. It was assumed that there was no difference in the changes in average working time between the two scenarios, although an acceleration of productivity growth would create scope for additional working time reduction in all its forms. Only in the upper simulation was the level of employment in the year 2000 higher (by 260 000) than in the reference scenario. In the medium simulation 360 000 people fewer were in employment in the year 2000.

Unemployment, however, does not change to the same extent as employment. This is due to the buffer effect of hidden unemployment ("Stille Reserve"). Higher rates of unemployment will exacerbate the discouraged worker effect leading many to give up the search for a job. When unemployment is falling job-seekers will be encouraged to register at employment offices. Changes in the numbers unemployed and in the "stille Reserve" are assumed to be roughly proportional.

Occasionally the hypothesis is put forward that as employment rises the "Stille Reserve" is reduced more quickly than registered unemployment because it is characterized by a more favorable skill structure. It has also been claimed that participation rates may develop in different ways in conditions of intensive technological change than in the reference scenario, independently of changes in overall levels of (un)employment. Changes in skill requirements and working time practices may, for example, lead to a more rapid increase in the participation rates of female workers. Neither argument was taken any further in this study. They would both require the effects of technological change on skill structures to be taken into consideration with reference to detailed studies in the relevant areas.

3.5 Analysis of the Sensitivity of the Results

The remaining areas of uncertainty associated with the assumptions which lie behind the innovation scenario can only be taken into account with the use of alternative scenarios. As mentioned above, the major sources of uncertainty were the volume of additional investment required in order to induce an acceleration of productivity growth of the dimension assumed and the extent of the effect of the improvement in competitive position on the level of exports.

In the upper simulation of the innovation scenario it is assumed that DM 27 000 Mill. (instead of DM 10 000 Mill. in the medium simulation) of additional investment would be necessary (1980 prices) in order to induce the same acceleration of productivity growth. Including indirect effects, the rate of increase of investment is 1.8 percentage points higher. In the lower simulation it is assumed that the acceleration of productivity growth is not embodied in additional autonomous investment but arises out of the "normal" process of investment (which, however, is accelerated to some extent by the higher rate of GDP growth compared with the reference scenario). In the upper simulation the elasticity of goods exports with respect to the volume of world exports is taken to be 1.2. In the lower simulation it is 1.0., i.e. Germany's share of world exports does not change due to structural effects.

At the same productivity growth acceleration as in the medium simulation the increase in GDP growth rates are lower (higher) in the lower (upper) simulation. In the year 2000 the level of employment in the lower simulation is 860 000 lower than in the reference scenario. In the upper simulation it is higher by 260 000. This means that under unfavorable conditions some two-thirds of job losses would be compensated for, while under favorable conditions the compensation rate reaches 110%.

4. Reduced Innovative Activity - An Illustrative Scenario

To a greater extent than with the other scenarios the simulation presented below, in which innovative activity actually declines compared with the reference scenario, is of an illustrative nature. While economic development along a path such as in the innovation scenario is feasible, it is most unlikely that those responsible for economic policy would tolerate a development such as is assumed in this scenario for long without taking counter measures.

The possibilities for the Federal Republic to convert an improvement in their competitive position vis a vis other countries into higher export revenues are limited. Such limits do not apply in the other direction, i.e. if the competitiveness of West German industry declines because technological developments are lagging behind those of their competitors. This can quite easily occur if other countries intensify their innovative activity while the Federal Republic fails to make additional efforts in this area.

In order to illustrate the consequences of such a development a slowdown in productivity growth is assumed compared with the reference scenario. The consequent deterioration in competitive position is reflected in the following developments:

- German exporters lose market shares (in the year 2000 West Germany's share of world exports is only 8.7% compared with 11.4% in the reference scenario);
- imported goods are substituted for domestically produced goods, as reflected in increased import elasticities;
- exporters cannot achieve the price increases they enjoyed in the reference scenario, while importers are able to demand higher prices. The terms of trade deteriorate rapidly, falling to 95 index points in the year 2000 as against 103 in the reference scenario.

The decrease in the annual rate of productivity growth by 0.8 percentage points would, by the year 2000 and in the absence of all other factors demand an increase in employment of more than 2.8 million to produce the same product. However, output growth is depressed by the effects of the deterioration in competitive position. A deterioration in competitiveness is subject to far less restrictions from the outside world than an improvement. If it assumed that in the year 2000 a trade balance results which is more or less in equilibrium, and taking all interdependencies into account, the gross increase in employment of almost 3 million would be more than offset by job losses of some 4 million, i.e. the net employment effect would be approx. 1 million less jobs than in the reference scenario. In conclusion the consequences for employment must be considered even graver than in the lower simulation of the innovation scenarios.

The employment effects occur in the context of a dramatic fall in growth rates: GDP growth is only half that in the reference scenario. The problems which this brings with it can be seen clearly by a glance at the increase of the budget deficit. Reduced innovative activity severely narrows the scope for solving conflicts over income distribution, and offers no hope of solving employment problems through a slowdown in productivity growth.

5. Conclusions

Our main conclusion is that an economic policy which relies entirely on accelerated innovative activity is in no position to solve the labor market problems foreseeable up to the year 2000 (DIW 1989). This is true even under favorable conditions. At the same time, innovative activity is indispensable as can be clearly seen from the scenario of reduced innovation. Therefore the opportunities offered by a strategy of intensified innovative activity via an improvement in the perspectives for economic growth must be exploited by other elements of an economic policy which are aimed explicitly at increasing employment (DIW 1987).

References

Abele, E. et al. (1984): Einsatzmöglichkeiten von flexibel automatisierten Montagesystemen in der industriellen Produktion: Montagestudie, Schriftenreihe 'Humanisierung des Arbeitslebens', vol. 61, Düsseldorf.

Abraham, K. (1983): Structural/Frictional vs. Deficient Demand Unemployment, in: The American Economic Review, vol. 73, pp. 708-724.

Abramovitz, M. (1956): Resource and Output Trends in the United States since 1870, in: The American Economic Review, vol. 46, Papers and Proceedings, pp. 5-23.

Adorno, T.W. (ed.) (1969): Spätkapitalismus oder Industriegesellschaft?, Stuttgart.

Akerlof, G.A./Main, B.G.M. (1981): An Experience-Weighted Measure of Employment and Unemployment Durations, in: The American Economic Review, vol. 71, no. 5, pp. 1003-1011.

Appelbaum, E./Schettkat, R. (eds.) (1990): The Impacts of Structural Change and Technological Progress on the Labor Market, New York (forthcoming).

Aquino, A. (1981): Changes over Time in the Pattern of Comparative Advantage on Manufacturing Goods - An Empirical Analysis for the Period 1962-1974, in: European Economic Review, vol. 15, pp. 41-62.

Arrow, K. (1971): Economic Welfare and the Allocation of Resources for Invention, in: Rosenberg, N. (ed.): The Economics of Technological Change. Selected Readings, Middlesex (England), pp. 164-181.

Bade, F.J. (1987): Die wachstumspolitische Bedeutung kleiner und mittlerer Unternehmen, in: Fritsch, M./Hull, C.J. (eds.) (1987a), pp. 73 ff.

Baethge, M. (1988): Neue Technologien, berufliche Perspektiven und kulturelles Selbstverständnis, in: Gewerkschaftliche Bildungspolitik, no. 1, pp. 15-23.

Baethge, M./Overbeck, H. (1986): Zukunft der Angestellten, Frankfurt/Main.

Bartenschlager, H.P. et al. (1982): Industrierobotereinsatz - Stand und Entwicklungstendenzen, Schriftenreihe 'Humanisierung des Arbeitslebens', vol. 23, Düsseldorf.

Battelle (1978): Soziale Implikationen der Einführung von Industrierobotern im Fertigungsbereich, unveröffentlichter Bericht, Frankfurt/Main.

Bauer, G. (1985): Schätzung von ökonometrischen Ungleichgewichtsmodellen, Idstein.

Baumol, W.J. (1967): Macroeconomics of Unbalanced Growth: The Anatomy of Urban Crisis, American Economic Review, vol. 57.

Baumol, W./Blackman, S./Wolff, E. (1985): Unbalanced Growth Revisited: Asymptotic Stagnancy and New Evidence, in: American Economic Review, vol. 75.

Benz-Overhage, K. (1985): Automatisierung bei der Arbeitsvorbereitung, Arbeitssteuerung und Konstruktion, in: Warneke, P. (ed.) (1985), pp. 237 ff.

Berglind, H. (1981): Age and Marginality in the Swedish Labour Market, in: Gardell, B./Johansson, G. (eds.), Working Life (John Wiley and Sons), pp. 145-158.

Berndt, E. (1981): Modelling the Simultaneous Demand for Factors of Production, in: Hornstein, Z./Grice, J./Webb, A. (eds.), The Economics of the Labour Market, London, pp. 127-142.

Berndt, E./Wood, D. (1975): Technology, Prices, and the Derived Demand for Energy, in: Review of Economics and Statistics, vol. 56, pp. 259-268.

BIBB/IAB (eds.) (1987): Neue Technologien: Verbreitungsgrad, Qualifikation und Arbeitsbedingungen, Analysen aus der BIBB/IAB-Erhebung 1985/86, Beiträge aus der Arbeitsmarkt- und Berufsforschung 118, Nürnberg.

Bispinck, R./Helfert, M. (1987): Technischer Wandel und gewerkschaftliche Arbeitspolitik, WSI-Mitteilungen, vol. 40, pp. 314 ff.

Blattner, N. (1986): Technischer Wandel und Beschäftigung: Zum Stand der Diskussion, in: Krupp, H.-J./Rohwer, B./Rothschild, K.W. (eds.), Wege zur Vollbeschäftigung, Freiburg, pp. 67-76.

Blazejczak, J. (1987): Simulation gesamtwirtschaftlicher Perspektiven mit einem ökonometrischen Modell für die Bundesrepublik Deutschland, Beiträge zur Strukturforschung des DIW, no. 100.

Blazejczak, J. (1989): Gesamtwirtschaftliche Auswirkungen verstärkter Innovationsanstrengungen. Szenarien der wirtschaftlichen Entwicklung in der Bundesrepublik bis zum Jahr 2000, in: Meyer-Krahmer, F. (ed.) (1989).

Bleicher, S. (1987): Die Zukunft der Fabrik - eine Aufgabe gewerkschaftlicher Technikgestaltung. Der Mensch steht im Mittelpunkt einer sozialverantwortlichen CIM-Entwicklung, in: CIM oder die Zukunft der Arbeit in rechnerintegrierten Fabrikstrukturen, Ergebnisse einer Fachtagung der IG Metall, Frankfurt, pp. 7 ff.

Bombach, G./Blattner, N. et al. (1976): Technischer Fortschritt. Kritische Beurteilung von Meß- und Prognosekonzepten, Göttingen.

Bombach, G./Gahlen, B./Ott, A. (eds.) (1980): Neuere Entwicklungen in der Investitionstheorie und -politik, Tübingen.

Bombach, G./Gahlen, B./Ott, A. (eds.) (1986): Technologischer Wandel - Analyse und Fakten, Schriftenreihe des wirtschaftswissenschaftlichen Seminars Ottobeuren, vol. 15, Tübingen.

Bombach, G./Gahlen, B./Ott, A. (eds.) (1987): Arbeitsmärkte und Beschäftigung. Fakten, Analysen, Perspektiven, Schriftenreihe des wirtschaftswissenschaftlichen Seminars Ottobeuren, vol. 16, Tübingen.

Börnsen, O./Glismann, H.H./Horn, E.-J. (1985): Der Technologietransfer zwischen den USA und der Bundesrepublik, Kieler Studien 192, Institut für Weltwirtschaft an der Universität Kiel, Tübingen.

Boskin, M.J. (1986): Macroeconomics, Technology and Economic Growth: An Introduction to some Important Issues, in: Landau, R./Rosenberg, N. (eds.), The Positive Sum Strategy, Washington D.C., pp. 33-56.

Boyer, R. (1988): Assessing the Impact of R&D on Employment. Puzzle or Consensus?, A Comment upon: Sectoral and Macroeconomic Impacts of Research and Development on Employment by J. Blazejczak, G. Erber, and G.A. Horn, Paris (mimeo).

Braverman, H. (1977): Die Arbeit im modernen Produktionsprozeß, Frankfurt/New York.

Brehm, H./Pohl, G. (eds.) (1986): Gewerkschaftliche Gegenwehr, Köln.

Briefs, U. (1986): Informationstechnologien und gewerkschaftliche Gegenwehr, in: Brehm, H./Pohl, G. (eds.), Gewerkschaftliche Gegenwehr, Köln, pp. 24 ff.

Brötz, R. (1983): Flexible Arbeitszeitregelungen am Beispiel eines multinationalen Mischkonzerns, WSI-Mitteilungen, vol. 36, pp. 480 ff.

Brown, M. (1966): On the Theory and Measurement of Technological Change, Cambridge.

Brown, M./de Cani, J.S. (1963): A Measure of Technological Employment, in: The International Economic Review, vol. 4, pp. 289-309.

Bundesministerium für Wissenschaft und Forschung (eds.) (1981): Mikroelektronik: Anwendungen, Verbreitung und Auswirkungen am Beispiel Österreichs, Wien.

Buttler, F./Kühl, J./Rahmann, B. (eds.) (1985): Staat und Beschäftigung, Beiträge zur Arbeitsmarkt- und Berufsforschung 88, Nürnberg.

Buttler, G. (1987): Zur mittleren Dauer der Arbeitslosigkeit, in: Mitteilungen aus der Arbeitsmarkt- und Berufsforschung 2/87, Nürnberg.

Christensen, L./Jorgenson, D./Lau, L. (1973): Transcendental Logarithmic Production Frontiers, in: Review of Economic and Statistics, vol. 55, pp. 28-45.

Clark, K.B./Summers, L.H. (1979): Labour Market Dynamics and Unemployment: A Reconsideration, in: Brookings Papers on Economic Activity, pp. 13-61.

Clauß, T. (1988): Beteiligung, Akzeptanz und Arbeitszufriedenheit - Die Einführung programmgesteuerter Arbeitsmittel aus der Sicht der Betroffenen, in: BIBB/IAB (eds.), pp. 437 ff.

Cramer, U. (1987): Klein- und Mittelbetriebe: Hoffnungsträger der Beschäftigungspolitik?, in: Mitteilungen aus der Arbeitsmarkt- und Berufsforschung, 1/87, Nürnberg.

Cramer, U./Egle, F. (1976): Zur durchschnittlichen Dauer der Arbeitslosigkeit, in: Mitteilungen aus der Arbeitsmarkt- und Berufsforschung, 4/76, Nürnberg.

Crusius, R./Stebani, J. (eds.) (1984): Neue Technologien und menschliche Arbeit, Berlin.

Cyert, R.M./Mowery, D.C. (eds.) (1987): Technology and Employment. Innovation and Growth in the U.S. Economy, Washington D.C.

d'Alcantara, G. et al. (1986): Technology and Macroeconomics, FAST-Occasional Papers no. 75 (International Document), Brüssel.

Dahrendorf, R. (1963): Bildung ist Bürgerrecht, Hamburg.

Dasgupta, P./Stiglitz, J.E. (1980): Uncertainty, Industrial Structure and the Speed of R&D, in: The Bell Journal of Economics 11/80, pp. 1-29.

Dasgupta, P./Stiglitz, J.E. (1988): Potential Competition, Actual Competition and Economic Welfare, in: European Economic Review 32/88, pp. 569-577.

David, P.A. (1975): Technical Choice and Economic Growth, London/New York.

Denison, E.F. (1974): Accounting for United States Economic Growth 1929-1969, Washington D.C.

Deutsch, E. (1988): Faktornachfrage auf strukturierten Arbeitsmärkten, Frankfurt/ Main.

Deutsches Institut für Wirtschaftsforschung (DIW) (1987a): Perspektiven der wirtschaftlichen Entwicklung in der Bundesrepublik Deutschland bis zum Jahr 2000. Bearbeiter: Blazejczak, J./Kirner, W./Krupp, H.-J., in: Wochenbericht des DIW Nr. 26/87.

Deutsches Institut für Wirtschaftsforschung (DIW) (1987b): Industriepolitik im westlichen Ausland - Rahmenbedingungen, Strategien, Außenhandelsaspekte, Beiträge zur Strukturforschung des DIW 92/1 (Allgemeiner Teil) und 92/2 (Länderberichte), Berlin.

Deutsches Institut für Wirtschaftsforschung (DIW) (1988): Exportgetriebener Strukturwandel bei schwachem Wachstum. Analyse der strukturellen Entwicklung der deutschen Wirtschaft, Strukturberichterstattung des DIW 1987, Beiträge zur Strukturforschung des DIW, no. 103, Berlin.

Deutsches Institut für Wirtschaftsforschung (DIW) (1989): Beschäftigungspolitik durch verstärkte Innovationsanstrengungen?, Bearb.: J. Blazejczak, in: Wochenbericht des DIW, Nr. 4/89.

DGB-Abteilung Angestellte und WSI-Arbeitsgruppe HDA (1984): Protokoll der Fachtagung für Betriebs- und Personalräte "Textverarbeitung und Mitbestimmung", Düsseldorf.

Diewert, W. (1974): Applications of Duality Theory, in: Intriligator, M./Kendrick, D. (eds.), Frontiers of Quantitative Economics, vol. II, Amsterdam.

Döbele-Berger, C./Berger, P./Kubicek, H. (1985): Handlungsmöglichkeiten bei der Einführung von Neuen Technologien in Büro und Verwaltung, hrsg. von der Arbeitskammer des Saarlandes, Saarbrücken.

Doeringer, P./Piore, M. (1971): Internal Labor Markets and Manpower Analysis, Massachussetts.

Dostal, W. (1982): Fünf Jahre Mikroelektronik-Diskussion, in: Mitteilungen aus der Arbeitsmarkt- und Berufsforschung, vol. 15, pp. 151 ff.

Duchin, F. (1986): Analyzing the Implications of Technological Change, Vortrag bei der "Eigth International Conference on Input-Output Techniques", Sapporo 1986.

Duchin, F./Szyld, D.B. (1985): A Dynamic Input-Output Model with Assured Positive Output, in: Metroeconomica, vol. XXXVII, pp. 269-282.

Eckart, W./v. Einem, E./Stahl, K. (1987): Dynamik der Arbeitsplatzentwicklung: Eine kritische Betrachtung der empirischen Forschung in den Vereinigten Staaten, in: Fritsch, M./Hull, C.J. (eds.) (1987a), pp. 21-48.

Egle, F. (1979): Ansätze für eine systematische Beobachtung und Analyse der Arbeitslosigkeit, in: Beiträge zur Arbeitsmarkt- und Berufsforschung 36/79.

Erber, G. (1986): Foreign Trade Modelling for the Federal Republic of Germany, in: Vierteljahrshefte 3/86, pp. 133-155.

Erber, G. (1989): Auswirkungen von Forschung und Entwicklung auf Außenhandel, internationale Wettbewerbsfähigkeit und Beschäftigung, in: Meyer-Krahmer, F., (ed.) (1989).

Erber, G./Horn, G.A. (1987): FuE-Aufwendungen in der Bundesrepublik Deutschland im Vergleich zu den USA und Japan, DIW Wochenbericht 23/87, pp. 385-391.

Ewers, H.-J./Becker, C./Fritsch, M. (1988): Wirkungen des Einsatzes computergestützter Techniken in Industriebetrieben, Institut für Stadtforschung und Strukturpolitik, Berlin.

Fachgemeinschaft Montage-Handhabung-Industrieroboter (MHI) im VDMA (1988): Das Portrait der Branche Montage, Handhabung, Industrieroboter (MHI).

Faust, K. (1987): Früherkennung technischer Entwicklungen auf der Basis von Patentdaten, Ifo-Studien zur Strukturforschung vol. 9/II, München.

Fishlow, A. (1966): Productivity and Technological Change in the Railroad Sector, in: Output, Employment and Productivity in the U.S. after 1800, Studies in Income and Wealth, no. 30, National Bureau of Economic Research.

Flaig G./Stadler M. (1988): Beschäftigungseffekte privater FuE-Aufwendungen. Eine Paneldaten-Analyse, in: Zeitschrift für Wirtschafts- und Sozialwissenschaften 108, pp. 43-61.

Fomby, T.B./Hill, R.C./Johnson, S.R. (1984): Advanced Econometric Methods, New York/Berlin.

Franz, W. (1987): Beschäftigungsprobleme aufgrund von Inflexibilitäten auf Arbeitsmärkten? Discussion Paper no. 340-87, Universität Stuttgart.

Freeman, Ch./Soete, L. (eds.) (1987): Technical Change and Full Employment, Oxford.

Freiburghaus, D. (1978): Dynamik der Arbeitslosigkeit, Meisenheim am Glan.

Friede, G. (1978): Untersuchungen zum Produzentenverhalten in der Bundesrepublik Deutschland mit Hilfe von Translog-Preisfunktionen, Dissertation, Universität Karlsruhe.

Fritsch, M./Hull, C.J. (eds.) (1987a): Arbeitsplatzdynamik und Regionalentwicklung, Beiträge zur beschäftigungspolitischen Bedeutung von Klein- und Großunternehmen, Berlin.

Fritsch, M./Hull, C.J. (1987b): Empirische Befunde zur Arbeitsplatzdynamik in großen und kleinen Unternehmen in der Bundesrepublik Deutschland - eine Zwischenbilanz, in: Fritsch, M./Hull, C.J. (eds.), Arbeitsplatzdynamik und Regionalentwicklung - Beiträge zur beschäftigungspolitischen Bedeutung von Groß- und Kleinunternehmen, Berlin, Sigma, pp. 149-172.

Fritsch, M. (1989): Die Arbeitsplatzentwicklung in Industriebetrieben, Berlin, Sigma.

Gahlen, B. (1972): Der Informationsgehalt der neoklassischen Wachstumstheorie für die Wirtschaftspolitik, Tübingen.

Gardell, B./Johansson, G. (eds.) (1981): Working Life (John Wiley and Sons).

Gerstenberger, W. (1986): Technische Neuerungen und Höhe der Investitionsausgaben. Analyse am Beispiel der Industrie, in: Mitteilungen aus der Arbeitsmarkt- und Berufsforschung 1/86, pp. 159-164.

Gerstenberger, W. et al. (1988): Wettbewerbsfähige Strukturen gestalten Expansionspolitik. Strukturberichterstattung 1987, Kernbericht, Schriftenreihe des Ifo-Instituts für Wirtschaftsforschung Nr. 120, Berlin/München.

Gerstenberger, W./Heinze, J./Vogler-Ludwig, K. (1984): Investitionen und Anlagevermögen der Wirtschaftsbereiche nach Eigentümer- und Nutzerkonzept, Ifo-Studien zur Strukturforschung, vol. 6, Berlin.

GEWIPLAN 1981 (Gesellschaft für Wirschaftsförderung und Marktplanung mbH): Marktpotentiale für Fertigungstechnologien 1990, Teil BX, Marktpotentiale für Industrieroboter in der BRD 1980-1990, Frankfurt/Main.

Giersch, H. (1985): Eurosclerosis, Discussion Paper 112, Kiel Institute for World Economics, Kiel.

Gizycki, R. v. (1980): Auswirkungen neuer Technologien - dargestellt am Beispiel von Industrierobotern, in: Beschäftigungspolitik für die achtziger Jahre, Sozialwissenschaftliche Reihe des Battelle-Instituts (ed.: L. Böckels).

Görzig, B. (1980): Der Einfluß des investitionsgebundenen Technischen Wandels auf die Arbeitsproduktivität, in: Bombach, G./Gahlen, B./Ott, A. (eds.) (1980), Tübingen.

Görzig, B./Blazejczak, J./Horn, G.A./Kirner, W./Schulz, E./Stille, F. (1988): Investitionen, Beschäftigung und Produktivität. Zu den Arbeitsplatzeffekten einer verstärkten Investitionstätigkeit vor dem Hintergrund sektoraler Entwicklungen, Beiträge zur Strukturforschung des DIW, no. 108.

Görzig, B./Schintke, J./Baumann, P. (1987): Produktionsvolumen und -potential, Produktionsfaktoren des Bergbaus und des Verarbeitenden Gewerbes in der Bundesrepublik Deutschland, Statistische Kennziffern, 29. Folge, 1970-1986, Berlin.

Griffin, J./Gregory, P. (1976): An Intercountry Model of Energy Substitution Responses, in: American Economic Review, vol. 66, pp. 845-857.

Griliches, Z. (1980): R&D and the Productivity Slowdown, in: The American Economic Review, vol. 70, pp. 343-348.

Griliches, Z. (ed.) (1984): R&D, Patents and Productivity, Chicago (University Press).

Griliches, Z./Intriligator, M. (eds.) (1986): Handbook of Econometrics, vol. 3, Amsterdam.

Gruber, W./Mehta, D./Vernon, R. (1967): The R&D Factor in International Trade and International Investment of United States Industries, in: Journal of Political Economy, vol. 75, no. 1, pp. 20-37.

Hagemann, H. (1985): Freisetzungs- und Kompensationseffekte neuer Technologien: Zur Gefahr einer technologischen Arbeitslosigkeit, in: Buttler, F./Kühl, J./ Rahmann, B. (eds.), Staat und Beschäftigung, Beiträge zur Arbeitsmarkt- und Berufsforschung 88, Nürnberg.

Hamermesh, D./Grant, J. (1979): Econometric Studies of Labor-Labor Substitution and their Implications for Policy, in: Journal of Human Resources, vol. 14, pp. 518-542.

Hanappi, G./Wagner, M. (1989): Ein makroökonomisches Modell der Innovationsdynamik, Wien (IWS).

Hansen, G. (1983): Faktorsubstitution in den Wirtschaftssektoren der Bundesrepublik, in: DIW-Vierteljahresheft 2/3, pp. 169-183.

Hansen, J.A. (1986): Innovation Characteristics of Industries in the United States, Papier für die Konferenz: Technology, Labour and Economics, Maastricht, The Netherlands, Oktober 1986.

Hansen, L. (1982): Large Sample Properties of Generalized Method of Moments Estimators, in: Econometrica, vol. 50, pp. 1029-1054.

Hansmann, K.-W./Roggon, A. (1984): Stand und Entwicklung des Industrieroboterinsatzes in der Deutschen Wirtschaft, in: Crusius, R./Stebani, J. (eds.), Neue Technologien und menschliche Arbeit, Berlin.

Hattry, W./Sydow, J. (1982): Information und Partizipation bei technisch-organisatorischen Änderungen der Arbeit, Office Management, vol. 1982, pp. 710 ff.

Helfert, M. (1987): Gewerkschaften und technische Entwicklung, WSI-Studie zur Wirtschafts- und Sozialforschung, Nr. 59, Köln.

Helberger, Ch. (1977): Ziele und Ergebnisse der Gesundheitspolitik, in: Zapf, W. (ed.), Lebensbedingungen in der Bundesrepublik, Frankfurt/New York.

Hickel, R. (1987): Ein neuer Typ der Akkumulation, Hamburg.

Hildebrandt, E. (1987): Das Handlungsdilemma von Betriebsräten bei der Einführung von computergestützten Produktionsplanungs- und -steuerungssystemen im Maschinenbau, in: CIM oder die Zukunft der Arbeit in rechnerintegrierten Fabrikstrukturen, Ergebnisse einer Fachtagung der IG Metall, Frankfurt, pp. 126 ff.

Hinz, H. (1986): Selbstverwirklichung durch Arbeit - Technologie-Abschätzung für den Arbeitnehmer, Blick durch die Wirtschaft vom 12.12.1986.

Hohmeyer, O./Rahner, H.-J. (1980): Untersuchung der Auswirkungen des Baus von ausgewählten Technologien zur rationellen Energieverwendung und zur Nutzung von regenerativen Energiequellen auf die Produktion und Beschäftigung in der Bundesrepublik, Diplom-Arbeit, Universität Bremen.

Horn, G.A. (1989): Auswirkungen von Forschung und Entwicklung auf Beschäftigung, Preise und Löhne, in: Meyer-Krahmer, F. (ed.) (1989).

Hornschild, K./Wessels, H. (1987): Beschäftigungseffekte von Fernmeldeinvestitionen und Breitbandverkabelung der Deutschen Bundespost, in: Wochenbericht des DIW, Nr. 42/87.

Hornstein, Z./Grice, J./Webb, A. (eds.) (1981): The Economics of the Labour Market, London.

Hughes, K. (1986): Exports and Technology, Cambridge.

Hyman, R./Streeck, W. (eds.) (1988): New Technology and Industrial Relations, Oxford (Basil Blackwell).

Ifo-Institut für Wirtschaftsforschung (1980): Technischer Fortschritt - Auswirkungen auf Wirtschaft und Arbeitsmarkt. Berlin.

Ifo-Institut für Wirtschaftsforschung (1987): Innovationsbericht. Ergebnisse des Ifo-Innovationstestes 1985/86, München.

Ifo-Institut für Wirtschaftsforschung (1988): Innovation, Wachstum und Beschäftigung. Einzelwirtschaftliche, sektorale und intersektorale Innovationsaktivitäten und ihre Auswirkungen auf die deutsche Wirtschaft in den achtziger Jahren, München.

Ifo/ISI/Infratest (1980): Technischer Fortschritt - Auswirkungen auf Wirtschaft und Arbeitsmarkt. Gutachten der Projektgemeinschaft Ifo-Institut für Wirtschaftsforschung, Fraunhofer-Institut für Systemtechnik und Innovationsforschung, Infratest Wirtschaftsforschung München, Schriftenreihe des Ifo-Instituts für Wirtschaftsforschung, Nr. 106, Berlin.

IG Metall (1983): "Maschinen wollen sie - uns Menschen nicht" - Rationalisierung in der Metallwirtschaft. Eine Bestandsaufnahme der IG Metall (Kurzfassung), o.O.

Infratest Sozialforschung (1988): Technikdiffusion und Beschäftigungswirkungen im privaten Dienstleistungssektor, München.

Institut für Sozialforschung und Gesellschaftspolitik (ISG) (1987): Innovation und Qualifikation in Klein- und Mittelbetrieben, Untersuchung im Auftrag des Rationalisierungskuratoriums der deutschen Wirtschaft (RKW), Köln.

Institut für Stadtforschung und Strukturpolitik (IfS) (1988): Wirkungen des Einsatzes computergestützter Techniken in Industriebetrieben, Berlin.

Institut für Stadtforschung und Strukturpolitik (IfS) (1988): Zwischenbericht des IfS zum META-II-Workshop in Berlin am 9./10, März 1988.

Intriligator, M./Kendrick, D. (eds.) (1974): Frontiers of Quantitative Economics, vol. II, Amsterdam.

Jäger, W. (1980): Die Struktur des Arbeitsmarktes in der Bundesrepublik Deutschland, Düsseldorf.

Johnston, J. (1984): Econometric Methods, 3rd edition, New York (McGraw-Hill).

Jorgenson, D.W. (1986): Econometric Methods for Modelling Producer Behavior, in: Griliches, Z./Intriligator, M. (eds.), Handbook of Econometrics, vol. 3, Amsterdam, pp. 1841-1915.

Jorgenson, D.W. (1988): Productivity and Economic Growth in Japan and the United States, in: American Economic Review, Papers and Proceedings 78/88, pp. 217-222.

Jorgenson, D.W/Laffont, J.-J. (1974): Efficient Estimation of Non-Linear Simultaneous Equations with Additive Disturbances, in: Annals of Social and Economic Measurement, vol. 3, pp. 615-640.

Kämpfer, S. (1984): Roboter: Die elektronische Hand des Menschen, Düsseldorf.

Keesing, D.B. (1967): The Impact of Research and Development of United States Trade, in: Journal of Political Economy, vol. 75, pp. 38-48.

Kern, H./Schumann, M. (1970): Industriearbeit und Arbeiterbewußtsein, 2 Bände, Studienausgabe, Frankfurt.

Kern, H./Schumann, M. (1985): Das Ende der Arbeitsteilung? Rationalisierung in der industriellen Produktion, 2. Auflage, München (Beck).

Kierzkowski, H. (1984): Monopolistic Competition and International Trade, Oxford.

Kleine, J. (1980): Innovation und Unternehmensgröße, Discussion Paper des Wissenschaftszentrum, Nr. IIM/LMP 80-9.

Klodt, H. (1984): Produktivitätsschwäche in der deutschen Wirtschaft, Tübingen.

Krelle, W. (1986): Technischer Fortschritt und Wachstum, in: Bombach, G./Gahlen, B./Ott, A.E. (eds.), Technologischer Wandel - Analyse und Fakten, Schriftenreihe des wirtschaftswissenschaftlichen Seminars Ottobeuren, vol. 15, Tübingen, pp. 87- 107.

Krengel, R. (ed.) (1982): Die Weiterentwicklung der Input-Output-Rechnung in der Bundesrepublik Deutschland. Sonderhefte zum Allgemeinen Statistischen Archiv, no. 18.

Krupp, H.-J. (1986): Innovation und Wettbewerbsfähigkeit der deutschen Volkswirtschaft, in: Bombach, G./Gahlen, B./Ott, A. (eds.) (1986), pp. 195-218.

Krupp, H.-J./Rohwer, B./Rothschild K.W. (eds.) (1986): Wege zur Vollbeschäftigung, Freiburg.

Kugler, P./Müller, U./Sheldon, G. (1987): Arbeitsmarktwirkungen moderner Technologien. Eine ökonometrische Untersuchung für die Bundesrepublik Deutschland. Meta-Studie Phase II. Abschlußbericht, Basel, Dezember 1987.

Landau, R./Rosenberg, N. (eds.) (1986): The Positive Sum Strategy, Washington.

Leontief, W./Duchin, F. (1986): The Future Impact of Automation on Workers, New York/Oxford.

Levin, R.C./Klevorick, A.K./Nelson, R./Winter, S.G. (1984): Survey Research on R&D Appropriability and Technological Opportunity. Part I: Appropriability. Yale.

Lutz, B. (1969): Produktionsprozeß und Berufsqualifikation, in: Adorno, T.W. (ed.), Spätkapitalismus oder Industriegesellschaft?, Stuttgart.

Lutz, B. (1986): Kann man Technik-Folgen abschätzen? in: Gewerkschaftliche Monatshefte, no. 9, pp. 561-570.

Lutz, B. (1987): Das Ende des Technik-Determinismus und die Folgen - Soziologische Technikforschung vor neuen Aufgaben und neuen Problemen, in: Lutz, B. (ed.), Technik und sozialer Wandel, Frankfurt/New York.

Lutz, B. (ed.) (1987): Technik und sozialer Wandel, Frankfurt/New York.

Lutz, B./Schultz-Wild, R. (1986): Aufklärung als Gestaltung - Zur Rolle der Sozialwissenschaften bei technisch-organisatorischen Innovationsvorhaben, WSI-Mitteilungen, vol. 39, pp. 669 ff.

Mansfield, E. (1984): R&D and Innovation: Some Empirical Findings, in: Griliches, Z. (ed.), R&D, Patents and Productivity. Chicago (University Press), pp. 127-154.

Mansfield, E./Romeo, A./Wagner, S. (1979): Foreign Trade and U.S. Research and Development, in: Review of Economics and Statistics, vol. 61, no. 1, pp. 49-57.

Meidner, R./Hedborg, A. (1984): Modell Schweden. Erfahrungen einer Wohlfahrtsgesellschaft, Frankfurt/New York.

Metcalfe, S. (1987): Technical Change, in: The New Palgrave, vol. 4, London (MacMillan), pp. 617-620.

Mettelsiefen, B./Barens, J. (1987): Direkte und indirekte Beschäftigungswirkungen technologischer Innovationen, Beiträge zur Arbeitsmarkt- und Berufsforschung 112, Nürnberg.

Meyer-Krahmer, F. (ed.) (1989): Sektorale und gesamtwirtschaftliche Beschäftigungswirkungen moderner Technologien. Gutachten des DIW im Rahmen der Meta-Studie II: Arbeitsmarktwirkungen moderner Technologien (im Erscheinen).

Meyer-Krahmer, F./Wessels, H. (1989): Intersektorale Technologiegeber- und Technologienehmerverflechtung in der Bundesrepublik Deutschland, in: Meyer-Krahmer, F. (ed.) (1989).

Müdespacher, A. (1987): Innovationen der Telematik: Adoptionsverhalten und regionalwirtschaftliche Effekte, in: Raumforschung und Raumordnung, 45. vol., pp. 71-79.

Müller, J./Wessels, H. (1982): Produktions- und Beschäftigungseffekte von Fernmeldeinvestitionen der Deutschen Bundespost, in: Wochenbericht des DIW, Nr. 34/82.

Naegele, G. (1983): Arbeitnehmer in der Spätphase ihrer Erwerbstätigkeit. Forschungsbericht Nr. 91, hrsg. vom Bundesministerium für Arbeit und Sozialordnung, Bonn.

Nakamura, S. (1984): An Inter-Industry Translog Model of Prices and Technical Change for the West German Economy, Berlin.

Nelson, R. (1981): Research on Productivity Growth and Productivity Differences: Dead Ends and New Departures, in: Journal of Economic Literature, vol. XIX, pp. 1029 - 1064.

Nelson, R./Winter, S.G. (1982): An Evolutionary Theory of Economic Change, Cambridge (Harvard University Press).

Nelson, R./Winter, S.G. (1977): In Search of Useful Theory of Innovation, in: Research Policy, vol. 6, no. 1, pp. 36-76.

Neubourg, C. de (1987): Unemployment, Labour Market Slack, and Labour Market Accounting, Groningen.

Oppenländer, K.H. (1987): Arbeitsmarktwirkungen moderner Technologien, in: Bombach, G./Gahlen, B,/Ott, A. (eds.), Arbeitsmärkte und Beschäftigung. Fakten, Analysen, Perspektiven, Schriftenreihe des wirtschaftswissenschaftlichen Seminars Ottobeuren. vol. 16. Tübingen.

Oppenländer, K.H. (1988): Wachstumstheorie und Wachstumspolitik, München.

Oppenländer, K.H. et al. (1971): Wirtschaftliche Auswirkungen des technischen Wandels in der Industrie, Wirtschaftliche und soziale Aspekte des technischen Wandels in der Bundesrepublik Deutschland, Dritter Band, Frankfurt/Main.

Ott, A. (1959): Technischer Fortschritt, in: Handwörterbuch der Sozialwissenschaften, Göttingen.

Ott, A. (1985): Neue Technologien in Klein- und Mittelbetrieben, Institut zur Erforschung sozialer Chancen, Arbeitsmaterialien Nr. 12, Köln.

Petersen, H.C. (1979): Estimation of Sector Sales and Employment Changes Associated with Solar Space and Water Heating Development, in: Solar Energy, vol. 22, pp. 175-182.

Pindyck, R. (1979): Interfuel Substitution and the Industrial Demand for Energy: An International Comparison, in: Review of Economics and Statistics, vol. 61, pp. 169-179.

Pindyck, R.S./Rubinfeld, D.L. (1981): Econometric Models and Economic Forecasts, 2nd edition, Auckland, Bogot (McGraw Hill).

Posner, M.V. (1961): Technical Change and International Trade, in: Oxford Economic Papers, vol. 13, pp. 323-341.

Prescott, E.C./Boyd, J.H. (1987): Dynamic Coalitions: Engines of Growth, in: The American Economic Review, vol. 77, pp. 63-67.

Reinhard, M./Scholz, L. unter Mitarbeit von W. v.d. Ohe (1979): Konzeption eines Innovationstests für die Bundesrepublik Deutschland, Forschungsprojekt, gefördert durch den BMFT, München.

Rempp, H./Lay, G. (1983): Herstellung und Anwendung von CNC-Werkzeugmaschinen in der Bundesrepublik Deutschland, in: RKW-Handbuch Mikroelektronik, Berlin (Erich Schmidt Verlag).

Reyher, L./Bach, H.-U. (1986): Sonderauswertung für die Methode, vgl.: Arbeitskräfte-Gesamtrechnung. Bestände und Bewegungen am Arbeitsmarkt, in: Mitteilungen aus der Arbeitsmarkt- und Berufsforschung 4/80.

Robak, B./Schlecht, M. (1983): Tarifliche Gestaltung des Einsatzes neuer Technologien - Erfahrungen und Konsequenzen aus der Anwendung des RTS-Tarifvertrages in der Druckindustrie, WSI-Mitteilungen, vol. 36, pp. 133 ff.

Rogers, E.M./Shoemaker, F.F. (1971): Communication of Innovations, New York.

Romer, P.M. (1987): Growth Based on Increasing Returns due to Specialization, in: The American Economic Review 77 (1987), pp. 56-62.

Rosenberg, N. (ed.) (1971): The Economics of Technological Change, Middlesex (Penguin).

Rosenberg, N. (1976): Perspectives on Technology, London/New York.

Rosenberg, N. (1979): Technological Interdependence in the American Economy, in: Technology and Culture, January 1979, pp. 25-50.

Rosenberg, N. (1982): Inside the Black Box, New York (Cambridge University Press).

Rosenberg, N. (1984): Inside the Black Box: Technology and Economics, New York (Cambridge University Press).

Rothkirch, C.v./Weidig, I. (1985): Die Zukunft der Arbeitslandschaft. Zum Arbeitskräftebedarf nach Umfang und Tätigkeiten bis zum Jahr 2000, Beiträge zur Arbeitsmarkt- und Berufsforschung 94.1 und 94.2, Nürnberg.

Schankerman, M./Nadiri, I.M. (1984): Investment in R&D, Costs of Adjustment and Expectations, in: Griliches, Z. (ed.), R&D, Patents, and Productivity, Chicago, pp. 315-338.

Scherer, F.M. (1980): Industrial Market Structure and Economic Performance, Chicago.

Scherer, F.M. (1982): Inter-Industry Technology Flows and Productivity Growth, in: Review of Economics and Statistics 1982.

Scherer, F.M. (1982): Interindustry Technology Flows in the United States, in: Research Policy 11 (August 1982), pp. 227-245, hier entnommen aus: Scherer, F.M., Innovation and Growth. Schumpeterian Perspectives (1984), Cambridge (MIT-Press).

Scherer, F.M. (1982): Innovation and Growth. Schumpeterian Perspectives (1984), Cambridge (MIT-Press).

Schettkat, R. (1987): Erwerbsbeteiligung und Politik, Berlin (edition sigma).

Schettkat, R. (1989): Innovation und Arbeitsmarktdynamik, Berlin (deGruyter, im Erscheinen).

Schmalholz, H./Scholz, L. (1985): Innovation in der Industrie, ifo-Studien zur Industriewirtschaft 28, München.

Schmalholz, H./Scholz, L. unter Mitarbeit von J. Gürtler (1985): Innovation in der Industrie, Ifo-Studien zur Industriewirtschaft, vol. 28, München.

Schmid, G. (1980): Strukturierte Arbeitslosigkeit und Arbeitsmarktpolitik, Königstein/Ts. (Athenäum).

Schmid, G. (1990): Institutions Regulating the Labour Market. Support or Impediment for Structural Change, in: Appelbaum, E./Schettkat, R. (eds.), The Impacts of Structural Change and Technological Progress on the Labor Market, New York (forthcoming).

Schmookler, J. (1971): Economic Sources of Inventive Activity, in: Rosenberg, N. (ed.), The Economics of Technological Change, Middlesex (Penguin), pp. 117-136.

Scholz, L. (1974): Technologie und Innovation in der industriellen Produktion, Schriftenreihe der Kommission für wirtschaftlichen und sozialen Wandel, vol. 21, Göttingen.

Scholz, L. (1977): Technik-Indikatoren. Ansätze zur Messung des Standes der Technik in der industriellen Produktion, Schriftenreihe des Ifo-Instituts für Wirtschaftsforschung, Nr. 91, Berlin/München.

Schott, K. (1978): The Relations between Industrial Research and Development and Factor Demands, in: The Economic Journal, no. 88, pp. 85-106.

Schraft, R.D. (1987): Dem Mechanisierungsgrad sind derzeit noch Grenzen gesetzt, in: Handelsblatt vom 24.06.1987, pp. B1.

Schraft, R.D. et al. (1984): Industrierobotertechnik - Einführung und Anwendung, Schriftenreihe Kontakt & Studium, vol. 115, Grafenau.

Schulte zur Surlage R. (1985): Qualifikationsstruktur der Arbeitsnachfrage, Frankfurt/Main.

Schumpeter, J. (1926): Theorie der wirtschaftlichen Entwicklung, München und Leipzig.

Schünemann, Th.M./Bruns, Th. (1986): Modellierung und Prognose der Diffusion von Industrierobotern in der Bundesrepublik Deutschland, in: Zeitschrift für Betriebswirtschaft, 56. vol., no. 10, pp. 953-988.

Schweizer, M. (1983): Flexible Automatisierung mit Industrierobotern - Heute und Morgen, in: RKW-Handbuch Mikroelektronik, Berlin (Erich Schmidt Verlag).

Sedeno-Andres, F. et al. (1981): Bildschirm-Arbeitsplätze: Vergleich geltender Betriebs- und Dienstvereinbarungen, Bericht II des Forschungsprojekts Rationalisierung und Folgewirkungen, Berlin.

Sengenberger, W. (1984): Zur Flexibilität im Beschäftigungssystem, SAMF-Arbeitspapier 1984-3.

Sengenberger, W. (1987): Struktur und Funktionsweise von Arbeitsmärkten, Frankfurt/New York.

Siebert, H. (1986): Technologischer Wandel, Beschäftigung und Wachstum, in: Bombach, G./Gahlen, B./Ott, A. (eds.) (1986), pp. 311-325.

Soete, L./Freeman, C. (1985): New Technologies, Investment and Employment Growth, in: Employment, Growth and Structural Change, OECD, Paris, pp. 52-82.

Solow, R.M. (1957): Technical Change and the Aggregate Production Function, in: Review of Economics and Statistics, vol. 39, pp. 312-320.

Soltwedel, R. (1984): Mehr Markt am Arbeitsmarkt. Ein Plädoyer für weniger Arbeitsmarktpolitik, München/Wien (Philosophia Verlag).

Sorge, A./Hartmann, G./Warner, M./Nicholas, I. (1982): Mikroelektronik und Arbeit in der Industrie, Frankfurt/New York.

Sorge, A./Streeck, W. (1987): Industrial Relations and Technical Change - The Case for an Extended Perspective, discussion paper Nr. IIM/LMP 87-1, Wissenschaftszentrum Berlin; wieder in: Hyman, R./Streeck, W. (eds.) (1988), New Technology and Industrial Relations, Oxford (Basil Blackwell).

Soziologisches Forschungsinstitut Göttingen (SOFI), Zentrale wissenschaftliche Einrichtung "Arbeit und Betrieb" der Universität Bremen (1981): Industrieroboter-Bedingungen und soziale Folgen des Einsatzes neuer Technologien in der Automobilproduktion, Schriftenreihe 'Humanisierung des Arbeitslebens', vol. 13, Frankfurt/Main.

Spahn, H.-P./Vobruba, G. (1986): Das Beschäftigungsproblem. Die ökonomische Sonderstellung des Arbeitsmarktes und die Grenzen der Wirtschaftspolitik, discussion paper IIM/LMP 86-14, Wissenschaftszentrum Berlin.

Spektrum der Wissenschaft in Zusammenarbeit mit dem DIW (1985): Input-Output-Struktur für die Wirtschaft der Bundesrepublik Deutschland, Input-Output-Wandtafel, Heidelberg und Berlin.

Springler, J.C. (1987): Rationalisierungschancen in der Serienmontage, in: Handelsblatt vom 18.2.1987, pp. 26.

Stäglin, R. (1982): Überblick über die Aktivitäten auf dem Gebiet der Input-Output-Rechnung in der Bundesrepublik Deutschland, in: Krengel, R. (ed.), Die Weiterentwicklung der Input-Output-Rechnung in der Bundesrepublik Deutschland. Sonderhefte zum Allgemeinen Statistischen Archiv, no. 18.

Statistisches Bundesamt: Fachserie 4, Produzierendes Gewerbe, Reihe 3.1, Produktion im Produzierenden Gewerbe des In- und Auslandes.

Staudt, E. (1988): Wenn die Investition steht, muß auch eine trainierte Mannschaft einsatzbereit sein, Handelsblatt vom 11.03.1988.

Stoneman, P. (1983): The Economic Analysis of Technological Change, Oxford.

Stooß, F. (1984): Berufsqualifikation: eine individuelle und gesellschaftliche Chance, in: Keine Arbeit - keine Zukunft? Die Bildungs- und Beschäftigungsperspektiven der geburtenstarken Jahrgänge, Frankfurt/Berlin/München (Diesterweg).

Stooß, F./Troll, L. (1988): Das "Arbeitsmittel"-Konzept als Instrumentarium zur Beobachtung des beruflichen Wandels, Mitteilungen aus der Arbeitsmarkt- und Berufsforschung, vol. 21, pp. 16 ff.

Storey, D.J./Johnson, S. (1987): Zur Arbeitsplatzdynamik in Großbritannien, in: Fritsch, M./Hull, C.J. (eds.) (1987a), pp. 49-70.

Tengler, H./Dahremöller, A., unter Mitarbeit von U. Cramer (1987): Der Dienstleistungssektor in der Bundesrepublik Deutschland und in Nordrhein-Westfalen: Strukturen, Entwicklungen, Unterschiede, IfmMaterialien, Nr. 57, Bonn.

Thwaites, A.T./Edwards, A./Gibbs, D.C. (1982): Interregional Diffusion of Production Innovations in Great Britain, Centre for Urban and Regional Development Studies, University of Newcastle upon Thyne.

Ulrich, E./ Bogdahn, J. (eds.) (1986): Auswirkungen neuer Technologien, Beiträge zur Arbeitsmarkt- und Berufsforschung 82, Nürnberg.

VDMA - Fachgemeinschaft Werkzeugmaschinen und Fertigungssysteme (1987): Sonderumfrage - Statistik über numerisch gesteuerte Werkzeugmaschinen 1986, (Juli 1987).

Vernon, R. (1966): International Investment and International Trade in the Product Cycle, in: Quarterly Journal of Economics, vol. 80, pp. 190-207.

Volkholz, V. (1982): Trends in the Use of Industrial Robots in the 80s - The Case of the Federal Republic of Germany, in: OECD, Micro-electronics, Robotics and Jobs, Information Computer Communication Policy, vol. 7, Paris.

Warnecke, H.-J./Schraft, R.-D. (1986): Industrieroboter, Katalog '86, Mainz.

Warneke, P. (ed.) (1985): Technischer Wandel und Einflußmöglichkeiten der Arbeitnehmer in Europa, Berlin.

Warnken, J./Ronning, G. (1988): Bestimmungsgründe betrieblicher Faktorsubstitution. Untersuchung im Rahmen der "Meta-Studie II" im Auftrag des Bundesministers für Forschung und Technologie, Köln, Juni 1988.

Weiß, J.-P. (1981): Importmatrix für die Bundesrepublik Deutschland 1976, in: Vierteljahrshefte zur Wirtschaftsforschung, no. 1, pp. 21-31.

Weißhuhn G./König, A./Sakkas, K./Seetzen, J. (1988): Betriebsgrößenspezifische Analyse der Beschäftigung 1980-1986 in der Bundesrepublik Deutschland. Schlußbericht zum Projekt des BMFT: Arbeitsmarktwirkungen moderner Technologien, Berlin.

Weißhuhn, G. (1986), Technischer Wandel, Beschäftigung und Qualifikation. Einige arbeitsmarktorientierte und bildungsökonomische Aspekte, in: Ulrich, E./Bogdahn, J. (eds.), Auswirkungen neuer Technologien. Beiträge zur Arbeitsmarkt- und Berufsforschung 82, Nürnberg, pp. 20-42.

Wermter, W. (1981): Die Beschäftigtenstatistik der Bundesanstalt für Arbeit, in: Mitteilungen aus der Arbeitsmarkt- und Berufsforschung, no. 4, pp. 428-435.

Wessels, H. (1976): Auswirkungen des Baus eines Kernkraftwerkes auf Produktion und Erwerbstätigenzahl, in: Wochenbericht des DIW 26-27/1976.

Zapf, W. (ed.) (1977): Lebensbedingungen in der Bundesrepublik, Frankfurt/New York.

The Authors

Bettina Bangel, 1961, Dipl.-Soziologin (sociologist), researcher Wissenschaftszentrum Berlin für Sozialforschung/Labor Market and Employment.

Carsten Becker, 1957, Dipl.-Volkswirt (economist), researcher Technische Universität Berlin and Institut für Stadtforschung und Strukturpolitik Berlin.

Jörg Beutel, 1942, Dipl.-Volkswirt (economist), Dr. rer. pol., researcher Ifo-Institut für Wirtschaftsforschung, currently advisor Saudi-Arabien Government, book publication: Input-Output-Analyse der Energieströme 1975, München 1984.

Jürgen Blazejczak, 1948, Dr. rer. oec. (economist), Dipl. Ing., researcher Deutsches Institut für Wirtschaftsforschung Berlin (DIW); main research topic: longterm scenarios of economic development; book publication: Simulation gesamtwirtschaftlicher Perspektiven mit einem ökonometrischen Modell für die Bundesrepublik Deutschland, Beiträge der Strukturforschung des DIW, no. 100, Berlin 1987.

Dietmar Edler, 1954, Dipl.-Volkswirt (economist), researcher Deutsches Institut für Wirtschaftsforschung Berlin (DIW); main research topics: labor market effects of new technologies, R&D and innovation in small and medium-size firms, input-output-analysis and econometrics.

Georg Erber, 1950, Dr. rer. pol. (economist), researcher Deutsches Institut für Wirtschaftsforschung (DIW), member of the project disaggregated forecasting model; main research topics: development and application of multisectoral econometric models, analysis of international competitiveness of industries, foreign trade, productivity trends, technological change; publications: Structural Change in World Trade: Analysis and Forecasts of the FIND-Model, in: The Future of the World Economy - Economic Growth and Structural Change, hrsg. von W. Krelle, S. 499-513, Berlin/ Heidelberg/New York/London/Paris/Tokyo 1989; The General Framework of a New Disaggregated Econometric Model for the Federal Republik of Germany, in: Vierteljahresheft 3/1986, pp. 108/109 Berlin.

Hans-Jürgen Ewers, 1942, Dipl.-Volkswirt (economist), Dr. rer. pol., Professor of economics Technische Universität Berlin and partner Institut für Stadtforschung und Strukturpolitik Berlin; main research topics: structural analysis, socio-economic research on technology, regional and industrial policy.

Renate Filip-Köhn, 1941, Dr. (economist), researcher Deutsches Institut für Wirtschaftsforschung Berlin (DIW); main research topics: input-output-analysis, labor market, natural resources, information technology, environmental research; publications: Wirtschaftsinformation in der Bundesrepublik Deutschland, Teil II: Angebotsseite und internationaler Vergleich des Online-Datenbankmarktes, Studie im Auftrag des Bundesministeriums für Wirtschaft, 1986 (jointly with W. Seufert).

Michael Fritsch, 1951, Dipl.-Volkswirt, Dr. rer. oec. (economist), researcher Technische Universität Berlin and Institut für Stadtforschung und Strukturpolitik Berlin; book publications: Arbeitsplatzdynamik und Regionalentwicklung (ed.), Berlin 1987; Arbeitsplatzentwicklung in Industriebetrieben (in print).

Werner Frühstück, 1957, Dipl.-Volkswirt (economist), Büro für empirische Sozial- und Wirtschaftsforschung Wien; book publication: Beschäftigung durch Umweltschutz, Linz (Trauner) 1988.

Hubertus Häbler, 1950, Dipl.-Soziologe (sociologist), researcher Infratest Sozialforschung München; book publications: Häbler/Kunz: Qualität der Arbeit und Verkürzung der Arbeitszeit in Schule und Hochschule, München 1985; Sozialräumliche Auswirkungen der neuen Informations- und Kommunikationstechniken, München 1986 (jointly with Sträter).

Lisa Höflich-Häberlein, 1955, Dipl.-Volkswirtin (economist), project leader Infratest Sozialforschung München; unpublished documents: Prognose des DV-Einsatzes im Bürobereich in der mittelständischen Wirtschaft, Infratest 1985; Prognose des DV-Einsatzes im Bürobereich in Unternehmen und Betrieben in der Bundesrepublik Deutschland, Infratest 1986.

Gustav A. Horn, 1954, Dipl.-Volkswirt (economist), researcher Deutsches Institut für Wirtschaftsforschung Berlin (DIW), member of the project disaggregated forecasting model; main research topics: econometric analysis of the development of industries, sectoral labor market analysis; publications: Konjunkturelle Wirkungen des technischen Wandels, in: Dose, N./Drexler, A.: Technologieparks, Opladen 1987; Die Bedeutung der Lohnentwicklung für eine kapitalintensive Produktion, in: Konjunkturpolitik, Drittes Heft 1988.

Andreas König, 1958, Dipl.-Volkswirt (economist), University of Vienna, researcher Technische Universität Berlin.

Peter Kugler, 1949, Prof. Dr. rer. pol. (economist), Professor of economics University Bern.

Frieder Meyer-Krahmer, 1949, Dr. rer. pol. (economist), Privatdozent, director of the department Industry and Technology Deutsches Institut für Wirtschaftsforschung Berlin (DIW); book publications: Politische Entscheidungsprozesse und Ökonomische Theorie der Politik Frankfurt/New York 1979; Innovationsförderung für kleine und mittlere Unternehmen Frankfurt/New York 1982; Technologiepolitik und Sozialwissenschaft (ed.) Frankfurt/New York 1986; Der Einfluß staatlicher Technologiepolitik auf industrielle Innovationen Frankfurt/New York 1989.

Urs Müller, 1957, Dr. rer. pol. (economist), assistant professor for statistics and econometrics University Basel.

Horst Penzkofer, 1959, Dipl.-Volkswirt (economist), researcher Ifo-Institut für Wirtschaftsforschung; main research topics: input-output-analysis, innovation research.

Gerd Ronning, 1939, Dipl.-Volkswirt (economist), Dr. rer. oec. pub., professor of econometrics and statistics Universität Konstanz; main research topics: econometrical and statistical market analysis (pharmaceuticals, insurances, tourism); econometric models based on business-cycle and investment survey micro-data.

Ronald Schettkat, 1954, Dr. rer. oec., Dipl.-Volkswirt (economist), Dipl.-Ingenieur, project leader Wissenschaftszentrum Berlin für Sozialforschung/ labor market and employment; book publications: Erwerbsbeteiligung und Politik, Berlin 1987 (Sigma); Beschäftigungsrisiko Innovation?, Berlin 1988 (Sigma) (jointly with E. Matzner and M. Wagner).

Heinz Schmalholz, 1946, Dipl.-Volkswirt (economist), researcher Ifo-Institut für Wirtschaftsforschung, main research topics: innovation research, technology assessment.

Lothar Scholz, 1940, Dipl.-Volkswirt (economist), Dr. rer. pol., director: Innovation and Market Analysis Ifo-Institut für Wirtschaftsforschung; book publications: Technischer Fortschritt - Auswirkungen auf Wirtschaft und Arbeitsmarkt, München 1980; Innovation in der Industrie, München 1985.

Jürgen Seetzen, 1929, Dr.-Ing., currently director of the program application and impact analysis VDI/VDE Technologiezentrum Informationstechnik in Berlin; book publications: Vorausschätzung gesamtwirtschaftlicher Rahmengrößen, Ergebnisbericht Band 4 zum Forschungsvorhaben "Nachfrageimpulse und Marktaussichten für Systeme und Endgeräte zur Vermittlung von Breitbandinformation", Heinrich-Hertz-Institut für Nachrichtentechnik, Berlin 1985; Die Anforderungen der politikberatenden Wissenschaft an die Statistik zur Erfassung der informationstechnischen Entwicklung, in: R. Stäglin/E. Südfeld (ed.): Informations- und Kommunikationstechnologien in Wirtschaft und Gesellschaft, Stuttgart 1988.

George Sheldon, 1948, PD Dr. rer. pol. (economist), deputy director Forschungsstelle für Arbeitsmarkt- und Industrieökonomik (FAI) Universität Basel; book publication: Die Dynamik der Arbeitslosigkeit in der Schweiz, Bern 1989.

Reiner Stäglin, 1938, Prof. Dr. (economist), researcher Deutsches Institut für Wirtschaftsforschung Berlin (DIW) and professor Freie Universität Berlin; main research topics: structural change, input-output-analysis, industry studies; publications: Anforderungen an die amtliche Statistik aus der sektoralen und einer regionalisierten Strukturberichterstattung, in: Jahrbücher für Statistik und Landeskunde von Baden-Württemberg (1987); Direkte und indirekte Exportabhängigkeit der Wirtschaftszweige in der Bundesrepublik Deutschland vom Europäischen Binnenmarkt, Wochenbericht des DIW, no. 47/1988.

Michael Wagner, 1945, Dr., Univ. Doz. (economist), director Institut für Wirtschafts- und Sozialforschung Wien; book publications: Beschäftigungsrisiko Innovation?, Berlin 1988 (Sigma) (jointly with E. Matzner and R. Schettkat); Wissen als Kapital, Regensburg 1989 (Transfer).

Jürgen Warnken, 1953, Dipl.-Volkswirt (economist), researcher Institut für Sozialforschung und Gesellschaftspolitik Köln; main research topics: labor market and occupational research, technology assessment.

Gernot Weißhuhn, 1943, Dr. rer. pol. (economist), professor of economics Technische Universität Berlin; main research topic: economics of education.

Hans Wessels, 1937, Dr. rer. oec. (economist), researcher Deutsches Institut für Wirtschaftsforschung Berlin (DIW); main research topics: sectoral analysis and input-output-analysis; publications: Struktur und Wettbewerbsfähigkeit der Schuhindustrie in der Bundesrepublik Deutschland, DIW-Beiträge zur Strukturforschung, Heft 104/1987 (jointly with G. Neckermann); Beschäftigungseffekte von Fernmeldeinvestitionen und Breitbandverkabelung der Deutschen Bundespost, in: Wochenbericht des DIW, no. 42/1987 (jointly with K. Hornschild).

Bei Fragen zur Produktsicherheit wenden Sie sich bitte an:
If you have any questions regarding product safety,
please contact:

Walter de Gruyter GmbH
Genthiner Straße 13
10785 Berlin
productsafety@degruyterbrill.com